Lecture Notes in Computer Science 13922

Founding Editors

Gerhard Goos
Juris Hartmanis

The series Lecture Notes in Computer Science (LNCS), including its subseries Lecture Notes in Artificial Intelligence (LNAI) and Lecture Notes in Bioinformatics (LNBI), has established itself as a medium for the publication of new developments in computer science and information technology research, teaching, and education.

LNCS enjoys close cooperation with the computer science R & D community, the series counts many renowned academics among its volume editors and paper authors, and collaborates with prestigious societies. Its mission is to serve this international community by providing an invaluable service, mainly focused on the publication of conference and workshop proceedings and postproceedings. LNCS commenced publication in 1973.

Amr El Abbadi · Gillian Dobbie · Zhiyong Feng ·
Lu Chen · Xiaohui Tao · Yingxia Shao ·
Hongzhi Yin
Editors

Database Systems for Advanced Applications

DASFAA 2023 International Workshops

BDMS 2023, BDQM 2023, GDMA 2023, BundleRS 2023
Tianjin, China, April 17–20, 2023
Proceedings

Editors
Amr El Abbadi
University of California, Santa Barbara
Santa Barbara, CA, USA

Gillian Dobbie
University of Auckland
Auckland, New Zealand

Zhiyong Feng
Tianjin University
Tianjin, China

Lu Chen
Zhejiang University
Hangzhou, China

Xiaohui Tao
The University of Southern Queensland
Queensland, Australia

Yingxia Shao
Beijing University of Posts
and Telecommunications
Beijing, China

Hongzhi Yin
The University of Queensland
Brisbane, QLD, Australia

ISSN 0302-9743 ISSN 1611-3349 (electronic)
Lecture Notes in Computer Science
ISBN 978-3-031-35414-4 ISBN 978-3-031-35415-1 (eBook)
https://doi.org/10.1007/978-3-031-35415-1

This Springer imprint is published by the registered company Springer Nature Switzerland AG
The registered company address is: Gewerbestrasse 11, 6330 Cham, Switzerland

Paper in this product is recyclable.

Preface

Along with the main conference, the workshops of the Database Systems for Advanced Applications (DASFAA) conference continue to provide valuable forums for researchers and practitioners to explore focused problem domains in the database area. This year, we are pleased to have hosted four successful workshops in conjunction with DASFAA 2023:

- The 9th International Workshop on Big Data Management and Service (BDMS 2023)
- The 8th International Workshop on Big Data Quality Management (BDQM 2023)
- The 7th International Workshop on Graph Data Management and Analysis (GDMA 2023)
- The 1st International Workshop on Bundle-based Recommendation Systems (BundleRS 2023)

These workshops were selected through a rigorous Call-for-Proposals process and were organized by their respective Workshop Organizing Committees and Program Committees. Each workshop focused on a specific area that contributed to the main themes of DASFAA 2023. Following the acceptance of proposals, the workshops conducted their own publicity to solicit contributions and a thorough review of submissions for academic merit. As a result, in total 23 papers were accepted, including eight for BDMS 2023, two for BDMQ 2023, four for BundleRS 2023 and nine for GDMA 2023.

We would like to express our sincere gratitude to all members of the Workshop Organizing Committees and Program Committees for their hard work and dedication in delivering such a great success to the DASFAA 2023 workshops. Our great gratitude also extends to the main conference organizers for their consistent support in making this year's workshops a valuable addition to the DASFAA conference series. Last but not least, we cannot thank enough the authors who contributed to the workshops: without them, the DASFAA 2023 workshops wouldn't be possible. We hope the workshops provided a stimulating and rewarding experience for all attendees, and we look forward to continued success and growth in the years to come.

April 2023

Lu Chen
Xiaohui Tao

Organization

BDMS 2023 Organization

Chair and Co-chairs

Yan Zhao	Aalborg University, Denmark
Xiaoling Wang	East China Normal University, China
Kai Zheng	University of Electronic Science and Technology of China, China

Program Committee

Chi Zhang	Brandeis University, USA
Dalin Zhang	Aalborg University, Denmark
Jin Chen	University of Electronic Science and Technology of China, China
Junhua Fang	Soochow University, China
Liwei Deng	University of Electronic Science and Technology of China, China
Kai Huang	Hong Kong University of Science and Technology, China
Minbo Ma	Southwest Jiaotong University, China
Hao Miao	Aalborg University, Denmark
Shuncheng Liu	University of Electronic Science and Technology of China, China
Yue Cui	Hong Kong University of Science and Technology, China

BDQM 2023 Organization

Chair and Co-chairs

Xiaoou Ding	Harbin Institute of Technology, China
Chengliang Chai	Beijing Institute of Technology, China
Zhixin Qi	Harbin Institute of Technology, China

Program Committee

Xueli Liu	Tianjin University, China
Shaoxu Song	Tsinghua University, China
Jiannan Wang	Simon Fraser University, Canada
Yajun Yang	Tianjin University, China
Chen Ye	Hangzhou Dianzi University, China
Feng Zhang	Renmin University of China, China
Kaiqi Zhang	Harbin Institute of Technology, China
Wenjie Zhang	University of New South Wales, Australia
Dongjing Miao	Harbin Institute of Technology, China

BundleRS 2023 Organization

Chair and Co-chairs

Lin Li	Wuhan University of Technology, China
Jianquan Liu	NEC Corporation, Japan

Program Committee

Kaixi Hu	University of Technology Sydney, Australia
Ru Wang	Shandong Normal University, China
Peipei Wang	Wuhan University of Technology, China
Xiaohua Wu	Ankang University, China
Yanhui Gu	Nanjing Normal University, China
Zilong Jiang	Guizhou University of Finance and Economics, China

GDMA 2023 Organization

General Chair

Lei Zou	Peking University, China

Program Chairs

Liang Hong	Wuhan University, China
Weiguo Zheng	Fudan University, China

Program Committee

Xiaowang Zhang	Tianjin University, China
Youhuan Li	Hunan University, China
Yuanyuan Zhu	Wuhan University, China
Yu Liu	Beijing Jiaotong University, China
Bo Xu	Donghua University, China
Kangfei Zhao	Beijing Institute of Technology, China
Peng Peng	Hunan University, China

Contents

BDQM

BundleRS

GDMA

Blood and Blood Products Management System Based on Blockchain, and NFT Technologies

Hong Khanh Vo[1], Bao Q. Tran[1], Hieu M. Doan[1], Kiet T. Le[1], Nguyen D. P. Trong[1], Hieu V. Le[1], Loc V. C. Phu[1], Duy N. T. Quoc[1], Nguyen H. Tran[1], Anh N. The[1], Huynh H. Nghia[1], Phuc N. Trong[1], Khoa T. Dang[1], Khiem H. Gia[1], Bang L. Khanh[1], Ngan N. T. Kim[2], and Luong Hoang Huong[1(✉)]

[1] FPT University, Can Tho City, Vietnam
khanhvh@fe.edu.vn, huonghoangluong@gmail.com
[2] FPT Polytecnic, Can Tho City, Vietnam

Abstract. Nowadays, the need to use blood and its products for the treatment and diagnosis of diseases is increasing. At present, no method can replace blood and its products in the treatment method - the only solution is to utilize the donors. On the one hand, the proposed processes only focus on solving the problem of preservation and transportation between medical facilities based on peer-to-peer and blockchain approaches. On the other hand, proposals to raise people's awareness of blood donation have not yet been incorporated into current systems. One of those methods is encouraging blood donors to continue participating in the future (i.e., each donor needs some time to recover depending on the blood and blood products they donate). In this paper, we propose a supply chain management model for blood and its products based on Blockchain and smart contracts. In addition, we also review the process of managing electronic certificates after each blood donation to replace the current paper certificates based on NFT technology. Therefore, our work contributes to four aspects: (a) propose a mechanism to manage the supply chain of blood and its products; (b) propose a model to generate electronic blood donation certificates; (c) implement the proposed model; and (d) deploy proof-of-concept on four platforms (i.e., BNB Smart Chain, Fantom, Polygon, and Celo) to choose the most suitable.

Keywords: Blood donation · Blockchain · Ethereum · blood supply chain · Smart contracts · NFT · Ethereum · Fantom · Polygon · BNB Smart Chain

1 Introduction

Today, the demand for supply chain management is extremely important. The above statement has been highlighted in the approaches to supporting delivery

A. El Abbadi et al. (Eds.): DASFAA 2023 Workshops, LNCS 13922, pp. 3–18, 2023.
https://doi.org/10.1007/978-3-031-35415-1_1

[10,11,14], payment [6,18,25], project management [20], transferring the products [10,23] as well as the waste disposal process [16]. Especially in the medical environment, supply chain management issues are an important factor in the diagnosis and treatment of disease, which is referred to as the therapeutic process (i.e., emergency [17,29]) medicine [37], or medical care (i.e., user-centric system [7,8]). One of the mandatory requirements in the treatment and diagnosis of diseases today is the process of managing the supply chain of blood and its products.

Current approaches focus on managing the supply chain of blood and its products. Specifically, from receiving blood from donors to transporting and giving blood to recipients. These approaches focus on exploiting traditional supply chain management problems to solve the problems of information storage and preservation. Regarding storage, blood and blood products have very different requirements in terms of temperature, humidity, and especially use time [2]. Therefore, approaches based on Blockchain technology and smart contract have exploited (a) transparency and (b) decentralized storage to bridge the gap between donor and recipient [20,37,38].

In the first aspect (a), the advantage of systems based on Blockchain technology is that it is easy for stakeholders to access data with high-reliability [7]. All information is validated before being stored on-chain - so donors easily see the status of their blood (i.e., transferred to the recipient), location and storage time - just like recipients know the source of the blood they received as well as the associated metadata (i.e., time, location of collection, storage period). Subject to the information protection policy of donors and recipients (i.e., access control [26,39]), personal data is intentionally accessed and must be verified by the owner, ie, only medical staff can access patient information [27,35].

However, the supply chain of blood and its products management systems has not focused on raising awareness of the importance of blood and blood banking in the treatment of disease. It is challenging to change people's awareness in the propaganda of blood donation [5]. One of the efforts of the governments of all countries is to offer as many benefits as possible, and honor donors who have made a significant contribution to the blood donation process and its products [4]. A small amount of donors is not enough to maintain because after about three months each blood donation [3] (i.e., a donor can only donate blood up to four times a year).

In addition, the incentives after donor blood donation have not been fully resolved. Specifically, how to encourage donors to donate blood or blood products in the future are unanswered questions. This is extremely important because the amount of health stored in the core is on a downward trend [9]. This comes from two reasons: i) there are no artificial products that can replace blood and blood products in the treatment of diseases; ii) more and more new treatments require blood and blood products during treatment [1]. It is for the above reasons that it is extremely important to raise people's understanding of blood donation and its products. Some countries are willing to buy blood at a high price (i.e., depending on the rarity of the blood, the blood value will change) to encourage their people to donate blood[1]. As an example, in developing countries (i.e., Vietnam), in addition to the initial items and money included, the blood donor will have a certificate documenting the time and place of donation; blood and the amount of blood; and blood group donated (eg, A, B, O, or AB). This information will help the donors when they need a corresponding amount of blood for the treatment process (i.e., free cost). However, these certificates are stored in paper form and are challenging to maintain (i.e., easy to lose or burnt).

For that reason, in addition to adopting blockchain and smart contract-based models, we exploit an NFT (non-fungible tokens) based approach to generate electronic blood donation certificates. Our solution builds on the blockchain technology of the Ethereum platform. We also mine NFT on EVM-supported platforms (i.e., ERC 721) to generate electronic blood donation certificates. In addition, we also deploy our smart contract protocols on those platforms that support EVM (Ethereum Virtual Machine), such as BNB Smart Chain, Fantom, Polygon, and Celo. Our main aim in this implementation was to identify the platform with the lowest fees (i.e., gas). Thereby, identifying a potential platform to implement our proposed model.[2]

2 Related Work

Nga et al. [24] presented the first attempts to manage the blood supply chain based on Blockchain technology. The authors implemented their proof-of-concept on the Hyperledger Fabric platform. The main idea of this approach is that they allow healthcare workers to enter information and store it on-chain. Besides, storage protocols are also moved from centralized to decentralized. To add attributes related to donor and recipient. Specifically, information related to donor and recipient are protected [15] under authorization mechanism - only authorized users can access corresponding personal information. This ensures system-wide privacy protection. In addition to the above two approaches, Kim et al. [12] to build a blood supply chain management system based on the Hyperledger Fabric platform. The authors have exploited the privacy of the Hyperledger Fabric platform to build a closed supply chain management system from blood collection to blood distribution to hospitals/medical centers. In addition, the above article also provides a method to identify donors to easily contact when there are any new blood donation requests. However, these articles have not provided a specific management procedure for different types of blood products. Specifically, for each type of product (i.e., red blood cells, white blood cells, plasma, platelets) there will be a different preservation process and shelf life.

Also exploiting the Hyperledger Fabric platform, Lakshminarayanan et al. [13] also applies a blockchain model to increase transparency in the transportation of blood from the donor to the receiving place. To help recipients access blood type information and corresponding metadata (eg, donor information,

[1] https://www.who.int/news-room/fact-sheets/detail/blood-safety-and-availability.

[2] We do not deploy smart contracts on ETH because the execution fee of smart contracts is too high.

time, location) Toyoda et al. [36] proposed a hybrid model of blockchain and RFID. Specifically, each blood unit after being donated will be assigned a corresponding RFID tag, which helps not only medical staff but also recipients to easily access information related to blood donation (eg, location, time) as well as the corresponding shipping process.

With blood supply chain management systems and its products on other platforms (i.e., Ethereum), [2] has proposed a decentralized solution based on Ethereum-based blockchain to empower administer blood and its products to a specific group of people (i.e., certified blood donation center (CBDC)) so that smart contracts can be deployed. Thereby, the entire blood donation system will be ensured to follow the correct procedure and limit possible loopholes in transportation and logistics (i.e., blood loss, expiry date). Donors can optionally access the system through an identifier such as their social security number and password. For a specific blood product, Peltoniemi et al. [22] discussed how the decentralized blockchain is for plasma monitoring and management. Donor-related information is recorded before plasma is separated from their blood. The system then conducts plasma analysis and determines the quality of the blood (i.e., good or bad).

Based on the detailed analysis of the above approaches to the problem of blood supply chain management based on Blockchain technology, we can conclude that the state-of-the-art has focused on solving all the problems. Vulnerabilities and difficulties of the traditional blood supply chain management system. However, the issue of encouraging blood donors to return remains unresolved. Specifically, this paper proposes a combination mechanism between Blockchain (i.e., Ethereum), Smart contract, and NFT to not only manage the supply chain of blood and its products, but also provide management electronic certifications (i.e., donor blood donation certificates).

3 Approach

In this section, we present blood donation models and create traditional blood donation certificates before proposing an approach based on Blockchain technology, smart contract, and NFT in the process of transporting and storing blood products as well as generate electronic blood donation certificates based on NFT.

3.1 Traditional Model of Blood Donation and Blood Management

Figure 1 presents the traditional blood donation model based on four methods. Depending on the geographical location of the donors, they can choose from one of four methods including: (a) medical center; (b) medical clinic; (c) hospital; and (d) mobile blood collection unit [19]. For the 4th method of blood donation (i.e., (d)), these locations are mobile (i.e., not fixed like the other three) and are short-lived (eg, weekends - holiday). The fourth method also aims to get people to donate blood to change their perception. For some countries with limited equipment to preserve blood and blood products, the fourth method is

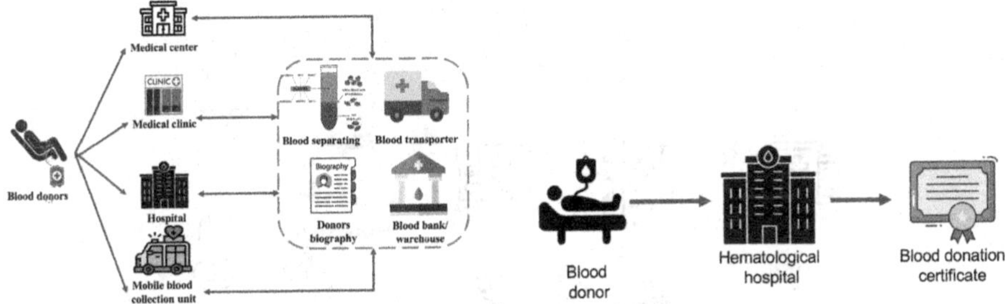

Fig. 1. Traditional model of blood donation and blood management (Color figure online)

Fig. 2. Traditional certificate of blood donation for donors

increasingly used to be able to collect blood from areas/areas that cannot be collected. Collect blood by the remaining three methods. After blood is received from donors, this blood is transported (i.e., blood transporter) to hematology hospitals or hospitals/medical centers with storage facilities (called blood banks). Here, the blood is separated into its respective components, including, red blood cells, platelets, white blood cells, and plasma. Donors' personal information is also entered (i.e., local) on the system server - this information is used for contact purposes only.

For the traditional blood donation certification process, Fig. 2 shows the process of getting a blood donation certificate from a hospital (eg, hematology hospital). These certifications have many benefits in boosting donor morale. In case donors suffer from blood-related diseases, they will receive blood in proportion to their donation. The analysis and assessment related to the limitations are presented in Introduction.[3] The next section presents our proposed model in building a blood donation and blood management model based on blockchain, smart contract and NFT technology.

3.2 Blood Donation and Blood Management Model Based on Blockchain Technology, Smart Contract and NFT

The main purpose of this paper is to build a system to optimize the use of blood volume between hospitals and medical centers as well as to propose a useful solution to encourage people to change their blood donation habits. surname. In this section we propose two blockchain-based models, smart contracts as well as NFT technology to solve the above limitations.

Figure 3 proposes a model of blood donation and blood management between hospitals in a given area (eg, city). Step 1 presents the doctor's examination and treatment process with their patients (i.e., blood recipient). In the event that a patient needs a sample for treatment, this request is sent to the hospital where the patient is being treated (step 2). There are two cases here: i) the hospital has

[3] Our studies highlighted the problems associated with traditional methods [15].

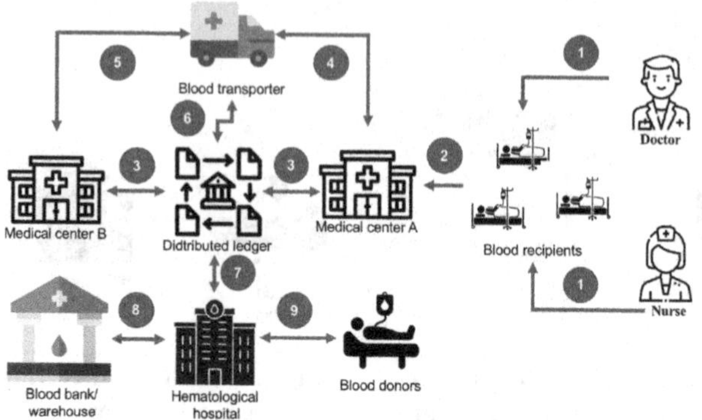

Fig. 3. Blood donation and blood management model based on blockchain technology, smart contract and NFT

a reserve of blood; ii) the hospital has no more blood for treatment. In case ii, the request is updated to the distributed ledger (step 3). The system retrieves other hospitals to find the corresponding blood type if the target hospital is identified - the hospital that contacts the blood carrier (step 4). Step 5 represents the carrier receiving blood from the target hospital. All data such as time, location are updated to the distributed ledger (step 6). In case the regional hospitals have no blood in reserve, the system sends a request to the hematology hospital and checks the blood bank of that region/city in two steps 7 and 8, respectively. If the amount of blood is not sufficient, the hematology hospital will call potential donors to collect them in step 9.

Figure 4 shows the steps to confirm and receive the NFT, ie, electronic blood donation certificate (left to right sequentially). The creation of NFTs for donors requires their consent as well as confirmation of the hospital - where the donor donates blood (eg, hematology hospital). We provide the respective services to both of these user groups before creating an electronic version of the blood donation certificate. The relevant information and documents are updated to the distributed ledger and the corresponding NFT is generated based on the functions defined in the smart contract.

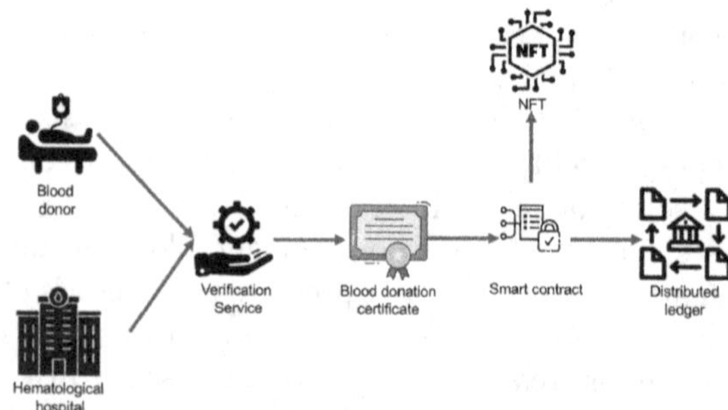

Fig. 4. Generating the electronic blood donation certificates based on NFT technology

4 Implementation

Our reality model focuses on two main purposes i) data manipulation (i.e., blood and blood product supply chain management) - creation, query and update - on blockchain platform and ii) create NFTs for each user's blood donation (i.e., donors) to encourage them to participate in the next blood donation.

4.1 Initialize Data/NFT

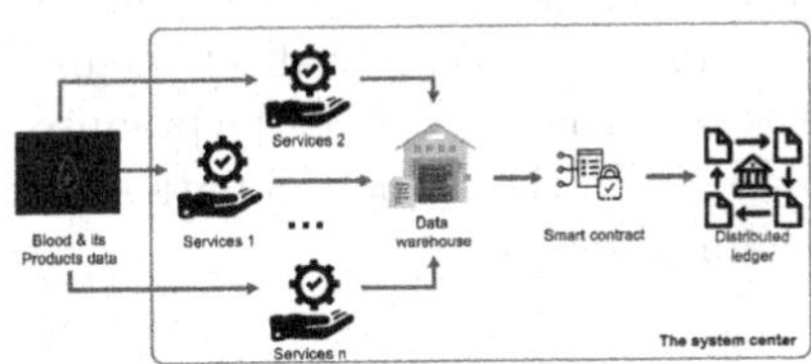

Fig. 5. Initialize data/NFT (Color figure online)

Figure 5 shows the steps to generate blood and blood product data. These data types include information regarding donors and contact information (i.e., address, ID code) or blood and blood products (eg, red blood cells, platelets, white blood cells, plasma) as well as shelf life (eg, shelf life of red blood cells). Blood products are required to be classified into many different products to facilitate preservation because each type of product has a different preservation process and shelf life. In addition, information regarding blood recipients and blood types are also important in the preservation and use of the recipients. In addition, information about medical staff (i.e., in which department, time, location) is also added to the metadata of blood and blood products. As for the storage process, services support concurrent storage (i.e., distributed processing as a peer-to-peer network) on a distributed ledger - supporting more than one user for concurrent storage. reduce system latency. In general, data on blood and blood products (eg, red blood cells) are organized as follows:

```
redBloodCellsObject = {
"donorID": donorID,
"medicalStaffID": medicalStaffID,
"type": type of blood,
"instituionID": instituionID,
"quantity": quantity,
"period": the remaining time of usage,
"packageID": packageID,
"time": time,
"location": location,
"state": null};
```

Specifically, in addition to information to extract content (i.e., place of origin, weight, blood type, etc.), we also store information related to the state of blood and blood products at the hospital. library (i.e., "state" - defaults to Null). Specifically, "state" changes to a value of 1 if the corresponding blood bag (i.e., "packageID") has been transported out of the medical center (i.e., taken to another medical center to assist in the treatment of the patient).); value 0 - is in storage at a hospital (i.e., instituionID). During transportation (i.e., "state": 1), based on "time" as well as "location" information, medical staff at the receiving location can retrieve the estimated time or location current of "packageID". Then the pre-designed constraints in the Smart Contract are called through the API (i.e., name of function) to sync them up the chain. This role of verification is extremely important because they directly affect the process of blood management and transport to the recipient, as well as the premise for disease treatment. For processes that initiate NFTs (i.e., blood donation certificates), the content of the NFT is defined as follows:

```
NFT BLOOD = {
"donorID": donorID,
"type": type of blood,
"times": times of donation,
"quantity": the total of quantity,
"date": the latest donation date,
"address": address};
```

All donor information related to the donation of blood and blood products is permanently stored on NFT, and the blood donation history is easily retrieved. This reduces the risk of losing information or certificates (i.e., paper versions). In addition, it is easy for health workers to track the information of donors.

4.2 Data Query

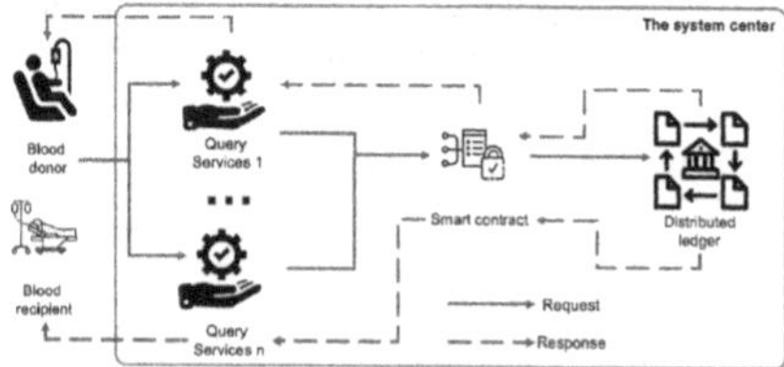

Fig. 6. Data query

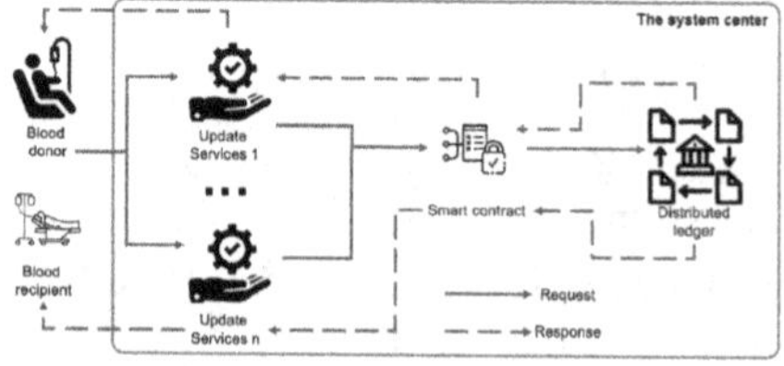

Fig. 7. Data updated

Similar to the data initialization steps, the data query process also supports many simultaneous participants in the system for access (i.e., distributed model). Support services receive requests from donors/recipients or nurses/doctors to access data. Depending on the query object we have different access purposes. Specifically, donors/recipients query for the purpose of checking the preservation and storage process for blood and blood products as well as information related to blood donation (i.e., number of blood donations, estimated time of for the next blood donation etc.). In contrast, health workers can query the data to find contact information related to donors to contact for the next blood donation. Figure 6 shows the steps to query blood and blood products data. These requests are sent as requests (i.e., pre-designed services as API calls) from the user to the smart contracts available in the system (i.e., name of function) before retrieving the data. from the distributed ledger. All retrieval requests are also saved as query history for each individual or organization. In case the corresponding information is not found (eg, wrong ID), the system will send a message not found results. For the NFT query process, all support services are provided as APIs.

4.3 Data Updated

The data update routine is invoked only after verifying that the data exists on the thread (i.e., after executing the corresponding data query procedure). In this section, we assume that the search data exists on the string. Where none exists, the system sends the same message to the user (see Sect. 4.2 for details). Similar to the two processes of query and data initialization, we support update services in the form of APIs to receive requests from users before passing them to smart contract (i.e., name of function) for processing. The purpose of this process is to update the time, blood volume and number of blood donations by donors. In addition, healthcare workers can still see which donors can donate blood in an emergency situation (i.e., meet the waiting time). Figure 7 shows the procedure for updating blood and blood product data. For NFTs (i.e., available) the update process includes only moving from the owner's address to the new (i.e., new owner). If any information is updated on an existing NFT, it will be stored as a new NFT.

5 Evaluation Scenarios

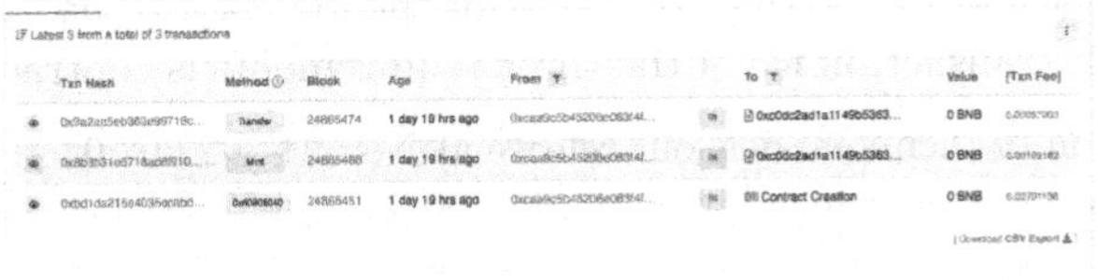

Fig. 8. The transaction info (e.g., BNB Smart Chain)

Since the proposed model generates blood donation certificates to incentivize donors for community service behavior, we implement the recommendation model on EVM-enabled blockchain platforms instead of mining platforms belonging to the Hyperledger eco-system because they are easily extensible (i.e., using existing platforms and systems). In addition, assessments based on system responsiveness (i.e., number of requests responded to successfully/failed, system latency - min, max, average) have been evaluated by us in the previous research paper. Therefore, in this paper, we determine the suitable platform for our proposed model. Specifically, we install a recommendation system on four popular blockchain platforms today, supporting Ethereum Virtual Machine (EVM), including Binance Smart Chain (BNB Smart Chain); Polygon; Fantom; and Celo. Our implementations on these four platforms are also shared as a contribution to the article to collect transaction fees corresponding to the four platforms' supporting coins[4], ie, BNB[5]; MATIC[6]; FTM[7]; and CELO[8]. For example, Fig. 8 details our three assessments of a successful installation on BNB Smart Chain (i.e., similar settings are shown for the other three platforms). Our implementations to evaluate the execution cost of smart contracts (i.e., designed based on Solidity language) run on testnet environments of four platforms in order to choose the most cost-effective platform to deploy. reality. Our detailed assessments focus on the cost of performing contract creation, NFT generation and NFT retrieval/transfer (i.e., updating NFT ownership address) presented in the respective subsections related to i) Transaction Fee; ii) Gas limit; iii) Gas Used by Transaction; and iv) Gas Price.

5.1 Transaction Fee

Table 1. Transaction fee

	Contract Creation	Create NFT	Transfer NFT
BNB Smart Chain	0.02731136 BNB ($8.37)	0.00109162 BNB ($0.33)	0.00057003 BNB ($0.18)
Fantom	0.009576826 FTM ($0.001860)	0.000405167 FTM ($0.000079)	0.0002380105 FTM ($0.000046)
Polygon	0.006840590024626124 MATIC($0.01)	0.00028940500115762 MATIC($0.00)	0.000170007500612027 MATIC($0.00)
Celo	0.0070973136 CELO ($0.004)	0.0002840812 CELO ($0.000)	0.0001554878 CELO ($0.000)

Table 1 shows the cost of creating contracts for the four platforms. It is easy to see that the highest transaction fee of the three requirements is contract creation for

[4] Implementation of theme models our release at Nov-24-2022 07:04:41 AM +UTC.

[5] https://testnet.bscscan.com/address/0xc0dc2ad1a1149b5363d7f58c2cf7231d83925c0c.

[6] https://mumbai.polygonscan.com/address/0xd20ae7123c4387d25d670a6fd74a6095f4dcaa56.

[7] https://testnet.ftmscan.com/address/0xd20ae7123c4387d25d670a6fd74a6095f4dcaa56.

[8] https://explorer.celo.org/alfajores/address/0xD20aE7123C4387d25d670a6fd74A6095F4dCaa56/transactions.

all four platforms. In which, the cost of BNB Smart Chain is the highest with the highest cost when creating a contract is 0.02731136 BNB ($8.37); whereas, the lowest cost recorded by the Fantom platform with the highest cost for contract initiation is less than 0.009576826 FTM ($0.001860). Meanwhile, the cost to enforce Celo's contract initiation requirement is lower than Polygon's with only $0.004 compared to $0.01. For the remaining two requirements (Create NFT and Transfer NFT), we note that the cost of implementing them for all three platforms, Polygon, Celo, and Fantom is very low (i.e., negligible) given the cost. trades close to $0.00. However, this cost is still very high when deployed on BNB Smart Chain with 0.00109162 BNB ($0.33) and 0.00057003 BNB ($0.18) for Create NFT and Transfer NFT, respectively.

Table 2. Gas limit

	Contract Creation	Create NFT	Transfer NFT
BNB Smart Chain	2,731,136	109,162	72,003
Fantom	2,736,236	115,762	72,803
Polygon	2,736,236	115,762	72,803
Celo	3,548,656	142,040	85,673

5.2 Gas Limit

Table 2 shows the gas limit for each transaction. Our observations show that the gas limits of the three platforms (i.e., BNB, Polygon, and Fantom) are roughly equivalent - where Polygon and Fantom are similar in all three transactions. The remaining platform (i.e., Celo) has the highest gas limit with 3,548,656; 142,040; and 85,673 for all three transaction types.

5.3 Gas Used by Transaction

Table 3. Gas Used by Transaction

	Contract Creation	Create NFT	Transfer NFT
BNB Smart Chain	2,731,136 (100%)	109.162 (100%)	57,003 (79.17%)
Fantom	2,736,236 (100%)	115,762 (100%)	68,003 (93.41%)
Polygon	2,736,236 (100%)	115,762 (100%)	68,003 (93.41%)
Celo	2,729,736 (76.92%)	109.262 (76.92%)	59,803 (69.8%)

Table 3 shows the amount of gas used when executing the transaction (i.e., what percentage of gas in total gas is shown in Table 2). Specifically, three platforms

BNB, Polygon, and Fantom use 100% of Gas Limit for two transactions Contract Creation and Create NFT. Meanwhile, Celo uses 76.92% of the Gas limit for the above two transactions. For the last transaction of Transfer NFT, the highest Gas level was recorded by Fantom and Polygon with 93.41% of Gas limit; while BNB and Celo use 79.17% and 69.8% of Gas limit.

5.4 Gas Price

Table 4. Gas Price

	Contract Creation	Create NFT	Transfer NFT
BNB Smart Chain	0.00000001 BNB (10 Gwei)	0.00000001 BNB (10 Gwei)	0.00000001 BNB (10 Gwei)
Fantom	0.0000000035 FTM (3.5 Gwei)	0.0000000035 FTM (3.5 Gwei)	0.0000000035 FTM (3.5 Gwei)
Polygon	0.000000002500000009 MATIC (2.500000009 Gwei)	0.00000000250000001 MATIC (2.50000001 Gwei)	0.000000002500000009 MATIC (2.500000009 Gwei)
Celo	0.0000000026 CELO (Max Fee per Gas: 2.7 Gwei)	0.0000000026 CELO (Max Fee per Gas: 2.7 Gwei)	0.0000000026 CELO (Max Fee per Gas: 2.7 Gwei)

Table 4 shows the value of Gas for all four platforms. Specifically, BNB, Fantom, and Celo have the same Gas value in all three transactions with values of 10 Gwei (i.e., the highest of the three platforms), 3.5 Gwei, and 2.7 Gwei, respectively. Meanwhile, the Gas value of the Polygon platform (i.e., MATIC) has the lowest value and fluctuates around 2.5 Gwei.

6 Discussion

According to our observation, the transaction value depends on the market capitalization of the respective coin. The total market capitalization of the 4 platforms used in our review (i.e., BNB (Binance Smart Chain); MATIC (Polygon); FTM (Fantom); and CELO (Celo)) are $50,959,673,206; $7,652,386,190; $486,510,485; and $244,775,762.[9] This directly affects the coin value of that platform - although the number of coins issued at the time of system implementation also plays a huge role. The total issuance of the four coins BNB, MATIC, FTM, and CELO is 163,276,974/163,276,974; 8,868,740,690/10,000, 000,000; 2,541,152,731/3,175,000,000 and 473, 376,178/1,000,000,000 coins. The coin's value is conventionally based on the amount of coins issued and the total market capitalization with a value of $314.98; $0.863099; $0.1909; and $0.528049 for BNB, MATIC, FTM, and CELO. Based on the measurements in Sect. 5 section, we have concluded that the proposed model deployed on Faltom brings

[9] Our observation time is 12:00PM - 11/26/2022.

many benefits related to system operating costs. In particular, generating and receiving NFTs has an almost zero (i.e., negligible) fee. Also, the cost of creating contracts with transaction execution value is also meager (i.e., less than $0.002).

In future work, we proceed to implement more complex methods/algorithms (i.e., encryption and decryption) as well as more complex data structures to observe the costs for the respective transactions. Deploying the proposed model in a real environment is also a possible approach (i.e., implementing the recommendation system on the FTM mainnet). In our current analysis, we have not considered issues related to the privacy policy of users (i.e., access control [27,28], dynamic policy [26,39]) - a possible approach would be implemented in upcoming research activities. Finally, infrastructure-based approaches (i.e., gRPC [21,32]; Microservices [30,33]; Dynamic transmission messages [34] and Brokerless [31]) can be integrated into the model of us to increase user interaction (i.e., API-call-based approach).

7 Conclusion

The article proposes a blood supply chain management model and its products based on blockchain technology, smart contracts, and NFT. Specifically, in the process of managing, transporting, and storing blood between medical facilities, we propose a supply chain management system based on blockchain technology and smart contract - supporting storage/processing centralization as well as increasing transparency for stored data. For NFT, we create electronic blood donation certificates (i.e., update blood donation count) to replace the current paper/centralized storage certificate of blood donation - encouraging people to donate blood. We have implemented proof-of-concept of the proposed model based on the Ethereum platform and Solidity language. We have deployed smart contracts on four popular EVM-enabled platforms, including BNB, FTM, MATIC, and CELO. The evaluation proved that our smart contract implementations (i.e., contract creation, NFT creation, NFT transfer) on the Fantom platform have the lowest cost. Causes and possible development directions are also suggested in the discussion.

Acknowledgement. This work was supported by Le Thanh Tuan and Dr. Ha Xuan Son during the process of brainstorming, implementation, and evaluation of the system.

References

1. Baek, E.J., Kim, H.O., Kim, S., Park, Q.E., Oh, D.J.: The trends for nationwide blood collection and the supply of blood in Korea during 2002–2006. Korean J. Blood Transfus. **19**(2), 83–90 (2008)
2. Çağlıyangil, M., Erdem, S., Özdağoğlu, G.: A blockchain based framework for blood distribution. In: Hacioglu, U. (ed.) Digital Business Strategies in Blockchain Ecosystems. CMS, pp. 63–82. Springer, Cham (2020). https://doi.org/10.1007/978-3-030-29739-8_4

3. Chapman, J.: Unlocking the essentials of effective blood inventory management. Transfusion **47**, 190S-196S (2007)
4. Chmielewski, D., Bove, L.L., Lei, J., Neville, B., Nagpal, A.: A new perspective on the incentive-blood donation relationship: partnership, congruency, and affirmation of competence. Transfusion **52**(9), 1889–1900 (2012)
5. Cooper, E., Jahoda, M.: The evasion of propaganda: how prejudiced people respond to anti-prejudice propaganda. J. Psychol. **23**(1), 15–25 (1947)
6. Duong-Trung, N., et al.: Multi-sessions mechanism for decentralized cash on delivery system. Int. J. Adv. Comput. Sci. Appl **10**(9) (2019)
7. Duong-Trung, N., Son, H.X., Le, H.T., Phan, T.T.: Smart care: integrating blockchain technology into the design of patient-centered healthcare systems. In: Proceedings of the 2020 4th International Conference on Cryptography, Security and Privacy, pp. 105–109 (2020)
8. Duong-Trung, N., Son, H.X., Le, H.T., Phan, T.T.: On components of a patient-centered healthcare system using smart contract. In: Proceedings of the 2020 4th International Conference on Cryptography, Security and Privacy, pp. 31–35 (2020)
9. Emmanuel, J.C.: The blood cold chain. WHO report (2017)
10. Ha, X.S., Le, H.T., Metoui, N., Duong-Trung, N.: DeM-CoD: novel access-control-based cash on delivery mechanism for decentralized marketplace. In: 2020 IEEE 19th International Conference on Trust, Security and Privacy in Computing and Communications (TrustCom), pp. 71–78. IEEE (2020)
11. Ha, X.S., Le, T.H., Phan, T.T., Nguyen, H.H.D., Vo, H.K., Duong-Trung, N.: Scrutinizing trust and transparency in cash on delivery systems. In: Wang, G., Chen, B., Li, W., Di Pietro, R., Yan, X., Han, H. (eds.) SpaCCS 2020. LNCS, vol. 12382, pp. 214–227. Springer, Cham (2021). https://doi.org/10.1007/978-3-030-68851-6_15
12. Kim, S., Kim, D.: Design of an innovative blood cold chain management system using blockchain technologies. ICIC Express Lett. Part B: Appl. **9**(10), 1067–1073 (2018)
13. Lakshminarayanan, S., Kumar, P.N., Dhanya, N.M.: Implementation of blockchain-based blood donation framework. In: Chandrabose, A., Furbach, U., Ghosh, A., Kumar M., A. (eds.) ICCIDS 2020. IAICT, vol. 578, pp. 276–290. Springer, Cham (2020). https://doi.org/10.1007/978-3-030-63467-4_22
14. Le, H.T., et al.: Introducing multi shippers mechanism for decentralized cash on delivery system. Int. J. Adv. Comput. Sci. Appl. **10**(6) (2019)
15. Le, H.T., et al.: Bloodchain: a blood donation network managed by blockchain technologies. Network **2**(1), 21–35 (2022)
16. Le, H.T., et al.: Medical-waste chain: a medical waste collection, classification and treatment management by blockchain technology. Computers **11**(7), 113 (2022)
17. Le, H.T., et al.: Patient-chain: patient-centered healthcare system a blockchain-based technology in dealing with emergencies. In: Shen, H., et al. (eds.) PDCAT 2021. LNCS, vol. 13148, pp. 576–583. Springer, Cham (2022). https://doi.org/10.1007/978-3-030-96772-7_54
18. Le, N.T.T., et al.: Assuring non-fraudulent transactions in cash on delivery by introducing double smart contracts. Int. J. Adv. Comput. Sci. Appl. **10**(5), 677–684 (2019)
19. Le Van, H., et al.: Blood management system based on blockchain approach: a research solution in Vietnam. IJACSA **13**(8) (2022)
20. Luong, H.H., Huynh, T.K.N., Dao, A.T., Nguyen, H.T.: An approach for project management system based on blockchain. In: Dang, T.K., Küng, J., Chung, T.M.,

Takizawa, M. (eds.) FDSE 2021. CCIS, vol. 1500, pp. 310–326. Springer, Singapore (2021). https://doi.org/10.1007/978-981-16-8062-5_21
21. Nguyen, L.T.T., et al.: BMDD: a novel approach for IoT platform (broker-less and microservice architecture, decentralized identity, and dynamic transmission messages). PeerJ Comput. Sci. **8**, e950 (2022)
22. Peltoniemi, T., Ihalainen, J.: Evaluating blockchain for the governance of the plasma derivatives supply chain: how distributed ledger technology can mitigate plasma supply chain risks. Blockchain in Healthcare Today (2019)
23. Quoc, K.L., et al.: SSSB: an approach to insurance for cross-border exchange by using smart contracts. In: Awan, I., Younas, M., Poniszewska-Marańda, A. (eds.) MobiWIS 2022. LNCS, vol. 13475, pp. 179–192. Springer, Cham (2022). https://doi.org/10.1007/978-3-031-14391-5_14
24. Quynh, N.T.T., et al.: Toward a design of blood donation management by blockchain technologies. In: Gervasi, O., et al. (eds.) ICCSA 2021. LNCS, vol. 12956, pp. 78–90. Springer, Cham (2021). https://doi.org/10.1007/978-3-030-87010-2_6
25. Son, H.X., Chen, E.: Towards a fine-grained access control mechanism for privacy protection and policy conflict resolution. Int. J. Adv. Comput. Sci. Appl. **10**(2) (2019)
26. Son, H.X., Dang, T.K., Massacci, F.: REW-SMT: a new approach for rewriting XACML request with dynamic big data security policies. In: Wang, G., Atiquzzaman, M., Yan, Z., Choo, K.-K.R. (eds.) SpaCCS 2017. LNCS, vol. 10656, pp. 501–515. Springer, Cham (2017). https://doi.org/10.1007/978-3-319-72389-1_40
27. Son, H.X., Hoang, N.M.: A novel attribute-based access control system for fine-grained privacy protection. In: Proceedings of the 3rd International Conference on Cryptography, Security and Privacy, pp. 76–80 (2019)
28. Son, H.X., Nguyen, M.H., Vo, H.K., Nguyen, T.P.: Toward an privacy protection based on access control model in hybrid cloud for healthcare systems. In: Martínez Álvarez, F., Troncoso Lora, A., Sáez Muñoz, J.A., Quintián, H., Corchado, E. (eds.) CISIS/ICEUTE -2019. AISC, vol. 951, pp. 77–86. Springer, Cham (2020). https://doi.org/10.1007/978-3-030-20005-3_8
29. Son, H.X., Le, T.H., Quynh, N.T.T., Huy, H.N.D., Duong-Trung, N., Luong, H.H.: Toward a blockchain-based technology in dealing with emergencies in patient-centered healthcare systems. In: Bouzefrane, S., Laurent, M., Boumerdassi, S., Renault, E. (eds.) MSPN 2020. LNCS, vol. 12605, pp. 44–56. Springer, Cham (2021). https://doi.org/10.1007/978-3-030-67550-9_4
30. Thanh, L.N.T., et al.: IoHT-MBA: an internet of healthcare things (IoHT) platform based on microservice and brokerless architecture. Int. J. Adv. Comput. Sci. Appl. **12**(7) (2021)
31. Thanh, L.N.T., et al.: SIP-MBA: a secure IoT platform with brokerless and microservice architecture (2021)
32. Thanh, L.N.T., et al.: Toward a security IoT platform with high rate transmission and low energy consumption. In: Gervasi, O., et al. (eds.) ICCSA 2021. LNCS, vol. 12949, pp. 647–662. Springer, Cham (2021). https://doi.org/10.1007/978-3-030-86653-2_47
33. Nguyen, T.T.L., et al.: Toward a unique IoT network via single sign-on protocol and message queue. In: Saeed, K., Dvorský, J. (eds.) CISIM 2021. LNCS, vol. 12883, pp. 270–284. Springer, Cham (2021). https://doi.org/10.1007/978-3-030-84340-3_22
34. Thanh, L.N.T., et al.: UIP2SOP: a unique IoT network applying single sign-on and message queue protocol. IJACSA **12**(6) (2021)

35. Thi, Q.N.T., Dang, T.K., Van, H.L., Son, H.X.: Using JSON to specify privacy preserving-enabled attribute-based access control policies. In: Wang, G., Atiquzzaman, M., Yan, Z., Choo, K.-K.R. (eds.) SpaCCS 2017. LNCS, vol. 10656, pp. 561–570. Springer, Cham (2017). https://doi.org/10.1007/978-3-319-72389-1_44
36. Toyoda, K., et al.: A novel blockchain-based product ownership management system (POMS) for anti-counterfeits in the post supply chain. IEEE access **5**, 17465–17477 (2017)
37. Huynh, T.K.N., Dao, T.A., Van Pham, T.N., Nguyen, K.H.V., Tran, N.C., Luong, H.H.: VBlock–blockchain-based traceability in medical products supply chain management: a case study in VietNam. In: Ibrahim, R., Porkumaran, K., Kannan, R., Mohd Nor, N., Prabakar, S. (eds.) International Conference on Artificial Intelligence for Smart Community. LNEE, vol. 758, pp. 429–441. Springer, Singapore (2022). https://doi.org/10.1007/978-981-16-2183-3_42
38. Nguyen, K.T.H., et al.: Domain name system resolution system with hyperledger fabric blockchain. In: Smys, S., Kamel, K.A., Palanisamy, R. (eds.) Inventive Computation and Information Technologies. LNNS, vol. 563, pp. 59–72. Springer, Singapore (2023). https://doi.org/10.1007/978-981-19-7402-1_5
39. Xuan, S.H., et al.: Rew-XAC: an approach to rewriting request for elastic ABAC enforcement with dynamic policies. In: 2016 International Conference on Advanced Computing and Applications (ACOMP), pp. 25–31. IEEE (2016)

ip2text: A Reasoning-Aware Dataset for Text Generation of Devices on the Internet

Yimo Ren[1,2], Zhi Li[1,2](✉), Hong Li[1,2], Peipei Liu[1,2], Jie Liu[1,2], Hongsong Zhu[1,2], and Limin Sun[1,2]

[1] School of Cyber Security, University of Chinese Academy of Sciences, Beijing, China

[2] Institute of Information Engineering, University of Chinese Academy of Sciences, Beijing, China

{renyimo,lizhi,lihong,liupeipei,liujie1,zhuhongsong,sunlimin}@iie.ac.cn

Abstract. Nowadays, Internet of Things (IoT) search engines are more and more popular for users to explore devices on the Internet. Table-to-text generation of devices is helpful for users to understand search results from IoT search engines. However, it has yet to be available, and difficult to obtain a good text description of the devices because of lacking quality data for this task. Also, the content is hidden in multiple attributes of the devices, and it takes work to mine them well and directly. Thus, this paper introduces ip2text, a challenging dataset for reasoning-aware table-to-text generation of devices on the Internet. The input data in ip2text are tables, which contain many attributes of devices collected from the Internet. And the output data is their corresponding descriptions. Generating descriptions of devices is costly, time-consuming, and does not scale to Internet data. To tackle this problem, this paper designs an annotation method based on active learning according to the characteristics of devices and studies the performance of existing and typical state-of-the-art models for table-to-text generation on ip2text. The automatic evaluation shows that existing pre-trained baselines could be challenging to perform satisfactorily on ip2text, with BLEU almost all less than 1. Further, the human evaluation shows that BART and T5 are prone to produce hallucinations when reasoning, and results show that Hallucination is more than 0.10. Therefore, it is not easy to achieve satisfactory performance using the existing and mainstream seq2seq models based on the reasoning-aware ip2text. So, continuous improvement is urgently needed for the models and datasets for the table-to-text generation of devices on the Internet.

Keywords: Internet of Things · Table-to-text generation · Reasoning-Aware

Supported by National Key Research and Development Projects (No. 2020YFB2103803) and National Natural Science Foundation of China (No. U1766215, No. 61931019).

A. El Abbadi et al. (Eds.): DASFAA 2023 Workshops, LNCS 13922, pp. 19–33, 2023.
https://doi.org/10.1007/978-3-031-35415-1_2

1 Introduction

Nowadays, search engines have gradually become the most convenient way for people to obtain massive amounts of information effectively. Among them, Internet of Things search engines, such as Shodan [1], Zoomeye, and Fofa [2], are used to identify computers and IoT devices connected to the Internet. IoT search engines collect information on online devices, including hardware, software, vulnerability, *etc.* Therefore, IoT search engines are essential and valuable for cyber security. For example, experts describe how they find online devices infected by malware using Shodan [3,4].

While the results of IoT search engines are usually highly professional, as Fig. 1 shows. Compared with existing search engines, IoT search engines lack explanations and make results challenging to understand. Figure 2 shows Google summarises results as a recommended answer for the user. The paper observes that the results of IoT search engines usually appear in tables. Therefore, the paper decides to use the methods of table-to-text to generate the summarization of search results from IoT search engines to make them easier to use and popularize.

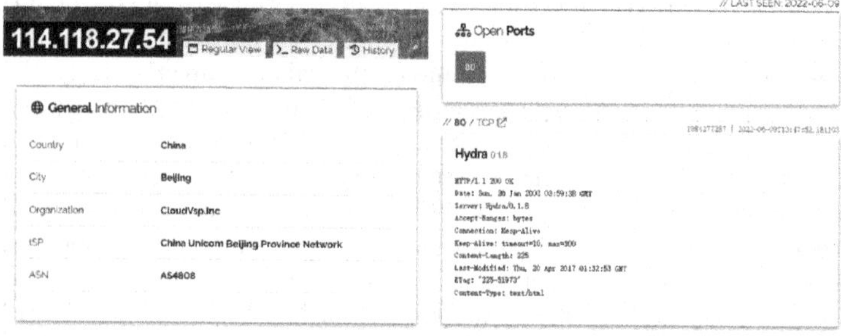

Fig. 1. An example of search results from Shodan. The results are usually displayed in the form of tables and are difficult to be understood.

Table-to-text generation is one of the established and practical tasks in NLP. Table-to-text generation makes data easier to be understood, in which the input is structured data, such as a table, and the output is a fluent and graceful text that describes the meaning of the input. Existing table-to-text datasets focus on weather reports [5], sports reports [6], restaurant descriptions [7], biography [8], science articles [9] or open domains [10] based on Wikipedia, except for IoT devices.

Thus, the paper introduces ip2text: A Reasoning-Aware Dataset for Text Generation of Devices on the Internet. ip2text is a semi-automated labeled dataset containing pairs of attribute tables of online devices from IoT search engines with corresponding descriptions. The descriptions in ip2text cover the attributes and include implicit and hidden information in tables. Most of the

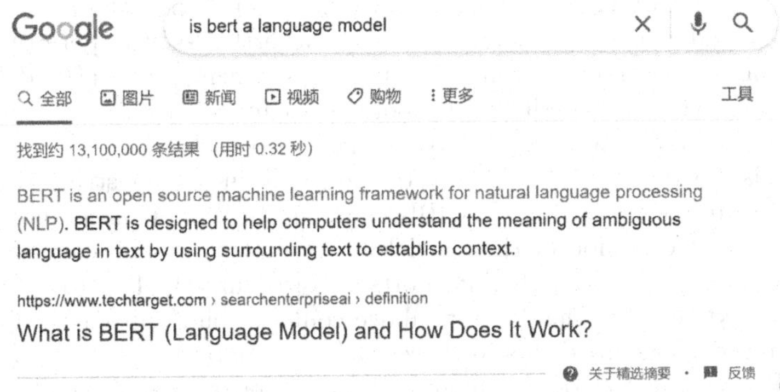

Fig. 2. An example of search results from Google. The summarization of results is regarded as the explanation.

existing table-to-text datasets are built by experts or extracted from a large amount of existing corpus. At the same time, the paper studies the performance of existing and typical state-of-the-art models for table-to-text on ip2text, including BART [11], T5 [12]. Moreover, the results show (1) the semi-automated annotation reduces as much as possible the cost of manual annotation, combined with the characteristics of IoT devices and based on ensuring data quality. (2) the generated texts of online devices based on existing methods are fluent and coherent but contain less implicit information and reasoning results. (3) ip2text can help non-expert people use IoT search engines easily and quickly.

Overall, the main contributions of this paper are: (1) This paper raises the demand for easy-to-use IoT search engines and proposes ip2text as a basis to realize text generation based on the devices' data. (2) To reduce the cost of labeling, This paper designs an annotation method based on active learning according to the data characteristics of devices. (3) This paper uses specific metrics for table-to-text generation of devices and studies the performance of existing and typical state-of-the-art models on ip2text.

2 Related Work

Recent researches have shown that fluent texts generated by table-to-text are beneficial for people to understand structure data in many actual scenarios, most of which are based on Deep Neutral Network. Some methods [5,6] use templates to generate texts of tables. Some methods build seq2seq models using structural features of tables [13,14] or hierarchically encoded embedding of tables [15]. Also, researches [9] convert tables to sequences to use seq2seq models directly. However, the output texts from most of the above models are mainly surface-level summaries of the data, which are hard to contain reasonable information from tables. KB-to-Language Generation [16] constructs a large knowledge graph to get reasonable texts from tables. logical2text [17] apply logical reasoning, and research [9] uses BART [11], T5 [12] for text generation to find the implicit information and reasoning results of tables. Nevertheless, the performance of those reasonable methods is still being improved when they are used for text generation of IoT devices. All the state-of-the-art models for table-to-text need a lot of good quality and specific data for the target tasks.

The existing datasets for table-to-text generation are constructed on text corpus from different domains. WikiBIO [8] collects Wikipedia infobox of persons and the introduction of their articles as samples automatically. Rotowire [10] uses the reports of basketball in sports websites directly. Due to the existing large-scale texts corpus, the datasets above could be built automatically. However, there are also some errors in those automatic datasets. At the same time, LogicNLG [18] and ToTTo [19] modify the table boxes and descriptions from Wikipedia manually, so their cost resources are relatively large. Thus, this paper proposes ip2text as a basis to realize text generation based on data from IoT devices.

3 Problem and Dataset

3.1 Problem Definition

ip2text is a dataset for generating descriptions for online devices by reasoning over their attributes in tables. An input in ip2text is a table T, collected from the Internet in many fields. Each input has a corresponding description C, namely a paragraph about the table T. $T = \{R_1, \ldots, R_n\}$ is a list of each sample representing a table row. The task is to generate a textual description D for the table that describes the most important findings of T by reasoning over its content. Figure 3 shows an example of the dataset.

3.2 Dataset Construction

For creating ip2text, the paper has collected data on IoT devices on the Internet that is accompanied by their corresponding descriptions.

Collection. For creating ip2text, the paper first collects data on IoT devices on the Internet. The example is as Fig. 3 shows, and the main parts are as follows:

- **IP** and **port** are generally the identity of devices [20] on the Internet. Users could control the devices by accessing their IP addresses and corresponding ports. The active IP addresses and ports could be purchased from IPIP (https://www.ipip.net/).
- **Protocol** is a set of rules established for the exchange of data between devices on the Internet. The type and content of protocol imply the information of devices or activities of users [21]. The content protocol could be extracted from the response data of devices when we build and send specific packets to the devices. The public protocol data could be downloaded from Rapid7 (https://www.rapid7.com/).

ip	219.142.226.34
port	554
protocol	rtsp
continent	Asia
country	China
city	Beijing
longitude	116.3889
latitude	39.9288
asn	4847
aso	China Networks Inter-Exchange
rdns	34.226.142.219.broad.bj.bj.dynamic .163data.com.cn
Whois description	Beijing Kuanjie Net communication technology Ltd
device	Video Device

(a) *table T*

219.142.226.34:554 is assigned to Beijing Kuanjie Net communication technology Ltd and operated by China Networks Inter-Exchange, locating in Beijing, China. The protocol deployed on the device is rtsp, therefore the device transfers multimedia data over networks, and the hardware of the device is Video Device. The users use the device by China Telecom because its rdns includes 163data.com.cn.

(b) *description D*

Fig. 3. An example of ip2text, including a table and corresponding description. The underlined texts in the description are the reasoning results by attributes from the table.

- **Location** includes the continent, country, and city of IP addresses. Also, some location of IP addresses is street-level, which exists in the database of AIWEN [22] (https://www.ipplus360.com/).
- **AS** [23] is a small group that autonomously decides which routing protocol should be used in the same system. ASN is the digit number of AS and ASO is the operator of AS. Moreover, **Whois** [24] is a database used to query whether a domain name has been registered and the detailed information of domains, such as register or owner of domains. The Whois and AS databases could be used in ICANN (https://lookup.icann.org/).
- **RDNS** [25] is a method of decomposing an IP address into a domain. A domain is the name of a device on the Internet, composed of a string of texts separated by dots. The RDNS (reversed domain name system) database could also be got in Rapid7 (https://www.rapid7.com/).

Annotation. Due to the insufficient table-to-text datasets about IoT but many IoT devices on the Internet, the paper proposes a semi-automated annotation based on active learning to generate descriptions for ip2text. The primary process of **annotation** is as Fig. 4 shows, which includes: **Templates Rules**, **Active Learning**, and **Experts Annotation**.

Firstly, for the unlabeled samples after data collection of IP addresses, **Templates Rules** builds several templates for each attribute of IP addresses to generate a preliminary description. Secondly, based on labeled and preliminary labeled samples, the table-to-text models, such as BART and T5, are trained, and **Activate Learning** selects the most valuable samples that need to be labeled from the preliminary labeled samples to reduce the labeling burden effectively. Thirdly, **Expert Annotation** further modifies the descriptions of the selected

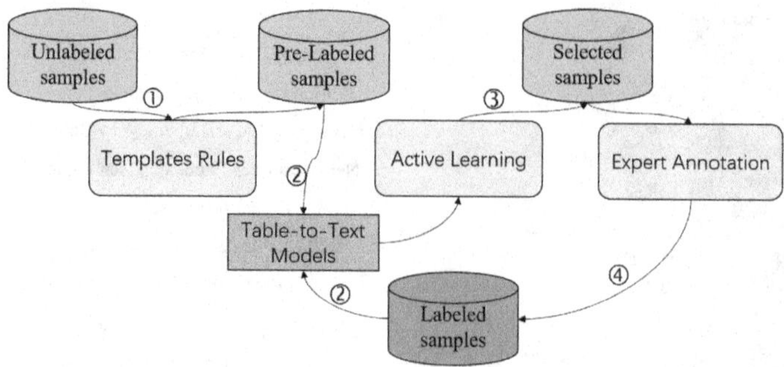

Fig. 4. The main annotation process of ip2text.

samples, paying particular attention to the overall fluency of the sentences and the coverage of reasoning content from the data of IP addresses.

Existing researches [26] claims that an open set of templates is crucial for enriching the phrase constructions and realizing varied generations. So, to get more diverse outputs, **Templates Rules** designs several templates for each attribute of tables about IP addresses. In this paper, we build 15 templates, and Table 1 shows examples of generating descriptions based on different templates. Different templates control different arrangements of descriptions to vary the text generation of IP addresses. There are also some defects in using templates to generate text descriptions directly. (1) Each template can only cover part of the table's contents. When the results of multiple templates are directly put together, the fluency of descriptions is always poor. However, the cost of modifying descriptions one by one is too high. (2) The results of templates hardly contain the reasoning content from attributes of tables. If we have to use templates to mine the reasoning content, a great many rules need to be designed carefully.

Table 1. Examples of generating descriptions based on different templates for IP addresses.

Table	**ip**[219.142.226.34], **port**[554], **protocol**[rtsp], **country**[China], **city**[Beijing]
Template1	[ip]:[port] is a device with [protocol], which locates in [city], [country]
Sentence1	219.142.226.34:554 is a device with RTSP, which locates in Beijing, China
Template2	[ip]:[port] locates in [city], [country]. The protocol deployed on it is [protocol]
Sentence2	219.142.226.34:554 locates in Beijing, China. The protocol deployed on it is RTSP
Table	**ip**[219.142.226.34], **port**[554], **asn**[4847], **aso**[China Network Inter-Exchange], **Whois description**[Beijing Kuanjie Net communication technology Ltd]
Template1	[ip]:[port] is assigned to [Whois description] and operated by [aso]
Sentence1	219.142.226.34:554 is assigned to Beijing Kuanjie Net communication technology Ltd and operated by China Network Inter-Exchange
Template2	Assigned to [Whois description], [ip]:[port] is operated by [aso]
Sentence2	Assigned to Beijing Kuanjie Net communication technology Ltd, 219.142.226.34:554 is operated by China Network Inter-Exchange

Activate Learning [27] attempts to maximize a model's performance gain while annotating the fewest samples. Generally, active learning selects samples that make the model perform poorly for manual annotation. So, this paper selects samples from two aspects:

- **Performance**. This paper chooses **BLEU** to measure the performance of pre-trained BART for the next generation of devices based on their tables. However, for table-to-texts tasks, existing automatic evaluation metrics could not measure the diversity and fluency of descriptions well [28] from multiple template rules. In other words, the description results with high BLEU scores could be the same and not smooth. Thus, this paper manually annotates description results with the highest and lowest BLEU scores, unlike ordinary active learning.
- **Confidence**. The output sentence generated by BART or T5 is predicted token by token. Each token has its confidence, which means certainty about the classification of the token. Further, the confidence of output descriptions of IP addresses is the average of their tokens' confidence. Thus, like ordinary active learning, the paper chooses the samples with less confidence to be labeled again by experts.

Expert Annotation modifies the descriptions of samples selected by **Activate Learning**. The descriptions are generated by table-to-text models trained by labeled samples or pre-trained models using data in other fields. The paper lets two persons label the data independently, and the third person chooses the best one of them as the final annotation result. In the annotation process, experts also add many reasoning contents that cannot be generated by rules easily. For example, "The users use the device by China Telecom because its RDNS includes 163data.com.cn" shown in Fig. 3 is the reasoning content based on the tables of IP addresses.

Statistics. Table 2 explains the characteristic of ip2text that ip2text is the first public table-to-text dataset, which includes 17K pairs and corresponding semi-automated descriptions from experts. Compared with other datasets, ip2text is built by the semi-automated method so that the number of ip2text is more significant and with reasoning content modified and generated by experts.

At the same time, the paper analysis the samples in ip2text in terms of locations and device types. Figure 5 shows the number of devices in different countries, and Fig. 6 shows the number of devices with different types. The figures show the diversity of samples in ip2text: (1) The devices are located in many countries, and there are more samples in China and the United States, accounting for 30% and 24%, respectively. (2) The type of devices is very different. However, Video devices and Routing devices are the top 2 devices in ip2text. Video covers 53% samples, and Routing covers 33% samples in ip2text.

Figure 7 shows the number of three different kinds of labeled samples, including:

Table 2. Comparison of ip2text to recent public table-to-texts datasets. **Pairs** show the number of annotated pairs in each dataset. $|Text|$ reports the average number of words in descriptions. $|Vocab|$ is the length of the corresponding vocabulary in each dataset.

Dataset	Pairs	$\|Text\|$	$\|Vocab\|$	Domain	Annotation	Reasoning
WikiBIO [8]	400K	97	400K	Biography	Automated	No
Rotowire [10]	11K	337	11.3K	Basketball	Automated	Few
LogicNLG [18]	37K	14	122K	Wikipedia	Semi-automated	Yes
ToTTo [19]	136K	17	136K	Wikipedia	Manual	Few
SciGen [9]	1.3K	116	11K	Scientific	Manual	Yes
	18K	124	54K		Semi-automated	Yes
	53K	133	127K		Semi-automated	Yes
ip2text	17K	56	132K	IoT	Semi-automated	Yes

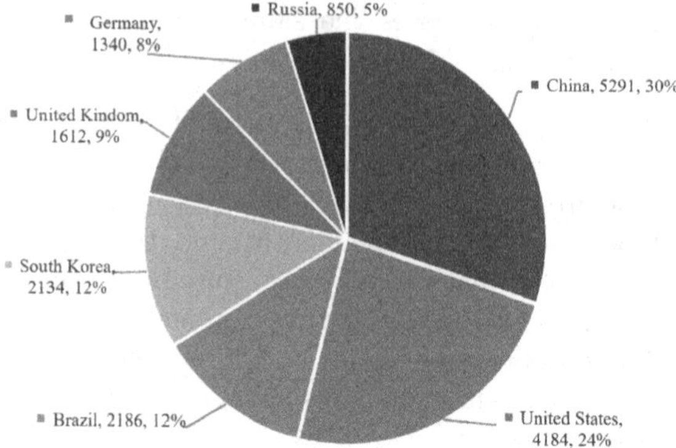

Fig. 5. The number of devices in different countries.

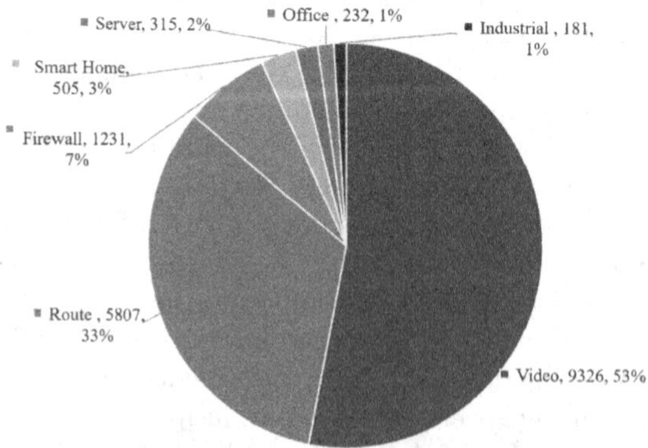

Fig. 6. The number of different device types.

- **manual label** means the samples are labeled by experts or accumulated in previous work.
- **automate label** means the samples are labeled by pre-trained and fine-tuned model BART. The annotation quality may be relatively poor, but also close to reality.
- **modified label** means the samples are labeled by models and selected by active learning. So, the annotation quality is the same as **manual label**, but the overall annotation work will be reduced by **modified label**.

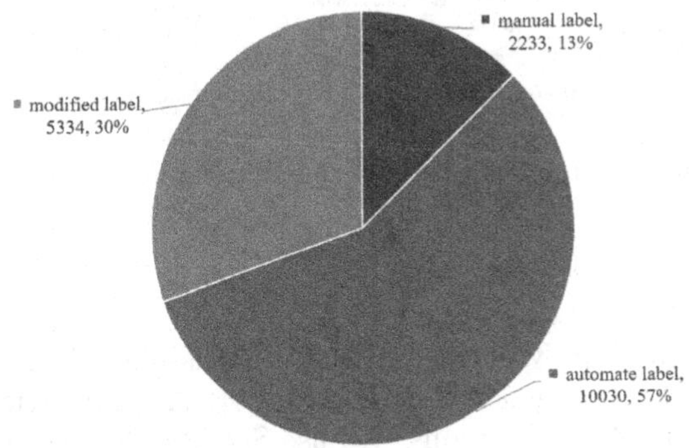

Fig. 7. The number of three different kinds of labeled samples.

4 Experiments

4.1 Settings

To use text-to-text generation baselines, the paper converts an input table T into a text sequence as representation R. In order to preserve the structure of the table, the paper uses unique tokens to specify the beginning of rows "⟨R⟩", cells "⟨C⟩", as defined in [29]. Figure 8 shows an input table T with its corresponding input representation R. After getting the representation of tables, the paper naturally transforms the problem of table-to-text generation of devices into a typical seq2seq task.

4.2 Baselines

Followed by the results of [30], there are two famous seq2seq models: BART [11] and T5 [12] pre-trained language models, consistently outperform recent specialized data-to-text models on various benchmarks. The paper studies the effectiveness of these two models on ip2text to evaluate the ability to generate descriptions of devices for IoT engines based on ip2text. For BART, the paper uses the model structures from HuggingFace, including Bart-base and Bart-large. For T5, the paper uses the t5-small, t5-base, and t5-large, respectively.

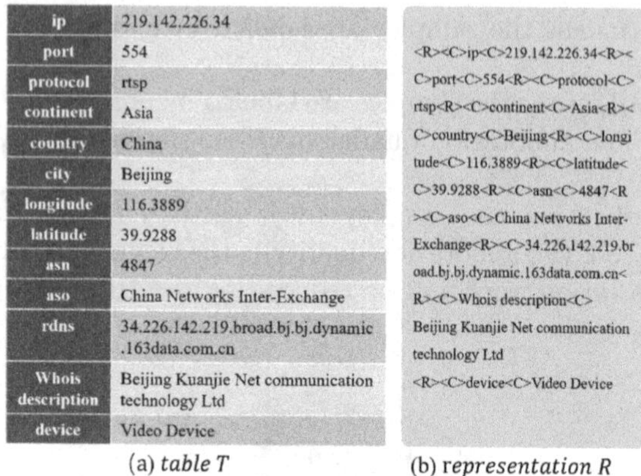

ip	219.142.226.34
port	554
protocol	rtsp
continent	Asia
country	China
city	Beijing
longitude	116.3889
latitude	39.9288
asn	4847
aso	China Networks Inter-Exchange
rdns	34.226.142.219.broad.bj.bj.dynamic .163data.com.cn
Whois description	Beijing Kuanjie Net communication technology Ltd
device	Video Device

(a) *table T*

(b) r*epresentation R*

Fig. 8. Sample table T with its corresponding input representation R.

4.3 Evaluation

Automatic Evaluation Metrics. BLEU [31] is one of the most common evaluation metrics for text generation. It computes the geometric average of the precision over output text's n-grams. We use SacreBLEU [32], which produces comparable and reproducible BLEU scores.

ROUGE [33] is also a frequently-used text generation evaluation metric like BLEU. However, ROUGE computes the geometric average of the precision over output texts' n-grams.

METEOR [34] aligns the output texts to the reference texts and calculates sentence-level similarity scores for the alignments.

BertScore [35] uses BERT embeddings and matches words in the output and reference sentences by cosine similarity.

MoverScore [36] computes the distance between the contextual representation of the output and reference texts. It captures the amount of shared content between two texts and how much the output texts deviate from the references. It uses BERT embeddings for computing contextualized representations.

BLEURT [37] is a learned evaluation metric based on BERT. It is first pre-trained on synthetic examples and then fine-tuned on human judgments for machine translation.

Human Evaluation Metrics. The Automatic Evaluation Metrics mostly measure the surface similarity of generated descriptions to labeled descriptions and could not evaluate the factual content of generated descriptions for corresponding input tables. For example, the Automatic Evaluation Metrics of Template1 and Template2 in Table 1 are relatively low. So, this paper evaluates the generated descriptions manually using Human Evaluation Metrics. Due to the table formats of devices, the paper designs specific metrics as follows:

- **Correct** is the ratio of words in output texts, which describes attributes in the table correctly.
- **Incorrect** is the ratio of words in output texts, which describes attributes in the table incorrectly.
- **Reasoning** is the ratio of words in output texts, which do not describe attributes in the table but show some facts inferred by the table.
- **Hallucination** is the ratio of words in output texts which do not describe attributes in the table and show the same facts not related to the table.

4.4 Results

Automatic Evaluation. Table 3 shows the results of baselines on ip2texts for table-to-text generation of devices using the evaluation metrics in Sect. 4.3. This paper reports the baselines' results from three aspects: (1) **Transfer learning** uses the pre-trained models directly to get the descriptions of devices. (2) **Fine tuning** uses ip2text to train the models based on the pre-trained parameters. (3) **Retraining** only uses ip2text to train the models.

Table 3. Automatic Evaluation of Baselines on ip2text

Settings	Models	BLEU	ROUGE	METEOR	BertScore	MoverScore	BLEURT
Transfer learning	Bart-base	0.63	0.11	0.07	0.71	0.31	−1.22
	Bart-large	1.06	0.23	0.12	0.73	0.32	−1.07
	t5-small	0.61	0.08	0.06	0.69	0.34	−1.25
	t5-base	0.86	0.13	0.09	0.71	0.36	−1.22
	t5-large	0.93	0.15	0.14	0.77	0.39	−1.27
Fine tuning	Bart-base	0.99	0.21	0.17	0.76	0.41	−1.19
	Bart-large	1.28	0.31	0.21	0.78	0.44	−1.08
	t5-small	1.02	0.25	0.20	0.75	0.41	−1.13
	t5-base	1.39	0.41	0.28	0.79	0.43	−1.01
	t5-large	1.43	0.49	0.33	0.81	0.46	−0.97
Retraining	Bart-base	2.55	0.49	0.33	0.83	0.55	−0.91
	Bart-large	3.02	0.53	0.37	0.85	0.59	−0.86
	t5-small	2.67	0.52	0.45	0.85	0.61	−0.88
	t5-base	2.99	0.52	0.47	0.87	0.65	−0.83
	t5-large	3.14	0.55	0.49	0.88	0.67	−0.79

The results shown in Table 3 reveal that:

- Existing pre-trained baselines could not be easy to perform satisfactorily on ip2text. The results of baselines in Transfer learning are much lower than those in Fine tuning and Retraining. For example, BLEU between generated descriptions and annotations is almost all less than 1. That may be because the characteristics of the corpus used by the existing pre-trained models are quite different from ip2text, resulting in poor migration of existing pre-trained models for table-to-text generation devices on the Internet.

- BertScore values are very high for all the experiments, while, as we could see in Table 3, other metrics are relatively low. Therefore, there needs to be an explicit agreement between the rankings of different metrics for the baseline models. In other words, some metrics, such as BLEU, are too strict, and other metrics, such as BertScore, are too loose for table-to-text generation of devices.
- According to automatic evaluation, the pre-trained Bart-large performs better than the other models, with BLEU 1.06. However, in the fine-tuning and retraining of baselines, the t5-large performs best, with BLEU 1.43 and 3.14, respectively. Thus, the familiar and public corpus used for pre-trained models is much different from ip2text, so in the current situation, fine-tuning and retraining on ip2text is essential for the table-to-text generation of devices.

Human Evaluation. For human evaluation, the paper randomly selects 1000 table-description pairs from ip2test and their corresponding generated descriptions from the retraining bart-large and t5-large. Then the manual metrics are calculated, and the results are as Table 4 shows.

Table 4. Human Evaluation of Baselines on ip2text.

Model	Correct	Incorrect	Reasoning	Hallucination
Bart-large	0.39	0.33	0.15	0.13
t5-large	0.42	0.31	0.12	0.15

The results show that t5-large receives a higher score, about 0.42 Correct, compared to bart-large based on all human evaluation metrics for the task of ip2text. At the same time, bart-large receives a higher score for the reasoning of ip2text, about 0.15 Reasoning, which means bart-large has a stronger reasoning ability to extract hidden content from the tables. However, the Incorrect of both models are not low, more than 0.30, and the Hallucination of both models are more than 0.10, making it difficult for the industrialized use of table-to-text generation devices on the Internet. The reasons may be as follows: (1) the distribution of ip2text is worldwide, located in many countries, as Fig. 5 shows. (2) the device type in ip2text are different and diversified, as Fig. 6 shows. (3) the descriptions of ip2text include some but not a few reasoning contents. (4) Bart and T5 need lots of data to train well and are prone to hallucinations while reasoning. Therefore, achieving state-of-the-art performance based on the existing and mainstream seq2seq models is difficult.

5 Conclusion

This paper introduces ip2text, a challenging dataset for reasoning-aware table-to-text generation of devices on the Internet. The input data in ip2text are tables, which contain attributes of devices collected from the Internet and their corresponding descriptions. Generating descriptions of devices is costly, time-consuming, and does not scale to Internet-scale data. To tackle this problem, this paper designs an annotation method based on active learning according to the data characteristics of devices and studies the performance of existing and typical state-of-the-art models for table-to-text generation on ip2text. The results show that existing pre-trained baselines, including BART and T5, could not be easy to get a satisfactory performance on ip2text, so in the current situation, fine-tuning and retraining on ip2text is essential for the table-to-text generation of devices. Further, the human evaluation shows Bart and T5 are prone to hallucinations while reasoning with Hallucination more than 0.10. Therefore, it is not easy to achieve satisfactory performance using the existing and mainstream seq2seq models based on the reasoning-aware ip2text. In the future, with the improvement of the seq2seq models to fit the ip2text proposed in this paper, we could achieve state-of-the-art performance and make it practical for the table-to-text generation of devices on the Internet.

References

1. Matherly, J.: Complete guide to Shodan. Shodan, LLC (2016–02-25), vol. 1 (2015)
2. Li, R., Shen, M., Yu, H., Li, C., Duan, P., Zhu, L.: A survey on cyberspace search engines. In: Lu, W., et al. (eds.) CNCERT 2020. CCIS, vol. 1299, pp. 206–214. Springer, Singapore (2020). https://doi.org/10.1007/978-981-33-4922-3_15
3. Ackley, D., Yang, H.: Exploration of smart grid device cybersecurity vulnerability using Shodan. In: 2020 IEEE Power & Energy Society General Meeting (PESGM) (2020)
4. Novianto, B., Suryanto, Y., Ramli, K.: Vulnerability analysis of internet devices from Indonesia based on exposure data in Shodan. In: IOP Conference Series: Materials Science and Engineering, vol. 1115, no. 1, p. 012045 (9pp) (2021)
5. Belz, A.: Automatic generation of weather forecast texts using comprehensive probabilistic generation-space models. Nat. Lang. Eng. **14**(4), 431–455 (2008)
6. Chen, D.L., Mooney, R.J.: Learning to sportscast: a test of grounded language acquisition. In: Proceedings of the 25th International Conference on Machine Learning, pp. 128–135 (2008)
7. Dušek, O., Novikova, J., Rieser, V.: Evaluating the state-of-the-art of end-to-end natural language generation: the E2E NLG challenge. Comput. Speech Lang. **59**, 123–156 (2020)
8. Lebret, R, Grangier, D., Auli, M.: Neural text generation from structured data with application to the biography domain. arXiv preprint arXiv:1603.07771 (2016)
9. Moosavi, N.S., Rücklé, A., Roth, D., Gurevych, I.: SciGen: a dataset for reasoning-aware text generation from scientific tables. In: Thirty-Fifth Conference on Neural Information Processing Systems Datasets and Benchmarks Track (Round 2) (2021)
10. Wiseman, S., Shieber, S.M., Rush, A.M.: Challenges in data-to-document generation. In: Proceedings of the 2017 Conference on Empirical Methods in Natural Language Processing (2017)
11. Lewis, M., et al.: BART: denoising sequence-to-sequence pre-training for natural language generation, translation, and comprehension. In: Meeting of the Association for Computational Linguistics (2020)

12. Raffel, C., et al.: Exploring the limits of transfer learning with a unified text-to-text transformer. J. Mach. Learn. Res. **21**(140), 1–67 (2020)
13. Liu, T., Wang, K., Sha, L., Chang, B, Sui, Z.: Table-to-text generation by structure-aware seq2seq learning. In: Thirty-Second AAAI Conference on Artificial Intelligence (2018)
14. Nan, L., et al.: DART: open-domain structured data record to text generation. In: Proceedings of the 2021 Conference of the North American Chapter of the Association for Computational Linguistics: Human Language Technologies, pp. 432–447. Association for Computational Linguistics, Online (2021)
15. Liu, T., Luo, F., Xia, Q., Ma, S., Chang, B., Sui, Z.: Hierarchical encoder with auxiliary supervision for neural table-to-text generation: learning better representation for tables. In: Proceedings of the Thirty-Third AAAI Conference on Artificial Intelligence and Thirty-First Innovative Applications of Artificial Intelligence Conference and Ninth AAAI Symposium on Educational Advances in Artificial Intelligence, pp. 6786–6793 (2019)
16. Wang, Q., et al.: Describing a knowledge base. arXiv preprint arXiv:1809.01797 (2018)
17. Chen, Z., et al.: Logic2text: high-fidelity natural language generation from logical forms. arXiv preprint arXiv:2004.14579 (2020)
18. Chen, W., et al.: TabFact: a large-scale dataset for table-based fact verification. arXiv preprint arXiv:1909.02164 (2019)
19. Parikh, A.P., et al.: ToTTo: a controlled table-to-text generation dataset. arXiv preprint arXiv:2004.14373 (2020)
20. Luo, Y., Chen, X., Ge, N., Lu, J.: Deep learning based device classification method for safeguarding internet of things. In: 2021 IEEE Global Communications Conference (GLOBECOM), pp. 1–6. IEEE (2021)
21. Wan, Y., Xu, K., Wang, F., Xue, G.: IoTMosaic: inferring user activities from IoT network traffic in smart homes. In: IEEE INFOCOM 2022-IEEE Conference on Computer Communications, pp. 370–379. IEEE (2022)
22. Wang, Y., Burgener, D., Flores, M., Kuzmanovic, A., Huang, C.: Towards street-level IP geolocation. In: 8th USENIX Symposium on Networked Systems Design and Implementation (NSDI 2011) (2011)
23. Paiva, T.B., Siqueira, Y., Batista, D.M., Hirata, R., Terada, R.: BGP anomalies classification using features based on as relationship graphs. In: 2021 IEEE Latin-American Conference on Communications (LATINCOM), pp. 1–6. IEEE (2021)
24. Lu, C., et al.: From WHOIS to WHOWAS: a large-scale measurement study of domain registration privacy under the GDPR. In: NDSS (2021)
25. Fiebig, T., Borgolte, K., Hao, S., Kruegel, C., Vigna, G., Feldmann, A.: In rDNS we trust: revisiting a common data-source's reliability. In: Beverly, R., Smaragdakis, G., Feldmann, A. (eds.) PAM 2018. LNCS, vol. 10771, pp. 131–145. Springer, Cham (2018). https://doi.org/10.1007/978-3-319-76481-8_10
26. Ye, R., Shi, W., Zhou, H., Wei, Z., Li, L.: Variational template machine for data-to-text generation. arXiv preprint arXiv:2002.01127 (2020)
27. Ren, P., et al.: A survey of deep active learning. ACM Comput. Surv. (CSUR) **54**(9), 1–40 (2021)
28. Alihosseini, D., Montahaei, E., Baghshah, M.S.: Jointly measuring diversity and quality in text generation models. In: Proceedings of the Workshop on Methods for Optimizing and Evaluating Neural Language Generation, pp. 90–98 (2019)

29. Voita, E., Sennrich, R., Titov, I.: When a good translation is wrong in context: context-aware machine translation improves on deixis, ellipsis, and lexical cohesion. In: Proceedings of the 57th Annual Meeting of the Association for Computational Linguistics, Florence, Italy, pp. 1198–1212. Association for Computational Linguistics (2019)
30. Ribeiro, L.F.R., Schmitt, M, Schütze, H., Gurevych, I.: Investigating pretrained language models for graph-to-text generation. arXiv preprint arXiv:2007.08426 (2020)
31. Papineni, K., Roukos, S., Ward, T., Zhu, W.-J.: Bleu: a method for automatic evaluation of machine translation. In: Proceedings of the 40th Annual Meeting of the Association for Computational Linguistics, pp. 311–318 (2002)
32. Post, M.: A call for clarity in reporting bleu scores. arXiv preprint arXiv:1804.08771 (2018)
33. Lin, C.-Y.: Rouge: a package for automatic evaluation of summaries. In: Text Summarization Branches Out, pp. 74–81 (2004)
34. Denkowski, M., Lavie, A.: Meteor universal: language specific translation evaluation for any target language. In: Proceedings of the Ninth Workshop on Statistical Machine Translation, pp. 376–380 (2014)
35. Zhang, T., Kishore, V., Wu, F., Weinberger, K.Q., Artzi, Y.: BERTScore: evaluating text generation with BERT. arXiv preprint arXiv:1904.09675 (2019)
36. Zhao, W., Peyrard, M., Liu, F., Gao, Y., Meyer, C.M., Eger, S.: MoverScore: text generation evaluating with contextualized embeddings and earth mover distance. arXiv preprint arXiv:1909.02622 (2019)
37. Sellam, T., Das, D., Parikh, A.P.: BLEURT: learning robust metrics for text generation. arXiv preprint arXiv:2004.04696 (2020)

Spatio-Textual Group Skyline Query

Mengmeng Sun, Yiping Teng(✉), Fanyou Zhao, Jiawei Qi, Dongyue Jiang, and Chunlong Fan

School of Computer, Large-Scale Distributed System Laboratory, Shenyang Aerospace University, Shenyang, China
{typ,fanchl}@sau.edu.cn,
{zhaofanyou,qijiawei,jiangdongyue}@stu.sau.edu.cn

Abstract. With the development of mobile Internet and location-based services, through spatial keyword queries users expect to retrieve the skyline groups of interested places based on their locations and individual preferences. Existing studies on spatial keyword queries focus on finding either single spatio-textual objects or top-k groups of spatio-textual objects, however few can directly solve the problem of spatio-textual group skyline queries. In this paper, we define and study the spatio-textual group skyline query problem and propose two spatio-textual group skyline query methods. In the basic method, we present a group dominance computation algorithm to calculate the spatio-textual group dominance of the generated candidate groups, with which the skyline groups w.r.t. the query group that cannot be dominated by other groups can be returned as the query results. To improve the efficiency of the basic method, we further propose an index-based method adapting the query processing in an IR-Tree index, and present a pruning algorithm based on the spatio-textual similarities over index nodes, which results in an efficient query processing without generating candidates by scanning the whole dataset. Analysis shows the validity and complexity of the proposed methods, and experimental results on the real and synthetic datasets illustrate the performance of our proposed methods.

Keywords: Spatio-textual data · Skyline groups · Spatio-textual group dominance

1 Introduction

With the rapid popularization of mobile devices and applications, the integration of geographical locations and textual descriptions motivates the users of mobile Internet applications to retrieve Points Of Interests (POIs) based on geographical proximity and textual relevance [1–3]. For example, users on Facebook can search interested places based on locations and textual description.

The skyline query [4–6] is a significant query type in the database field, which can be applied in preference analysis, multi-criteria decision making applications and so on. As an important variants among the skyline queries, taking a group in a given size of multi-dimensional tuples as the basic unit, the skyline groups query is enabled to find the groups from the tuples that cannot be dominated

A. El Abbadi et al. (Eds.): DASFAA 2023 Workshops, LNCS 13922, pp. 34–50, 2023.
https://doi.org/10.1007/978-3-031-35415-1_3

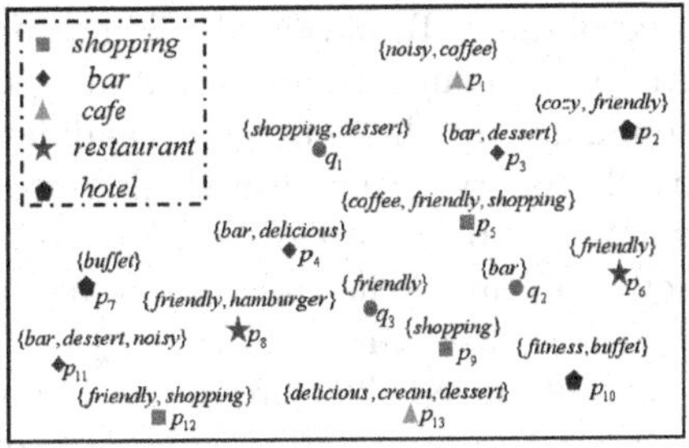

Fig. 1. Example of spatio-textual group skyline query

by any other groups in the same size, of which the problem was first defined in [7,8], and deeply studied in many existing works [7–10]. However, since the above group skyline queries are studied in the area of multi-dimensional data queries, in the location-based service scenario, with the locations and preferences, when a group of users expects to find a group of places which satisfies the constrains of the space and interests, the existing group skyline query method can hardly support such spatio-textual queries of groups in an efficient way.

Different from the traditional skyline query, the spatio-textual group skyline query is oriented to a set of query requests that expresses different query preferences. Many applications in real world scenarios focus on finding better groups of objects. Figure 1 illustrates an example of spatio-textual group skyline query, where a group of three users $\{q_1, q_2, q_3\}$ is decide go shopping, have a lunch and find a bar to have a drink. Each user has his/her location and the interests represented by a set of keywords such as {shopping, dessert}, {bar} and {friendly}. There is a set of spatio-textual points $\{p_1, p_2, p_3, ..., p_{13}\}$, of which each has a location and a set of keywords, e.g., {noisy, coffee}, {cozy, friendly} ,..., {fitness, buffer}. Considering the example in Fig. 1, the candidate groups can be $\{p_3, p_5, p_6\}$, $\{p_3, p_5, p_8\}$, $\{p_4, p_5, p_8\}$ and $\{p_8, p_{11}, p_{12}\}$. The spatio-textual skyline groups are that can not be spatio-textually dominated by other groups among all the candidate groups. Existing studies on spatial keyword queries focus on retrieving either single spatio-textual objects [11–15] or top-k groups of spatio-textual objects [16–18] based on the spatial proximity and textual similarity, of which few can directly solve the querying problem of spatio-textual skyline groups, and thus it calls for an effective group query method to the search for skyline groups of spatio-textual objects.

In this paper, we first define and study the spatio-textual group skyline query problem and propose the spatio-textual group skyline query methods. To address this problem, we first propose a straight-forward query processing method as the Basic Spatio-Textual Group Skyline Query (BSTGSQ) method to find the skyline groups. In BSTGSQ method, we first preprocess the spatio-textual data to filter out the objects containing the query keywords and generate a processed dataset. To generate candidate groups, we guarantee that each candidate group covers all query keywords and the group size is equal to the number of the query locations in the processed dataset. To obtain the dominance relations between groups, we present a group dominance computation algorithm to calculate the spatio-textual group dominance of the generated candidate groups. With the group dominance w.r.t. the query group, the skyline groups that cannot be dominated by other groups can be returned to the users as the query results. To address the efficiency problem in the basic query method, we further propose an Index-based Spatio-Textual Group Skyline Query (ISTGSQ) method. In ISTGSQ method, we adapt the STGSQ processing in an IR-Tree index [11], which is widely applied in existing studies on spatial keyword queries [1,13,19]. To effectively narrow down the candidates groups, we further propose a pruning algorithm based on the spatio-textual similarities based on index nodes to remove objects within the nodes that cannot be involved in the skyline groups. Through the pruning algorithm, the candidate group can be generated without scanning the whole dataset, and thus the skyline groups can be retrieved more efficiently compared to BSTGSQ method. Thorough analysis shows the validity and complexity of the proposed query methods. Besides, extensive experimental results on real datasets and synthetic datasets illustrate the performance of our proposed methods. We summarize our contributions of this paper as follows.

- To the best of our knowledge, this is the first attempt to define and study the spatio-textual group skyline query, in which dynamic queries are supported with respect to a given query group.
- To address this problem, we first propose a Basic Spatio-Textual Group Skyline Query (BSTGSQ) method, in which a group dominance computation algorithm is presented to calculate the spatio-textual group dominance of the generated candidate groups.
- To improve the query efficiency, we further propose an Index-based Spatio-Textual Group Skyline Query (ISTGSQ) method, in which the query processing is adapted in an IR-Tree index and a pruning algorithm based on the spatio-textual similarities over index nodes is presented.
- We provide thorough analysis to show the validity and complexity of the proposed query methods, and conduct extensive experiments on real and synthetic datasets to evaluate the performance of our proposed methods.

The rest of the paper is organized as follows. Section 2 introduces the related work. Section 3 states the problem definition and preliminaries in this work. Section 4 introduces the two proposed spatio-textual group skyline query method and analyzes validity and computational complexity of them. In Sect. 5 reports the experimental results and Sect. 6 concludes the paper.

2 Related Work

Spatio-textual data queries have been widely studied these years, where rich query semantics with various indices are developed including top-k spatial keyword queries [11,12,14,15], joint top-k spatial keyword queries [20], moving spatial keyword queries [21,22], etc. Shi et al. [13], as the first attempt, presented a spatio-textual skyline query processing scheme by defining the spatio-textual dominance, which, however, cannot directly solve the spatio-textual group skyline querying problem. In recent years, the skyline groups queries have been studied in many previous works [7–10], while few of them can retrieve the skyline groups over spatial-textual data. As the related work, we survey the studies on skyline groups queries and spatial keyword queries as follows.

2.1 Skyline Groups Query

In order to query the optimal groups, two types of skyline group query methods are proposed. In the first type, the dominance of groups is defined using the traditional dominance relationships between aggregate points of all the groups [7,8,23]. By selecting one of aggregate functions, such as SUM, MIN and MAX, the aggregate points of all the groups are calculated, and traditional dominance tests are conducted on these aggregate points, such that if an aggregate point is a skyline point, the corresponding group is a skyline group. As stated in [9], it is hard to choose an appropriate aggregate function for users. More importantly, it may neglect some representative groups, which cannot be captured by the aggregate function. Thus, in the another type, Liu et al. [9,10] proposed the definition of G-Skyline aiming to find all pareto optimal groups which are not dominated by any other group with the same size. They developed a two-step algorithm to accelerate G-Skyline query processing by constructing the directed skyline graph (DSG) to maintain the dominance relationships among the first k skyline layers and presenting two heuristic algorithms, point-wise and unit group-wise algorithms, to identify all G-Skyline groups by employing DSG. Zhu et al. [24,25] proposed a top-k dominance query on skyline groups to return k skyline groups that dominate the maximum number of points in a given dataset. The efficiency of G-Skyline computation has been further improved in [26,27].

2.2 Spatial Keyword Query

Combining the spatial query and keyword search, the spatio keyword query is to find the results of the objects that satisfy both spatial proximity and textual similarity w.r.t. the query locations and query keywords respectively. Zhou et al. [28] fused the inverted list and R^*-tree to construct a hybrid index structure to implement spatial and text query algorithms. De Pelipe et al. [29] proposed a hybrid index structure called IR^2-tree, which implemented Boolean spatio-textual query based on R-tree and text signature technology. Cao et al. [30] studied the collective spatial keyword querying problem, which aimed to find a set of spatial data points that satisfied the textual constrains with the smallest internal distance. Li et al. [19] proposed a spatio-textual data query method based on an IR-tree index, which associates each R-tree node with its corresponding inverted linked list. Based on the weighted calculation of the spatial distance and textual similarity, the query method can support the ranking query on spatio-textual data. Wu et al. [1] proposed a new indexing framework for top-k spatio-textual retrieval. The framework leverages the inverted file for text retrieval and the R-tree for spatial proximity querying. Wu et al. [20] studied the joint processing of multiple top-k spatial keyword queries, which aimed to find the object close to the query location that covers all the keywords specified in the query. They proposed an index structure called WIR-tree, which partitions objects into multiple groups such that each group shares a few keywords as possible. Qian et al. [14] studied the problem of semantics-based spatial keyword queries and returned top-k objects most similar to the query. Zhang et al. [31] studied answering why-not questions on top-k spatial keyword queries over moving objects. The query returned the top-k objects, moving or static, based on a ranking function that considered the spatial distance and textual similarity between the query and objects. There is also a lot of research focused on collective keyword query. Cao et al. [16] defined the problem of retrieving a group of spatio-textual objects such that the group's keywords cover the query's keywords and such that the objects are nearest to the query location and have the smallest introbject distances. Jin et al. [32] proposed a novel way of searching on a spatial knowledge, namely collective spatial keyword query on a knowledge base (CoSKQ-KB). Zhao et al. [33] proposed a social-aware spatial keyword top-k group query problem which aimed to retrieve a set of k groups of POI objects that satisfy the preferences of multiple users, taking into account spatial proximity, social relevance, and keyword constraints.

3 Problem Definitions and Preliminaries

In this section, we first introduce the definitions of spatio-textual data, spatio-textual similarity and spatio-textual dominance respectively. Then, we define the spatio-textual group dominance and spatio-textual group skyline query. At last, we state some necessary preliminaries in this paper.

3.1 Problem Definitions

Definition 1 **(Spatio-Textual Data).** Let P be a spatio-textual dataset, denoted as $P = \{p_1, p_2, ..., p_n\}$, which contains n spatio-textual objects. Each spatio-textual object p_i in P is represented as $p_i = (p_i.loc, p_i.doc)$, where $p_i.loc$ is the spatial descriptor (generally expressed as coordinates $\langle x_i, y_i \rangle$), and $p_i.doc$ is the textual description. $p_i.doc$ can be usually denoted by a set of keywords $\{t_1, t_2, ..., t_a\}$.

Definition 2 **(Spatio-Textual Group Skyline Query Request).** Let Q be an spatio-textual group skyline query request denoted as $Q = \{q_1, q_2, ..., q_m\}$, which contains m sub-requests. Each sub-request q_j in Q is represented as $q_j = (q_j.loc, q_j.doc)$, where $q_j.loc$ is the location of q_j, and $q_j.doc$ can be denoted as a set of query keywords $\{u_1, u_2, ..., u_b\}$. We need to satisfy all sub-requests, which means the group size is equal to the number of the sub-request locations and the skyline group contains all keywords.

Definition 3 **(Spatio-Textual Similarity)** [13]. Given a sub-request q, let $st(p,q)$ be the spatio-textual similarity between p and q, which can be evaluated through combining the spatial distance and textual relevancy are generally defined as follows,

$$st(p,q) = \alpha d(p.loc, q.loc) + (1-\alpha)(1 - w(p.doc, q.doc)), \quad (1)$$

where α is a balance factor to express users' preference on spatial distances and text relevancy for $\alpha \in [0,1]$. $d(p.loc, q.loc))$ represents a normalized Euclidean distance between $p.loc$ and $q.loc$ in two-dimensional space denoted as follows,

$$d(p.loc, q.loc) = \frac{\sqrt{(x_p - x_q)^2 + (y_p - y_q)^2}}{d_{max}}, \quad (2)$$

where the Euclidean distance is normalized by the maximum distance d_{max} between objects in P. $\omega(p.doc, q.doc)$ represents a normalized score of text relevancy between $p.doc$ and $q.doc$, which can be defined as follows,

$$\omega(p.doc, q.doc) = \frac{\Pi_{\varepsilon_i \in Q} \omega(\varepsilon_i, p_i.doc)}{\omega_{max}}, \quad (3)$$

where $\omega(p.doc, q.doc)$ stands for textual relevance between $p.doc$ and $q.doc$, which can be calculated following the language model in [34], w_{max} is a normalization factor for textual relevance, $\omega(\varepsilon_i, p_i.doc)$ shows that, if ε_i appears in $p_i.doc$, then $\omega(\varepsilon_i, p_i.doc)$ is the weight of ε_i in $p_i.doc$, otherwise, $w(p.doc, q.doc)$ is a smoothing factor (for example, 0.02).

Definition 4 **(Spatio-Textual Dominance)** [13]. Given two spatio-textual objects p_i and p_j with query request Q, p_j is dominated by p_i spatio-textually, if $st(p_j, q_k) \geq st(p_i, q_k)$,for all q_k in Q, and there exists at least one q_k in Q satisfying $st(p_j, q_k) > st(p_i, q_k)$.

Definition 5 **(Spatio-Textual Group Dominance).** Given spatio-textual dataset P and query request Q, Let $G = \{p_1, p_2, ..., p_s\}$ and $G' = \{p'_1, p'_2, ..., p'_s\}$ be two different groups with s points of P, calculate the spatio-textual dominance relationship between each data object and the query point in the group separately, such that $st(p_i, q_j) \geq st(p'_i, q_j)$, for all i $(1 \leq i \leq s)$, j $(1 \leq j \leq m)$ and $st(p_i, q_j) > st(p'_i, q_j)$ for at least one i, we say group G is dominated by G', denoted by $G' \prec G$.

Definition 6 **(Spatio-Textual Group Skyline Query).** Given spatio-textual dataset P and query request Q, a spatio-textual group skyline query (STGSQ) is to retrieve a set of groups consisting of spatio-textual objects, which cannot be dominated by any other groups.

3.2 Preliminaries

IR-Tree. Facilitating efficient search based on both spatial distance and textual relevance, the IR-tree [11] is an effective hierarchical spatio-textual index.

Figure 2 illustrates the basic structure of an exemplary IR-tree. Each leaf node in the IR-tree contains entries of the form $(p, p.\lambda, p.di)$, where p represents objects p in the dataset P, $p.\lambda$ represents the bounding rectangle of p, and $p.di$ is an identifier of the description of p. Each leaf node also contains a pointer to an inverted file, and the object keywords for the inverted file is stored in the node. An inverted file has two main components: a vocabulary of all the different terms that appear in the object description and a posting list for each term. Each non-leaf node R in the IR-tree contains entries of the form $(cp, rect, cp.di)$, where cp is the address of a child node of R, $rect$ is the Minimum Bounding Rectangle (MBR) of all rectangles for child node entries and $cp.di$ is a pseudotext description identifier which is the union of all text descriptions of the child node. For the limitation of space, the details of IR-Tree can be referred to [11].

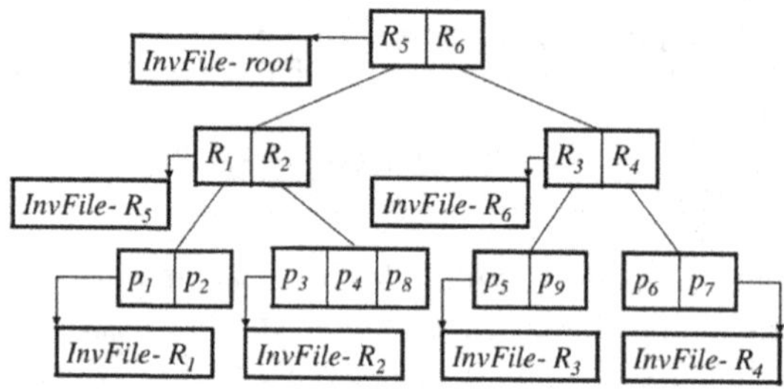

Fig. 2. IR-Tree

4 Approach

In this section, we first present a straight-forward query processing method for STGSQ named Basic Spatio-Textual Group Skyline Query (BSTGSQ) method. Then, to improve the efficiency of STGSQ, we further propose index-based query processing method named Index-based Spatio-Textual Group Skyline Query (ISTGSQ) method over the IR-Tree index. At last, we analyze the validity and computational complexity of the proposed methods.

4.1 Basic Spatio-Textual Group Skyline Query

Data Processing Based on Query Keywords. In order to narrow down the candidates, based on the query keywords, we first need to filter out the spatio-textual data objects p containing no query keywords from the large dataset which cannot be the objects of the skyline groups. Accordingly, since each q in the STGSQ request Q is independent, we can construct the union of the query keywords for all q in Q and filter the objects in P of which the keyword set and the union of the query keywords have an empty intersection. Next, using the filtered dataset, we use tf-idf weighing scheme [35] to calculate the weight of each keyword of objects, and then we can get the textual relevance between p and q. After the processing, the results are shown in Table 1 as an example. Finally, For the processed dataset, the C_n^m candidate groups can be enumerated and the spatio-textual group dominance of the candidate groups w.r.t. the query request can be computed.

Table 1. Weight of Each Keyword of Objects

ID	noisy	coffee	cozy	friendly	bar	dessert	delicious	...
p_1	0.349	0.349	0	0	0	0	0	
p_2	0	0	0.500	0.151	0	0	0	
p_3	0	0	0	0	0.261	0.261	0	
p_4	0	0	0	0	0.261	0	0.349	
p_5	0	0.233	0	0.100	0	0	0	
p_6	0	0	0	0.301	0	0	0	
...								

Group Dominance Computation. To find the skyline groups, we propose a group dominance computation algorithm. For two groups G_1 and G_2 with m objects respectively, calculate $st(p, q)$ with each p in each group (G_1 and G_2) and each $q \in Q$, then choose the maximum similarity denoted by st_{max} and the minimum similarity denoted by st_{min} for each group separately. We use $markset$ to store the minimum value of $\Sigma_{p \in G, q \in Q}(st(p, q)_{max} + st(p, q)_{min})$ so far. If the st_{max} in G_1 is smaller than st_{min} in G_2, that is $st_{max} < st_{min}$, output G_1 to the result list, otherwise, delete G_1 from the candidate groups. The algorithm is shown in Algorithm 1.

Algorithm 1: Group Dominance Comparison Algorithm

Input: Candidate Skyline Group G_i (i=1, ..., C_n^m)
Output: ResultSet of Skyline Group List

List=∅;
for $i = 1$ **to** C_n^m **do**
    Ascending by $\Sigma_{p \in G_i, q \in Q}(st(p, q)_{max} + st(p, q)_{min})$;
    Choose the first $\Sigma_{p \in G_i, q \in Q}(st(p, q)_{max} + st(p, q)_{min})$ as a skyline group;
    Add corresponding group to List;
    Delete the group dominated groups in candidate group;
    Delete $\Sigma_{p \in G_i, q \in Q}(st(p, q)_{max} + st(p, q)_{min})$ from $markset$;
Return List;

Query Processing. In the query processing step, we use the already processed dataset which contains the query keywords of STGSQ request to enumerate $C_{n'}^{m}$ candidate groups. To satisfy all keywords of the STGSQ request for the data objects within the candidate groups, we delete groups that do not contain the union of the query keywords. In order to compute the dominance between groups, the Group Dominance Computation Algorithm can be performed which is shown in Algorithm 1. Finally, output the results of skyline groups. Algorithm 2 shows the pseudocode of BSTGSQ method.

Algorithm 2: Basic Spatio-Textual Group Skyline Query

Input: *Dataset*:$P=\{p_1, p_2, ..., p_n\}$; *Queryrequest*:$Q=\{q_1, q_2, ..., q_m\}$;
Output: ResultSet of Skyline Group List

1 Data preprocessing:
2 List=∅,List1=∅;
3 **for** $i = 1$ **to** n **do**
4   **if** p_i.*keywords contains Q.keywords* **then**
5   Add p_i to the List1 ;
6   **else**
7   **delete** p_i;
8 Generate a new dataset:$D=\{p_1, p_2, ..., p_{n'}\}$;
9 Query processing:
10 Generate $C_{n'}^{m}$ candidate skyline groups from List1;
11 Delete groups that do not contain the union of the query keywords;
12 Use Algorithm 1 to calculate the dominance relationship between groups;
13 Add do not dominated groups to List;
14 Return List;

In Algorithm 2, we first filter out the spatio-textual data objects containing no query keywords from the large dataset which cannot be the objects of the skyline groups, and add their to the *list*1 which storage the candidate objects (Line 1–7). And then we generate a new dataset which objects in list1 (Line 8). Then to generate $C_{n'}^{m}$ candidate skyline groups which the objects in from list1 (Line 10). To satisfy all STGSQ request keywords in the candidate skyline group, we delete groups that do not contain the union of the query keywords (Line 11). Next, in order to compute the dominance between groups, we deal with it in Algorithm 1 (Line 11). After that, we add corresponding skyline groups to the list (Line 13). Finally, output the results of skyline groups (Line 14).

4.2 Index-Based Spatio-Textual Group Skyline Query

To improve the efficiency, on the basis of BSTGSQ method, we further propose an Index-based Spatio-Textual Group Skyline Query (ISTGSQ) method to avoid traversing the entire dataset. The main idea of ISTGSQ method is to create a query unit for the query request Q by construct an MBR for the query locations, then we could use it to prune the search space. Specifically, starting from the root of the IR-tree, calculate the spatio-textual similarity between index node and the query unit and visit the index nodes that may contain candidate objects. Then, output these candidate objects to generate candidate groups iteratively. Finally, the final results are obtained according to the group dominance relationship. We first introduce the necessary definitions that can facilitate pruning.

Pruning Method Based on Spatio-Textual Similarity over Index
The main idea of this pruning method is to prune off the index nodes of which the distances are far away from the query locations and the texts are not relevant. Therefore, we define the spatio-textual similarity over the index, which aims to calculate the spatio-textual similarity based on index nodes.

In the following, we use U to represent the query unit of Q with its MBR, D to represent an index node with the MBR and inverted file. Firstly, we give the definitions of spatio-textual similarity based on index nodes and then give the definitions of spatial distance $d(D, U)$ and textual relevance $\omega(D, U)$ between D and U.

Definition 7 **(Spatio-Textual Similarity based on Index Node).** Let $st(D, U)$ be the spatio-textual similarity between D and U, which can be evaluated through combining the spatial distance and textual relevance are generally defined as follows,

$$st(D, U) = \alpha d(D, U) + (1 - \alpha)(1 - \omega(D, U)), \tag{4}$$

where α is a balance factor which is the same with that in Definition 3. $d(D, U)$ represents the minimum spatial distance between D and U, $\omega(D, U)$ represents the maximum textual relevance between D and U denoted as follows,

$$d(D, U) = MIN\{\frac{d(N.l, U.l)}{d_{max}}\}, \tag{5}$$

where $N.l$ is the MBR of D and $U.l$ is the MBR of Q. $d(N.l, U.l)$ is the distance between $N.l$ and $U.l$, and d_{max} is the maximum distance in the location space.

$$\omega(D, U) = MAX\{\omega(p.doc, q.doc)\}. \tag{6}$$

That is, $\omega(D, U)$ is the maximum textual relevance value of p with q according to Eq. 3, which p in D, and $q \in U$.

We calculate the $st(D, U)$, that is, selecting the index node with the minimum spatial distance and the maximum textual relevance. Given two index nodes D_1 and D_2, we calculate the minimum spatial distances $d(D_1, U)$ and $d(D_2, U)$, which represent the minimum spatial distance between D_1 and U and between D_2 and U respectively. As introduced in the previous section, we have calculated the $st(p, q)$ between p and q, and we choose the maximum textual relevance value $\omega(D_1, U)$ and $\omega(D_2, U)$. Finally, we calculate $st(D_1, U)$ and $st(D_2, U)$. If $st(D_2, U) > st(D_1, U)$, we prune the objects in D_2 and output those within D_1.

Algorithm 3: Index-based Spatio-Textual Group Skyline Query

```
Input: Dataset:P={p_1,p_2,...,p_n};QueryRequest:Q={q_1,q_2,...,q_m};Index
       IR-Tree;
Output: ResultSet of Skyline Group List
1  List=∅, List2=∅, Queue H=∅;
2  Add root of IR-Tree to H;
3  while H is not empty do
4      e=H.pop();
5      if e is leaf node then
6          Add e to List2 ;
7      else
8          ListNode=∅;
9          Obtain the children node and add to ListNode;
10         Calculate st(D,U) and in descending order by st(D,U);
11         for w = 1 to the size of ListNode do
12             if st(D_{w+1},U)>st(D_w,U) then
13                 Add objects in D_w to H;
14             else
15                 Add objects in D_{w+1} to H;
16 Generate C^m_{n''} candidate skyline groups from List2;
17 Use Algorithm 1 to calculate the dominance relationship between groups;
18 Add do not dominated groups to List;
19 Return List;
```

Query Processing. ISTGSQ method utilizes the IR-tree and the pruning algorithm introduced in the previous section, of which the pseudocode is shown in Algorithm 3. In Algorithm 3, firstly initialize the skyline group result list, queue H, minimum heap *ListNode* to store the child node and list2 to store the objects that have not been pruned. Next, calculate the spatio-textual similarities based on MBRs and perform the above pruning method to prune the objects in dominated MBRs, then output objects that can not be pruned to list2. Then, by generating the candidate groups based on the non-pruned objects, to satisfy all the query keywords to be covered by the candidate groups, remove the groups that do not contain the union of the query keywords. Finally, to compute the dominance between groups, the Group Dominance Computation Algorithm (Algorithm 1) can be performed, and the results of skyline groups can be output.

In Algorithm 3, the result list and queue H are initialized (Line 1) and the root of IR-tree is pushed into the queue (Line 2). *ListNode* is used to store the child node. We traverse the entire IR-tree, if it is a leaf node, add it to list2, otherwise to obtain its' children node and add to *ListNode* (Line 4–9). Then $st(D, U)$ is calculated and ordered in a descending order (Line 10–11). According to the pruning method, we prune off the index nodes which spatial distance is farther and textual relevance is lower than other MBRs, and output the non-pruned objects (Line 13–15). Next, generate the candidate groups and compute the dominance between groups using Algorithm 1 (Line 16–17). After that, we add corresponding skyline groups to the list (Line 18). Finally, output the results of skyline groups (Line 19).

4.3 Analysis

Validity Analysis. In BSTGSQ method, we use the enumeration to get all candidate groups. Next, we compute the group dominance relationship between each candidate groups and then obtain the skyline groups which are not dominated by other groups. Since all the candidate groups are constructed after the filtering based on the query keywords, the skyline group results will be correctly returned. In ISTGSQ method, we first use IR-tree to find the objects in MBRs that cannot be pruned off. Then, generate candidate groups which are ordered by $\Sigma_{q_j \in Q} st(p_i, q_j)$ ascendingly. Hence, the groups accessed later cannot dominate the groups found earlier. Thus, we judge whether the groups are dominated or not pruned. The method discovers skyline groups progressively and in the end returns the correct results.

Complexity Analysis. In BSTGSQ method, in the data processing period, n times of processing are required and a processed dataset is generated. Then C_n^m candidate groups are generated from the processed dataset. The complexity of generating candidate groups is O^2 in the worst case. Although the computational costs are expensive, use query keyword filtering and the group dominance comparison algorithm can effectively reduce the query response time to find the skyline groups that are not dominated. In ISTGSQ method, with the IR-tree index, in the query processing, we query from the root node and access the query index layer by layer, which requires $O(\log n)$ complexity. According to the pruning method, we remove the MBRs of which the spatial distance is larger and textual relevance is lower than those of other MBRs. The candidate groups are generated from the remaining data objects after pruning. Although the method can prune MBRs ($m \times n$ pruning), the costs of group dominance tests have to be conducted.

5 Experiments

5.1 Experimental Setups

Setup. We implement all the above algorithms in Java and perform experiments on a Tower Server (Dell Precision 7920) with two 40-core Intel(R) Xeon(R) Bronze 3204 1.90 GHz CPUs and 256 GB RAM running Ubuntu 20.04.

Datasets. We use four datasets in the experiments. The detailed of the datasets are shown in Table 2. We use two real datasets named POI[1] and SINA[2]. For POI,

[1] http://www.poilist.cn/.
[2] https://hub.hku.hk/.

we combine *category* and *subcategory* of several real Points of Interest(POI) as the textual contents, while keep the original locations of the data objects. For SINA with real locations, we tokenize the original textual data, remove the stop words and keep the keywords with high frequency. Besides, we synthesize two datasets from SINA named TF and FT. The locations in TF are real, and the texts are randomly generated. On the contrary, the locations of FT are randomly generated, and the texts are combined from real microblogs.

Table 2. Spatio-Textual Datasets

Dataset	POI	SINA	TF	FT
# of objects	30,000	30,000	10,000	10,000
Total # of keywords	100	300	500	500
Avg. # of keywords	53	6	277	274

5.2 Experimental Evaluation

Evaluation on Index Construction

The costs of index construction are evaluated by varying the number of objects. Figure 3 shows the running time and storage costs of the index construction, respectively. As observed from the figures, both the time and storage costs increase closely linearly following the growing number of objects. This is because to create an index, the data objects are accessed linearly. When the number of objects getting 10,000, the index size reaches about 800 MB, and the construction time is about 200 s at most.

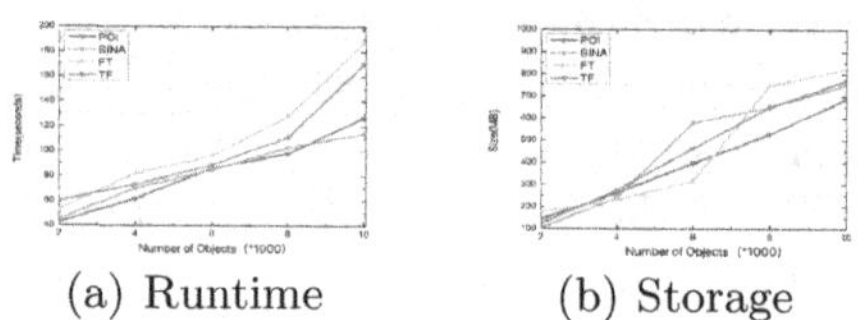

(a) Runtime (b) Storage

Fig. 3. Index construction costs, varying the number of objects

Evaluation on Query Processing

Impact of Number of Objects. Figure 4 compares the runtimes of the proposed methods on four datasets respectively. The runtime increases linearly, as the size of datasets increases and the algorithms maintain the difference of quality. The results show that the query response time of BSTGSQ method grows rapidly following the rising number of objects, while that of ISTGSQ method performs more stable. The query performance of ISTGSQ method is 3 to 4 times faster than that of BSTGSQ method.

Impact of Number of Query Locations. Figure 5 shows the query response time for different number of query locations on FT, TF, POI and SINA, respectively. The results show that the query response time of BSTGSQ method grows following the rising number of query locations, while that of ISTGSQ method performs gentle. From the comparisons of different datasets, the query response time of TF is the longest.

Impact of Number of Query Keywords. We test the query performance when varying the number of query keywords in the query request. Figure 6 illustrates the query response time for different numbers of query keywords on four datasets respectively. We observe that the query response time of BSTGSQ method increases following the rising number of query group keywords in an approximate linear fashion. The query response time of ISTGSQ method is 2 to 3 times faster than that of BSTGSQ method. The main reason is that more textually relevant objects involved into the query processing following the increasing number of query keywords result in higher computation overheads.

Evaluation on Dominance Tests

Figure 7 shows the number of dominance tests when varying the number of objects on FT, TF, POI and SINA, respectively. In this experiment, we set 5 query locations with 50 query keywords. The results show that the number of dominance tests of all methods increases as the increasing size of the datasets.

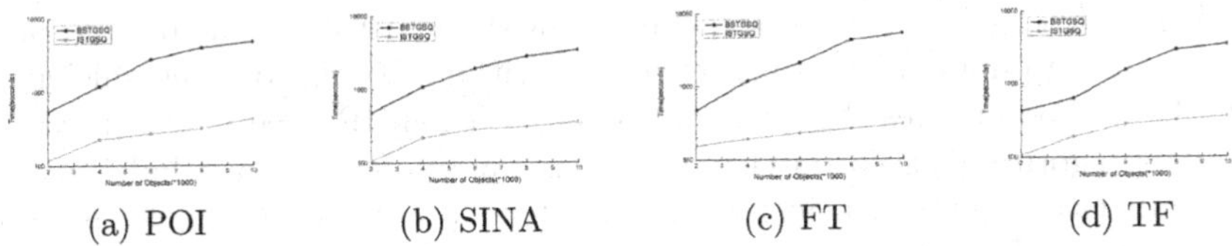

(a) POI (b) SINA (c) FT (d) TF

Fig. 4. Query response time, varying the number of objects

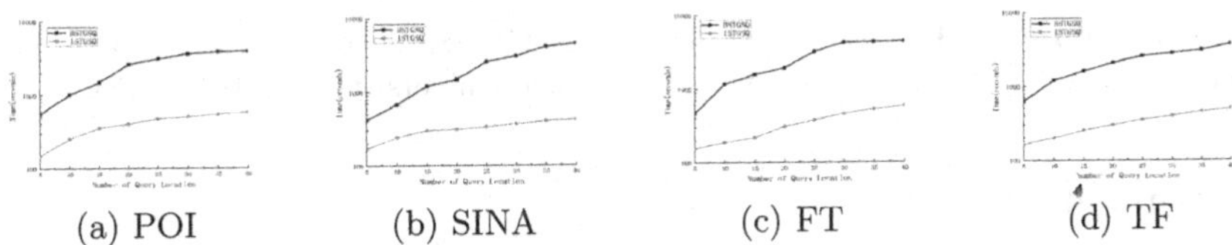

(a) POI (b) SINA (c) FT (d) TF

Fig. 5. Query response time, varying the number of query locations

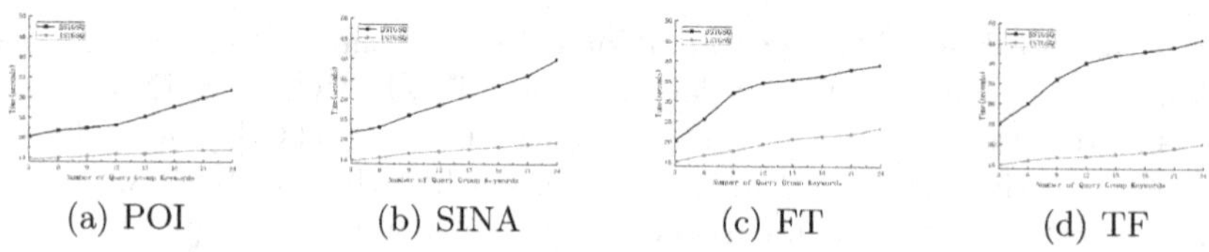

(a) POI (b) SINA (c) FT (d) TF

Fig. 6. Query response time, varying the number of query keywords

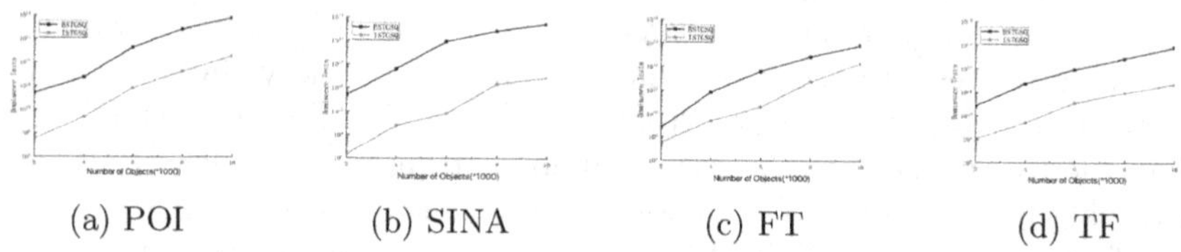

(a) POI (b) SINA (c) FT (d) TF

Fig. 7. Number of dominance tests, varying the number of objects

6 Conclusion

In this paper, we propose two spatio-textual group skyline query methods, denoted as Basic Spatio-Textual Group Skyline Query (BSTGSQ) method and Index-based Spatio-Textual Group Skyline Query (ISTGSQ) method. In BSTGSQ method, the skyline groups w.r.t. the query group that cannot be dominated by other groups can be retrieved from the generated candidate groups through a group dominance computation algorithm. To improve the efficiency, ISTGSQ method enables the query processing over an IR-Tree index, in which a pruning algorithm based on the spatio-textual similarities over index nodes is presented to support efficient queries without scanning the whole dataset. The validity and complexity of the proposed methods are analyzed, and the performance of our proposed methods is evaluated by the experiments on the real and synthetic datasets. For the future work, we plan to further study the spatio-textual group skyline queries in a dynamic programming manner without generating candidate groups.

Acknowledgement. The work is supported by College Students' Innovative Entrepreneurial Training Project of Shenyang Aerospace University (202210143013).

References

1. Wu, D., Cong, G., Jensen, C.S.: A framework for efficient spatial web object retrieval. VLDB J. **21**(6), 797–822 (2012)
2. Chen, L., Cong, G., Jensen, C.S., Wu, D.: Spatial keyword query processing: an experimental evaluation. Proc. VLDB Endow. **6**(3), 217–228 (2013)
3. Cao, X., et al.: Spatial keyword querying. In: Atzeni, P., Cheung, D., Ram, S. (eds.) ER 2012. LNCS, vol. 7532, pp. 16–29. Springer, Heidelberg (2012). https://doi.org/10.1007/978-3-642-34002-4_2

4. Börzsönyi, S., Kossmann, D., Stocker, K.: The skyline operator. In: ICDE, pp. 421–430. IEEE Computer Society (2001)
5. Su, I.-F., Chung, Y.-C., Lee, C.: Top-k combinatorial skyline queries. In: Kitagawa, H., Ishikawa, Y., Li, Q., Watanabe, C. (eds.) DASFAA 2010. LNCS, vol. 5982, pp. 79–93. Springer, Heidelberg (2010). https://doi.org/10.1007/978-3-642-12098-5_6
6. Dehaki, G.B., Ibrahim, H., Sidi, F., Udzir, N.I., Alwan, A.A., Gulzar, Y.: Efficient computation of skyline queries over a dynamic and incomplete database. IEEE Access **8**, 141523–141546 (2020)
7. Li, C., Zhang, N., Hassan, N., Rajasekaran, S., Das, G.: On skyline groups. In: CIKM, pp. 2119–2123. ACM (2012)
8. Zhang, N., Li, C., Hassan, N., Rajasekaran, S., Das, G.: On skyline groups. IEEE Trans. Knowl. Data Eng. **26**(4), 942–956 (2014)
9. Liu, J., Xiong, L., Pei, J., Luo, J., Zhang, H.: Finding pareto optimal groups: group-based skyline. Proc. VLDB Endow. **8**(13), 2086–2097 (2015)
10. Liu, J., Xiong, L., Pei, J., Luo, J., Zhang, H., Yu, W.: Group-based skyline for pareto optimal groups. IEEE Trans. Knowl. Data Eng. **33**(7), 2914–2929 (2021)
11. Cong, G., Jensen, C.S., Wu, D.: Efficient retrieval of the top-k most relevant spatial web objects. Proc. VLDB Endow. **2**(1), 337–348 (2009)
12. Rocha-Junior, J.B., Gkorgkas, O., Jonassen, S., Nørvåg, K.: Efficient processing of top-k spatial keyword queries. In: Pfoser, D., et al. (eds.) SSTD 2011. LNCS, vol. 6849, pp. 205–222. Springer, Heidelberg (2011). https://doi.org/10.1007/978-3-642-22922-0_13
13. Shi, J., Wu, D., Mamoulis, N.: Textually relevant spatial skylines. IEEE Trans. Knowl. Data Eng. **28**(1), 224–237 (2016)
14. Qian, Z., Xu, J., Zheng, K., Zhao, P., Zhou, X.: Semantic-aware top-k spatial keyword queries. World Wide Web **21**(3), 573–594 (2018)
15. Wu, D., Zhu, Y., Jensen, C.S.: In good company: efficient retrieval of the top-k most relevant event-partner pairs. In: Li, G., Yang, J., Gama, J., Natwichai, J., Tong, Y. (eds.) DASFAA 2019. LNCS, vol. 11447, pp. 519–535. Springer, Cham (2019). https://doi.org/10.1007/978-3-030-18579-4_31
16. Yao, K., Li, J., Li, G., Luo, C.: Efficient group top-k spatial keyword query processing. In: Li, F., Shim, K., Zheng, K., Liu, G. (eds.) APWeb 2016. LNCS, vol. 9931, pp. 153–165. Springer, Cham (2016). https://doi.org/10.1007/978-3-319-45814-4_13
17. Ahmad, S., Kamal, R., Ali, M.E., Qi, J., Scheuermann, P., Tanin, E.: The flexible group spatial keyword query. In: Huang, Z., Xiao, X., Cao, X. (eds.) ADC 2017. LNCS, vol. 10538, pp. 3–16. Springer, Cham (2017). https://doi.org/10.1007/978-3-319-68155-9_1
18. Apon, S.H., Ali, M.E., Ghosh, B., Sellis, T.: Social-spatial group queries with keywords. ACM Trans. Spatial Algorithms Syst. **8**(1), 1–32 (2022)
19. Li, Z., Lee, K.C.K., Zheng, B., Lee, W., Lee, D.L., Wang, X.: IR-tree: an efficient index for geographic document search. IEEE Trans. Knowl. Data Eng. **23**(4), 585–599 (2011)
20. Wu, D., Yiu, M.L., Cong, G., Jensen, C.S.: Joint top-k spatial keyword query processing. IEEE Trans. Knowl. Data Eng. **24**(10), 1889–1903 (2012)
21. Wu, D., Yiu, M.L., Jensen, C.S., Cong, G.: Efficient continuously moving top-k spatial keyword query processing. In: ICDE, pp. 541–552. IEEE Computer Society (2011)
22. Huang, W., Li, G., Tan, K., Feng, J.: Efficient safe-region construction for moving top-k spatial keyword queries. In: CIKM, pp. 932–941. ACM (2012)

23. Im, H., Park, S.: Group skyline computation. Inf. Sci. **188**, 151–169 (2012)
24. Zhu, H., Zhu, P., Li, X., Liu, Q.: Top-k skyline groups queries. In: EDBT, pp. 442–445. OpenProceedings.org (2017)
25. Zhu, H., Li, X., Liu, Q., Xu, Z.: Top-k dominating queries on skyline groups. IEEE Trans. Knowl. Data Eng. **32**(7), 1431–1444 (2020)
26. Yu, W., Qin, Z., Liu, J., Xiong, L., Chen, X., Zhang, H.: Fast algorithms for pareto optimal group-based skyline. In: CIKM, pp. 417–426. ACM (2017)
27. Wang, C., Wang, C., Guo, G., Ye, X., Yu, P.S.: Efficient computation of G-skyline groups. IEEE Trans. Knowl. Data Eng. **30**(4), 674–688 (2018)
28. Zhou, Y., Xie, X., Wang, C., Gong, Y., Ma, W.: Hybrid index structures for location-based web search. In: CIKM, pp. 155–162. ACM (2005)
29. Felipe, I.D., Hristidis, V., Rishe, N.: Keyword search on spatial databases. In: ICDE, pp. 656–665. IEEE Computer Society (2008)
30. Cao, X., Cong, G., Jensen, C.S., Ooi, B.C.: Collective spatial keyword querying. In: SIGMOD Conference, pp. 373–384. ACM (2011)
31. Zhang, W., Li, Y., Shu, L., Luo, C., Li, J.: Shadow: answering why-not questions on top-k spatial keyword queries over moving objects. In: Jensen, C.S., et al. (eds.) DASFAA 2021. LNCS, vol. 12682, pp. 738–760. Springer, Cham (2021). https://doi.org/10.1007/978-3-030-73197-7_51
32. Jin, X., Shin, S., Jo, E., Lee, K.: Collective keyword query on a spatial knowledge base. IEEE Trans. Knowl. Data Eng. **31**(11), 2051–2062 (2019)
33. Zhao, X., Zhang, Z., Huang, H., Bi, X.: Social-aware spatial keyword top-k group query. Distrib. Parallel Databases **38**(3), 601–623 (2020)
34. Ponte, J.M., Croft, W.B.: A language modeling approach to information retrieval. SIGIR Forum **51**(2), 202–208 (2017)
35. Manning, C.D., Raghavan, P., Schütze, H.: Introduction to Information Retrieval. Cambridge University Press, Cambridge (2008)

Towards a Cash-on-Delivery Management Solution Based on Blockchain, Smart Contracts, and NFT

Luong Hoang Huong[1], Bao Q. Tran[1], Hieu M. Doan[1], Nguyen D. P. Trong[1], Hieu V. Le[1], Loc V. C. Phu[1], Duy N. T. Quoc[1], Kiet T. Le[1], Nguyen H. Tran[1], Anh N. The[1], Huynh H. Nghia[1], Phuc N. Trong[1], Khoa T. Dang[1], Khiem H. Gia[1], Bang L. Khanh[1], Ngan N. T. Kim[2], and Hong Khanh Vo[1(✉)]

[1] FPT University, Can Tho City, Vietnam
khanhvh@fe.edu.vn

[2] FPT Polytecnic, Can Tho City, Vietnam

Abstract. The demand for transporting goods is increasingly diversified, it is associated with the economic development needs of a country. For domestic (eg, Cash-on-delivery - CoD) or international (eg, Letter-of-Credit) shipping models, they require trust with a third party between the seller and buyer, ie, carrier or bank. These two traditional models have a lot of holes that affect the interests of sellers and buyers, so there are many approaches to applying blockchain technology as well as smart contracts. However, some blockchain application models have revealed many shortcomings when removing the role of the shipper. Although many models have been implemented in practice and proven effective, some problems still require the intervention of a trusted third party to resolve the conflict. In this paper, we propose a model based on Blockchain technology, smart contracts to build a CoD system between seller - shipper - buyer, and NFT technology to create electronic invoices (i.e., proof-of-delivery) - dispute resolution if something goes wrong. Our research direction is towards eliminating the role of trusted third parties.

Keywords: Letter-of-Credit · Cash-on-delivery · Blockchain · Smart contracts · NFT · Ethereum · Fantom · Polygon · Binance Smart Chain

1 Introduction

Current shipping processes are based on mutual trust (i.e., between seller and buyer) or shipping company (i.e., shipper) [13]. Indeed, for traditional shipping models, there have been many unfortunate incidents related to the problem of unreliable freight. A good example of international trade is based on the Letter-of-Credit (LoC) model between a place of sale (i.e., a cashew nut export company in Vietnam) and an importer of the product on the other. products (i.e., cashew nut importers in Italy).[1] All exchanges between the two parties

[1] For some information security reasons, we do not disclose information about victims of export/import companies in Vietnam and Italy, respectively.

A. El Abbadi et al. (Eds.): DASFAA 2023 Workshops, LNCS 13922, pp. 51–66, 2023.
https://doi.org/10.1007/978-3-031-35415-1_4

are recorded and authenticated by a trusted third party (i.e., the bank). In case either party loses original documents (i.e., letter), the possibility of loss of goods/money is very high. A specific example happened in 2021. Specifically, four out of 100 containers of cashews exported from Vietnam to Italy are at risk of losing everything because exporting companies in Vietnam cannot present documents. original.[2] Luckily this issue has been resolved with the participation of the Ministry of Foreign Affairs of Vietnam and the Consulate General of Vietnam in Italy. However, this is also a warning to the traditional shipping process.

For the Cash-on-delivery (CoD) model, the seller must accept the risk of trusting and authorizing the shipping company (i.e., money, goods). Specifically, the amount of goods will be returned to the seller by the shipping company after a fixed period of time (eg, monthly, quarterly) or a certain amount is reached. For this type of shipping, the risk for the seller is very high because the shipping companies can use their money for something other than sending the money back to the seller. Most of today's processing is based on an agreement between two parties (i.e., the seller and the carrier). An example of a seller whose money was stolen by GNN Express occurred in Vietnam in 2017 and 2018.[3]. Specifically, the entire amount of goods worth about $154,900 was not transferred to the seller but was used by GNN Express for other purposes.

The two examples above are just one of many risks faced by traditional shipping models (i.e., LoC, CoD), where risks for both seller and buyer. Because of the above reasons, a series of models were born, applying the advances of science and technology in supply chain management to transport goods from sellers and buyers. One of those efforts is to apply blockchain technology and smart contracts to maintain the constraints between the seller and the buyer as well as to try to eliminate the reliance on a trusted third party (i.e., the bank). One of the first attempts at this was introduced in 2009 as Bitcoin [15]. The recommendation model is based on a Peer-to-Peer network with a combination of consensus protocol (i.e., Proof-of-Work PoW) to increase transparency for peer-to-peer networks. However, the method suffers from some limitations when it ignores an important group (i.e., shippers). This can affect the arbitration process when there is any conflict between seller and buyer [21]. This has prompted a number of theoretical studies that combine the role of shippers using blockchain technology and smart contracts [2]. Nevertheless, these approaches still present a challenge when transporting packages between shippers or delivery companies. Specifically, packages may be damaged during transit [25]. Therefore, in this study, we aim to determine the contents of the package and related information as it moves from the seller to the buyer and the shipper. Specifically, we apply NFT technology to generate information related to packages (i.e., sender, recipient, order content, weight, estimated time of receipt, etc). When the buyer receives the package, he can check the information sent from the seller.

---

[2] https://vietnamnet.vn/en/100-containers-of-cashew-nuts-exported-to-italy-suspected-of-being-scammed-821553.html.

[3] https://vir.com.vn/gnn-scandal-rocks-delivery-segment-62710.html.

Therefore, our contribution covers four aspects (a) proposing a shipping model based on blockchain technology and smart contract; (b) proposing a package information storage model based on Ethereum's NFT technology (i.e., ERC721); (c) implementing the proposed model by designing smart contracts (i.e., proof-of-concept); (d) deploying smart contracts on four EVM-enabled platforms including BNB Smartchain, Fantom, Celo, and Polygon to find a suitable platform for the proposed model.[4]

2 Related Work

Previous approaches exploiting Blockchain technology to increase interaction among stakeholders do not only apply to delivery problems such as, Cash-on-delivery (eg, [8]), Letter- of-Credit (eg, [18]), traditional delivery (eg, [25]), but also other systems that require high transparency such as healthcare systems such as, traditional healthcare (eg, [3,5]), supply chain management systems for blood and its products (eg, [10,19]), patient information access systems (eg, [11]) or in emergency [26] and many other systems [12,33,34].

Bitcoin is considered the first system to support trusted third-party payments introduced in early 2009 [16]. Although there are many limitations related to operating costs (i.e., largely guaranteed by the Proof-of-Work - PoW model), where a lot of miners are needed and operating costs are high (i.e., consume a lot of resources to maintain the system). Specifically, Bitcoin requires high fees to create new blocks as well as verify the correctness of transactions [20]. Aside from the limitations associated with operating the Bitcoin system, support for an exchange protocol with limited validation time remains an open question. In addition to trying to solve the aforementioned problems, several other approaches have introduced a new platform to replace Bitcoin's role in leading the current blockchain trend. The most famous of them is Ethereum [2], which introduced the first definitions of Smart Contracts as well as the Solidity programming language to assist developers in defining smarter transactions. In addition, the performance of the system is also more guaranteed when compared to Bitcoin.

One of the prominent examples of introducing a reliable exchange of goods between sellers and buyers is Local Ethereum introduced by [6]. This was one of the first attempts to build a traditional barter system between a seller (i.e., deliver the item and receive money) and a buyer (i.e., receive the item and transfer the money). Transactions are completely secured using a unique wallet address between transferring and receiving funds. Similar in concept, Open Bazaar is an open platform that supports a single channel between sellers and buyers [17] built on top of the `localEthereum` extension. The roles of the seller and the buyer are not clear between the two systems mentioned above, so depending on each specific transaction, their roles are changed. However, members of the same system can still monitor ongoing transactions even though they do not play a role. This has increased transparency for the Blockchain-based money

[4] We do not deploy smart contracts on ETH because the execution fee of smart contracts is too high.

transfer system. Comparing two stub systems for shipping and receiving (i.e., Local Ethereum and Open Bazaar), `OpenBazaar` involved three parties: the supplier, the requestor, and the moderator (i.e. has a new controlling role) instead of being limited by the two traditional user groups, sellers and buyers. This means that in the previously defined Smart Contracts, the seller and the buyer are required to follow the predefined rules (i.e., cannot be changed).

However, the disadvantage of systems using default smart contracts is that these systems still have to depend on a trusted third party. Specifically, if any problems arise between the seller and the buyer, there should be an arbitration (i.e., trust) to resolve the dispute between the seller and the buyer. Because of the above reason, a number of other approaches have supported a self-defining smart contract solution - called middleman [18]. Middleman creates pre-existing smart contracts with corresponding penalties for violations between sellers and buyers. In addition to the cost to pay for the system (i.e., gas), the parties involved will also provide a surcharge to the middleman. On a larger scale, the connection between the producer (i.e., the place where the product is made) and the consumer (i.e., the consumer of the product). Ethereum introduces a new protocol under the COD/LOC mechanism called decentralize application (Dapps) [1]. These transactions are still based on Ethereum's existing platform to connect stakeholders. Specifically, instead of focusing on the traditional transaction between the seller and the buyer, the new COD models consider new actors (i.e., the shipper or the shipping company). [21] argues that the shipper's role is so important that this component's behavior is reliable. This is a limitation for models that support shippers. On the other hand, these models require arbitration when there is a conflict of interest between the parties to determine the subject of the penalty. Specifically, each member has to pay a specific deposit (eg, 50% the value [14]).

To solve the above problems, we propose a freight model based on blockchain technology combined with NFT certificates. Specifically, NFTs will be generated when the item is shipped from the seller.

3 Approach

3.1 The Traditional Model of Freight Transport

Delivery and shipping are authorized by a shipping company. Depending on the commitment between the parties (i.e., seller, buyer, carrier) the role of the shipping company is different (i.e., shipping or transporting and receiving money). The transfer of funds from the buyer to the seller does not involve the shipping company. Figure 1 shows the steps of shipping goods from seller to buyer through a shipping company. In this model the buyer has to pay the full invoice cost to the seller before receiving the goods (step 1) - shipping costs are covered by the seller. Step 2 includes the preparation steps to ship the goods to the buyer (i.e., the seller aggregates all the invoices available for the day to send to many different buyers). In step 3, the seller packages the products for shipment to each respective buyer, who then forwards them to the shipping company (step 4).

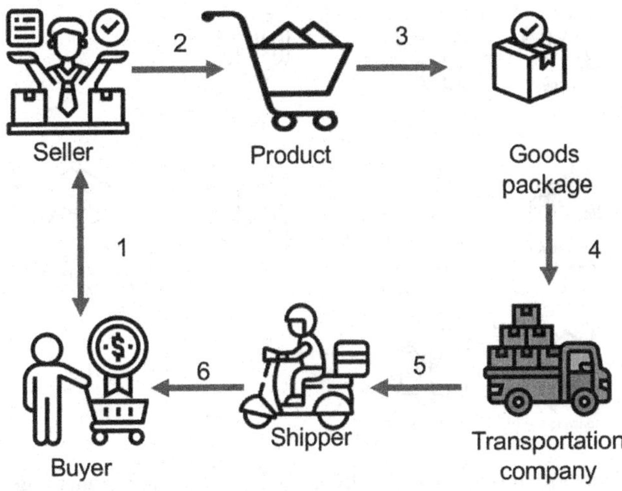

Fig. 1. The traditional model w.r.t transportation company of freight transport

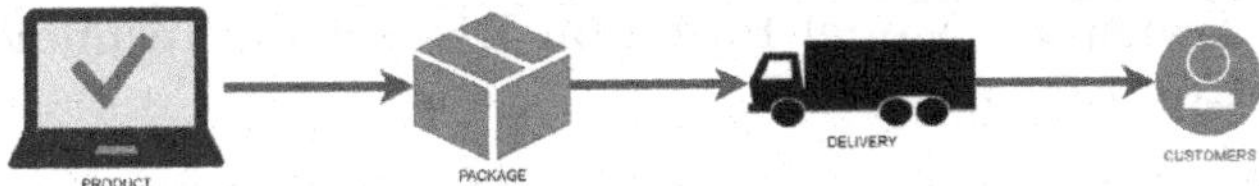

Fig. 2. The traditional model w.r.t cash-on-delivery of freight transport

The shipping company splits the packages for each shipper (i.e., responsible for the distribution of goods for each specific region/region) - step 5. Finally, the buyer receives the goods from the shipper.

Another example that is riskier for merchants and is used in many countries is Cash-on-Delivery (i.e., CoD). In this model, the seller authorizes the shipping company to deliver the goods and receive money from the customer. Specifically, Fig. 2 presents the traditional CoD transport model. The seller and the buyer will agree on the item and the form of payment. Assuming that the buyer selects a laptop, the seller packages the product and selects a shipping company to deliver the product to the buyer.

The risks related to the transportation of goods based on the traditional model have been analyzed by us in the Introduction and Related work sections. Our contribution in this paper is to propose a model based on blockchain technology, NFT and smart contract to propose a model of goods transportation and identify possible violations between relevant parties (ie, seller, buyer, or shipper). The next section presents our proposed model.

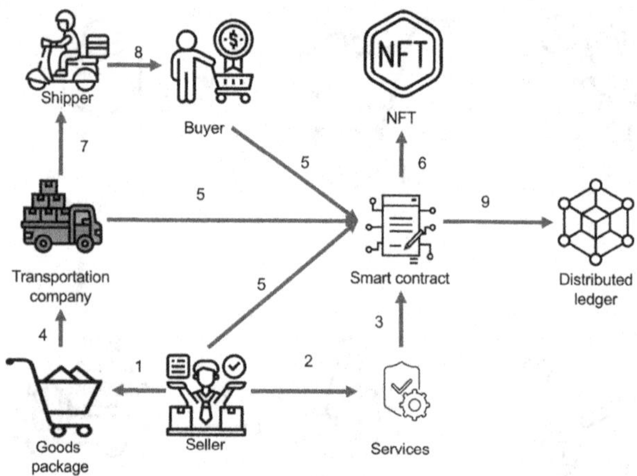

Fig. 3. Cargo transport model based on Blockchain technology, smart contract & NFT

3.2 Cargo Transport Model Based on Blockchain Technology, Smart Contract and NFT

Figure 3 shows our proposal process based on blockchain, NFT and smart contract technologies. The role of the blockchain (i.e., consensus protocol, distributed ledger) is to store transactions (i.e., in case of breach/compliance with the constraints on the contract); as well as increasing transparency for data stored on the handle - relevant parties can easily retrieve and confirm the information. The role of NFT in our proposed model is to store the information agreed between the parties involved and as proof of handling when any errors occur during the transfer. Meanwhile, the role of the smart contract is to enforce the agreed-upon constraints between the parties involved automatically. Our model attempts to eliminate the role of trusted third parties, including in the case of conflict resolution between parties.

Before showing the implementation steps of the system, let us clarify the role of each participant in the process (step 5 - Fig. 3). In which, the seller still has to enter an amount of money to ensure the correct delivery of the goods and the right quality (i.e., if the seller intentionally violates the information provided in the smart contract - step 2 - this amount is used to compensation to the carrier and the buyer). For shippers, their management company must pay a guarantee fee to avoid the shipper losing the goods or in case the company goes bankrupt before the time to refund the buyer (i.e., the amount Their deposit depends on the exchange between seller and shipping company). In the event of a conflict (i.e., shippers lose or damage the goods or the shipping company goes bankrupt) the smart contract automatically transfers the company's deposit to the seller through their address. For the buyer's deposit, this includes the shipping fee and part of the deposit of the product (i.e., depends on the agreement between the seller and the buyer). In case the buyer refuses to purchase the product, the shipping fee and deposit of the product are automatically transferred to the address of the shipping company and the seller, respectively.

Figure 3 suggests a procedure based on the step itself. Specifically, in step 1, the seller prepares the product and enters the information related to the package (i.e., weight, unit price, item type, etc.) into the system in step 2. These services call the corresponding functions in the smart contract to create a shipping contract between the seller and the buyer (step 3) before selecting the corresponding shipping company (step 4). Step 5 presents the deposit confirmation process for each participating group. The above conventions are monitored through protocols designed on smart contracts and stored on NFTs with the consent of all three parties (i.e., seller, buyer, and shipping company).) - step 6. Step 7 presents the shipping process between shippers (i.e., depending on the distance, the number of shippers is 1 person or many). This process is managed and operated by the shipping companies. Step 8 presents the final shipment (i.e., the buyer receives the item) - confirmation from the buyer that the package is correct and that the information is stored in the NFT. The transactions are conducted specifically, the buyer pays the remaining amount of the product; the shipping company receives the deposit and shipping; The seller receives the money for the sale of the product. Risks and contract breaches are resolved based on the cases designed in step 5. Finally, the smart contract updates the transactions to the distributed ledger and prepares for a new shipping process.

4 Implementation

Our implementation model focuses on two main purposes i) data manipulation (i.e., package) - initialization, query and update - on blockchain platform and ii) generation of NFTs for each order goods for easy retrieval by sellers and buyers (i.e., product reviews before and after delivery).

4.1 Initialize Data/NFT

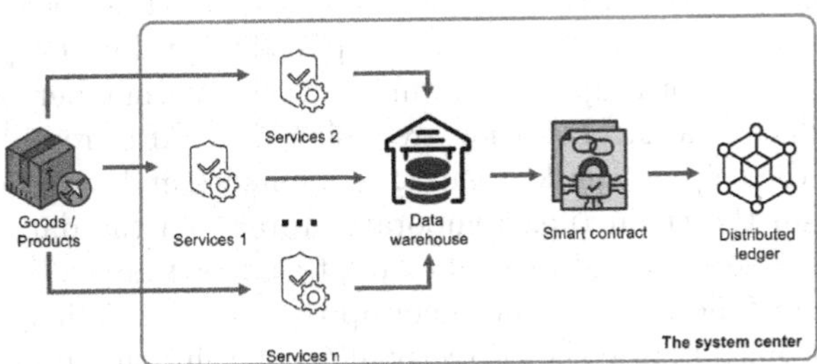

Fig. 4. Initialize data/NFT

Figure 4 shows the steps to initialize the package data. These package types include information related to the sender (i.e., receiving address, weight, type of item), the recipient (i.e., receiving address, estimated time of receipt). In addition, the delivery and receipt of goods also requires a deposit of all three parties depending on the purpose and transactions between the parties to ensure automatic conflict resolution on the smart contract. In addition, information about which carrier belongs to which company, time, and place of delivery and receipt is also added to the metadata of the package. This is extremely important in cases where more than one shippers are involved in the shipment (i.e., the same or different shipping company). As for the storage process, services support concurrent storage (i.e., distributed processing as a peer-to-peer network) on a distributed ledger - supporting more than one user for concurrent storage reduce system latency. In general, the package data is organized as follows:[5]

```
goodsObject = {
"goodsID": goodsID,
"deliveryCompanyID": deliveryCompanyID,
"shipperID": shipperID,
"type": type of goods,
"buyerID": buyerID,
"sellerID": sellerID,
"quantity": quantity,
"unit": unit,
"packageID": packageID,
"addressReceived": received address,
"addressDelivery": delivery address,
"time": time,
"location": location,
"state": Null};
```

Specifically, in addition to information for content extraction (i.e., place of origin, weight, item type, etc), we also store information regarding the status of the package at "addressReceived" (ie., "state" - default value is Null). Specifically, "state" changes to 1 if the corresponding package has been received and shipped by the shipping company (i.e., "shipperID"); value 0 - pending (i.e., waiting for shipper to pick up). Also, "unit" stores the number of orders (eg, 10) as well as which "packageID" they are assigned to. After receiving packages from the seller, the shipper checks them for compliance and waits for validation before syncing up the chain (i.e., temporary storage on the data warehouse). Then the pre-designed constraints in the Smart Contract are called through the API (i.e., name of function) to sync them up the chain. This inspection role is extremely important because they directly affect the shipping process of goods, as well as the premise for conflict resolution when any problems arise (eg, goods damage, lost packages).

[5] The information related to the system participants is not listed in the article. Readers can refer to the group's previous studies at [4,7,24].

The information on the NFT contributes to conflict resolution (eg, delivery delays). Definitions related to stakeholder deposits have been defined in our previous articles.

4.2 Data Query

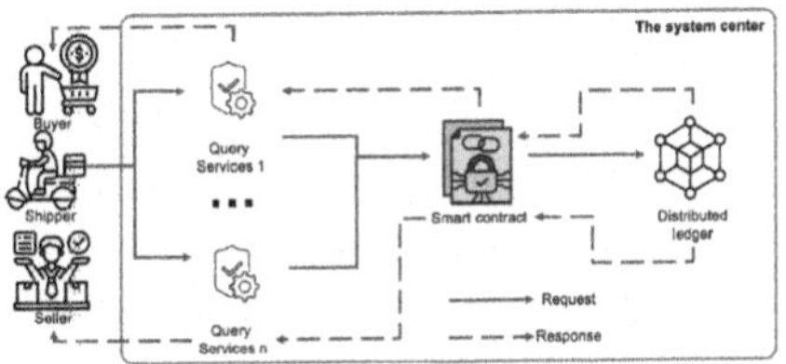

Fig. 5. Data query

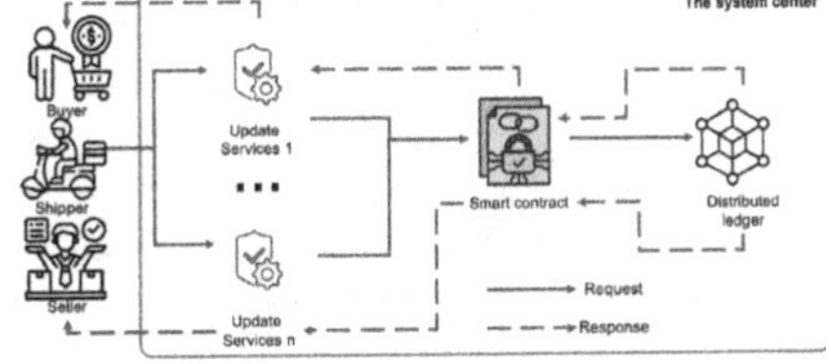

Fig. 6. Data updated

Similar to the data initialization steps, the data query process also supports many simultaneous participants in the system for access (i.e., distributed model). Support services receive requests from shippers or sellers/buyers to access data (i.e., respective packages). Depending on the query object we have different access purposes. Specifically, shippers query for the purpose of identifying consignee information and their addresses. In contrast, sellers/buyers view the status of their orders (i.e., after being delivered and received) as well as handle conflicts when something goes wrong. Figure 5 shows the steps to query the order data. These requests are sent as requests (i.e., pre-designed services as API calls) from the user to the smart contracts available in the system (i.e., name of function) before retrieving the data from the distributed ledger. All retrieval requests are also saved as query history for each individual or organization. For a shipping process that involves multiple discounts (i.e., multiple shippers deliver and receive the goods before reaching the buyer's address), NFTs are similarly created between shippers (i.e., within or different shipping companies). If the corresponding info is not found (eg, wrong ID), the system will send a message not found results. For the NFT query process, all support services are provided as APIs.

4.3 Data Updated

The data update routine is invoked only after verifying that the data exists on the thread (i.e., after executing the corresponding data query procedure). In this section, we assume that the search data exists on the string. Where none exists, the system sends the same message to the user (see Sect. 4.2 for details). Similar to the two processes of query and data initialization, we support update services in the form of APIs to receive requests from users before passing them to smart contract (i.e., name of function) for processing. The purpose of this process is to update the status of the package during transit as well as handle conflicts when something goes wrong (i.e., a combination of smart contract and NFT). Figure 6 shows the process of updating order data. For NFTs (i.e., available) the update process includes only moving from the owner's address to the new (i.e., new owner). If any information is updated on an existing NFT, it will be stored as a new NFT (see Sect. 4.1 for details).

Fig. 7. The transaction info (e.g., BNB Smart Chain)

5 Evaluation

Because of the connection model between seller - shipper - buyer and support for payment currency (i.e., consensus protocol), we implement the proposed model on blockchain platforms that support EVM instead of mining platforms. belonging to Hyperledger eco-system. In addition, assessments based on system responsiveness (i.e., number of requests responded successfully/failed, system latency - min, max, average) have been evaluated by us in the previous research paper. Therefore, in this paper we determine the suitable platform for our proposed model. Specifically, we install a recommendation system on four popular blockchain platforms today, supporting Ethereum Virtual Machine (EVM), including Binance Smart Chain (BNB Smart Chain); Polygon; Fantom; and Celo. Our implementations on these four platforms are also shared as a contribution to the article to collect transaction fees corresponding to the four platforms' supporting coins[6], ie, BNB[7]; MATIC[8]; FTM[9]; and CELO[10]. For example, Fig. 7 details our three assessments of a successful installation on BNB Smart Chain (i.e., similar settings are shown for the other three platforms). Our

6 Implementation of theme models our release at Nov-24-2022 07:42:20 AM +UTC.

7 https://testnet.bscscan.com/address/0x2ec70f233d91ade867259ff20c75f5c54e1ff008.

8 https://mumbai.polygonscan.com/address/0xd9ee80d850ef3c4978dd0b099a45a559fd7c5ef4.

9 https://testnet.ftmscan.com/address/0xd9ee80d850ef3c4978dd0b099a45a559fd7c5ef4.

10 https://explorer.celo.org/alfajores/address/0xD9Ee80D850eF3C4978Dd0B099A45a559fD7c5EF4/transactions.

implementations to evaluate the execution cost of smart contracts (i.e., designed based on Solidity language) run on testnet environments of four platforms in order to choose the most cost-effective platform to deploy. reality. Our detailed assessments focus on the cost of performing contract creation, NFT generation and NFT retrieval/transfer (i.e., updating NFT ownership address) presented in the respective subsections related to i) Transaction Fee; ii) Gas limit; iii) Gas Used by Transaction; and iv) Gas Price.

Table 1. Transaction fee

	Contract Creation	Create NFT	Transfer NFT
BNB Smart Chain	0.02731184 BNB ($8.32)	0.00109162 BNB ($0.33)	0.00057003 BNB ($0.17)
Fantom	0.009576994 FTM ($0.001850)	0.000405167 FTM ($0.000078)	0.0002380105 FTM ($0.000046)
Polygon	0.006840710032835408 MATIC($0.01)	0.000289405001852192 MATIC($0.00)	0.000170007501088048 MATIC($0.00)
Celo	0.0070974384 CELO ($0.004)	0.0002840812 CELO ($0.000)	0.0001554878 CELO ($0.000)

5.1 Transaction Fee

Table 1 shows the cost of creating contracts for the four platforms. It is easy to see that the highest transaction fee of the three requirements is contract creation for all four platforms. In which, the cost of BNB Smart Chain is the highest with the highest cost when creating a contract is 0.02731184 BNB ($8.32); whereas, the lowest cost recorded by the Fantom platform with the highest cost for contract initiation is less than 0.009576994 FTM ($0.001850). Meanwhile, the cost to enforce Celo's contract initiation requirement is lower than Polygon's with only $0.004 compared to $0.01. For the remaining two requirements (Create NFT and Transfer NFT), we note that the cost of implementing them for all three platforms, Polygon, Celo, and Fantom is very low (i.e., negligible) given the cost. trades close to $0.00. However, this cost is still very high when deployed on BNB Smart Chain with "0.00109162 BNB ($0.33)" and 0.00057003 BNB ($0.17) for Create NFT and Transfer NFT, respectively.

5.2 Gas Limit

Table 2 shows the gas limit for each transaction. Our observations show that the gas limits of the three platforms (i.e., BNB, Polygon, and Fantom) are roughly equivalent - where Polygon and Fantom are similar in the first two transactions. Particularly in the third transaction, BNB's gas limit was extremely high with 3,000,000. While the gas volume of Polygon and Fantom is equivalent to 72,803. The other platform (i.e., Celo) has the highest gas limit in the first two transactions with 3,548,719; 142,040.

Table 2. Gas limit

	Contract Creation	Create NFT	Transfer NFT
BNB Smart Chain	2,731,184	109,162	3,000,000
Fantom	2,736,284	115,762	72,803
Polygon	2,736,284	115,762	72,803
Celo	3,548,719	142,040	85,673

Table 3. Gas Used by Transaction

	Contract Creation	Create NFT	Transfer NFT
BNB Smart Chain	2,731,184 (100%)	109,162 (100%)	57,003 (1.9%)
Fantom	2,736,284 (100%)	115,762 (100%)	68,003 (93.41%)
Polygon	2,736,284 (100%)	115,762 (100%)	68,003 (93.41%)
Celo	2,729,784 (76.92%)	109,262 (76.92%)	59,803 (69.8%)

5.3 Gas Used by Transaction

Table 3 shows the amount of gas used when executing the transaction (i.e., what percentage of gas in total gas is shown in Table 2). Specifically, three platforms BNB, Polygon, and Fantom use 100% of Gas Limit for two transactions Contract Creation and Create NFT. Meanwhile, Celo uses 76.92% of the Gas limit for the above two transactions. For the last transaction of Transfer NFT, BNB's Gas level was only 1.9% with 57,003 (i.e., lowest) while the highest Gas level was recorded by Fantom and Polygon with 93.41% of Gas limit; while BNB and Celo use 79.17% and 69.8% of Gas limit.

5.4 Gas Price

Table 4. Gas Price

	Contract Creation	Create NFT	Transfer NFT
BNB Smart Chain	0.00000001 BNB (10 Gwei)	0.00000001 BNB (10 Gwei)	0.00000001 BNB (10 Gwei)
Fantom	0.0000000035 FTM (3.5 Gwei)	0.0000000035 FTM (3.5 Gwei)	0.0000000035 FTM (3.5 Gwei)
Polygon	0.000000002500000012 MATIC (2.500000012 Gwei)	0.000000002500000016 MATIC (2.500000016 Gwei)	0.000000002500000016 MATIC (2.500000016 Gwei)
Celo	0.0000000026 CELO (Max Fee per Gas: 2.7 Gwei)	0.0000000026 CELO (Max Fee per Gas: 2.7 Gwei)	0.0000000026 CELO (Max Fee per Gas: 2.7 Gwei)

Table 4 shows the value of Gas for all four platforms. Specifically, BNB, Fantom, and Celo have the same Gas value in all three transactions with values of 10 Gwei (i.e., the highest of the three platforms), 3.5 Gwei, and 2.7 Gwei, respectively. Meanwhile, the Gas value of the Polygon platform (i.e., MATIC) has the lowest value and fluctuates around 2.5 Gwei.

6 Discussion

According to our observation, the transaction value depends on the market capitalization of the respective coin. The total market capitalization of the 4 platforms used in our review (i.e., BNB (Binance Smart Chain); MATIC (Polygon); FTM (Fantom); and CELO (Celo)) are $50,959,673,206; $7,652,386,190; $486,510,485; and $244,775,762.[11] The total issuance of the four coins BNB, MATIC, FTM, and CELO is 163,276,974/163,276,974 coins, respectively; 8,868, 740,690/10,000, 000,000 coins; 2,541,152,731/3,175,000,000 coins and 473, 376,178/ 1,000,000,000 coins. The value of the coin is conventionally based on the amount of coins issued and the total market capitalization with a value of $314.98; $0.863099; $0.1909; and $0.528049 for BNB, MATIC, FTM, and CELO respectively. Based on the measurements and analysis in Sect. 5 section, we have concluded that the proposed model deployed on Faltom brings many benefits related to system operating costs. In particular, generating and receiving NFTs has an almost zero (i.e., negligible) fee. Also, the cost of creating contracts with transaction execution value is also meager (i.e., less than $0.002).

In future work, we proceed to implement more complex methods/algorithms (i.e., encryption and decryption) as well as more complex data structures to observe the costs for the respective transactions. Deploying the proposed model in a real environment is also a possible approach (i.e., implementing the recommendation system on the FTM mainnet). In our current analysis, we have not considered issues related to the privacy policy of users (i.e., access control [22,24,32], dynamic policy [23,35]) - a possible approach would be implemented in upcoming research activities. Finally, infrastructure-based approaches (i.e., gRPC [9,29]; Microservices [27,30]; Dynamic transmission messages [31] and Brokerless [28]) can be integrated into the model of us to increase user interaction.

7 Conclusion

The article focuses on the model of goods transportation (i.e., Cash-on-delivery) between seller - buyer - carrier combined with Blockchain technology, Smart Contract, NFT. We build a solution that is independent of a trusted third party (i.e., arbitration) and can be deployed in a developing country - limited to the infrastructure and IT background of the user people. The article also summarizes the development directions based on Blockchain technology and smart contract before proposing our solution. We exploited the Ethereum platform as well as the

[11] Our observation time is 12:00PM - 11/26/2022.

Solidity language to design smart contracts before deploying them on four EVM-enabled platforms (i.e., BNB, FTM, MATIC, and CELO). The three-transaction deployment fee assessments (i.e., contract creation, NFT create, NFT transfer) have proven the Fantom platform to have the lowest cost. In addition, in the discussion, we explained the reasons for this statement as well as possible development directions for our current system.

Acknowledgement. This work was supported by Le Thanh Tuan and Dr. Ha Xuan Son during the process of brainstorming, implementation, and evaluation of the system.

References

1. Two party contracts (2022). www.dappsforbeginners.wordpress.com/tutorials/two-party-contracts/
2. Buterin, V., et al.: A next-generation smart contract and decentralized application platform. White Pap. **3**(37), 2–1 (2014)
3. Duong-Trung, N., et al.: Smart care: integrating blockchain technology into the design of patient-centered healthcare systems. In: Proceedings of the 2020 4th International Conference on Cryptography, Security and Privacy, pp. 105–109 (2020)
4. Duong-Trung, N., et al.: Multi-sessions mechanism for decentralized cash on delivery system. Int. J. Adv. Comput. Sci. Appl. **10**(9) (2020)
5. Duong-Trung, N., et al.: On components of a patient-centered healthcare system using smart contract. In: Proceedings of the 2020 4th International Conference on Cryptography, Security and Privacy, pp. 31–35 (2020)
6. Ethereum: How our escrow smart contract works (2022). www.thenational.ae/business/technology/cash-on-delivery-the-biggest-obstacle-to-e-commerce-in-uae-and-region-1
7. Ha, X.S., et al.: DeM-CoD: novel access-control-based cash on delivery mechanism for decentralized marketplace. In: 2020 IEEE 19th International Conference on Trust, Security and Privacy in Computing and Communications (TrustCom), pp. 71–78. IEEE (2020)
8. Ha, X.S., Le, T.H., Phan, T.T., Nguyen, H.H.D., Vo, H.K., Duong-Trung, N.: Scrutinizing trust and transparency in cash on delivery systems. In: Wang, G., Chen, B., Li, W., Di Pietro, R., Yan, X., Han, H. (eds.) SpaCCS 2020. LNCS, vol. 12382, pp. 214–227. Springer, Cham (2021). https://doi.org/10.1007/978-3-030-68851-6_15
9. Lam, N.T.T., et al.: BMDD: a novel approach for IoT platform (broker-less and microservice architecture, decentralized identity, and dynamic transmission messages). PeerJ (2022)
10. Le, H.T., et al.: Bloodchain: a blood donation network managed by blockchain technologies. Network **2**(1), 21–35 (2022)
11. Le, H.T., et al.: Patient-chain: patient-centered healthcare system a blockchain-based technology in dealing with emergencies. In: Shen, H., et al. (eds.) PDCAT 2021. LNCS, vol. 13148, pp. 576–583. Springer, Cham (2022). https://doi.org/10.1007/978-3-030-96772-7_54
12. Luong, H.H., Huynh, T.K.N., Dao, A.T., Nguyen, H.T.: An approach for project management system based on blockchain. In: Dang, T.K., Küng, J., Chung, T.M., Takizawa, M. (eds.) FDSE 2021. CCIS, vol. 1500, pp. 310–326. Springer, Singapore (2021). https://doi.org/10.1007/978-981-16-8062-5_21

13. Madhwal, Y., et al.: Proof of delivery smart contract for performance measurements. IEEE Access **10**, 69147–69159 (2022)
14. Marzo, G.D., Pandolfelli, F., Servedio, V.D.: Modeling innovation in the cryptocurrency ecosystem. Sci. Rep. **12**(1), 1–12 (2022)
15. Nakamoto, S.: Bitcoin: a peer-to-peer electronic cash system. Decentralized Bus. Rev. 21260 (2008)
16. Nakamoto, S.: Bitcoin: a peer-to-peer electronic cash system bitcoin: a peer-to-peer electronic cash system. Bitcoin.org. Disponible en (2009). www.bitcoin.org/en/bitcoin-paper
17. OpenBazaar: Truly decentralized, peer-to-peer ecommerce features (2022). www.openbazaar.org/features/
18. Quoc, K.L., et al.: SSSB: an approach to insurance for cross-border exchange by using smart contracts. In: Awan, I., Younas, M., Poniszewska-Marańda, A. (eds.) MobiWIS 2022. LNCS, vol. 13475, pp. 179–192. Springer, Cham (2022). https://doi.org/10.1007/978-3-031-14391-5_14
19. Quynh, N.T.T., et al.: Toward a design of blood donation management by blockchain technologies. In: Gervasi, O., et al. (eds.) ICCSA 2021. LNCS, vol. 12956, pp. 78–90. Springer, Cham (2021). https://doi.org/10.1007/978-3-030-87010-2_6
20. Shang, G., et al.: Need for speed, but how much does it cost? Unpacking the fee-speed relationship in bitcoin transactions. J. Oper. Manag. **69**, 102–126 (2022)
21. Sinha, D., Chowdhury, S.R.: Blockchain-based smart contract for international business-a framework. J. Glob. Oper. Strateg. Sourcing **14**, 224–260 (2021)
22. Son, H.X., Hoang, N.M.: A novel attribute-based access control system for fine-grained privacy protection. In: Proceedings of the 3rd International Conference on Cryptography, Security and Privacy, pp. 76–80 (2019)
23. Son, H.X., Dang, T.K., Massacci, F.: REW-SMT: a new approach for rewriting XACML request with dynamic big data security policies. In: Wang, G., Atiquzzaman, M., Yan, Z., Choo, K.-K.R. (eds.) SpaCCS 2017. LNCS, vol. 10656, pp. 501–515. Springer, Cham (2017). https://doi.org/10.1007/978-3-319-72389-1_40
24. Son, H.X., Nguyen, M.H., Vo, H.K., Nguyen, T.P.: Toward an privacy protection based on access control model in hybrid cloud for healthcare systems. In: Martínez Álvarez, F., Troncoso Lora, A., Sáez Muñoz, J.A., Quintián, H., Corchado, E. (eds.) CISIS/ICEUTE -2019. AISC, vol. 951, pp. 77–86. Springer, Cham (2020). https://doi.org/10.1007/978-3-030-20005-3_8
25. Son, H.X., et al.: Towards a mechanism for protecting seller's interest of cash on delivery by using smart contract in hyperledger. Int. J. Adv. Comput. Sci. Appl. **10**(4), 45–50 (2019)
26. Son, H.X., Le, T.H., Quynh, N.T.T., Huy, H.N.D., Duong-Trung, N., Luong, H.H.: Toward a blockchain-based technology in dealing with emergencies in patient-centered healthcare systems. In: Bouzefrane, S., Laurent, M., Boumerdassi, S., Renault, E. (eds.) MSPN 2020. LNCS, vol. 12605, pp. 44–56. Springer, Cham (2021). https://doi.org/10.1007/978-3-030-67550-9_4
27. Thanh, L.N.T., et al.: IoHT-MBA: an internet of healthcare things (IoHT) platform based on microservice and brokerless architecture. Int. J. Adv. Comput. Sci. Appl. **12**(7) (2021)
28. Thanh, L.N.T., et al.: SIP-MBA: a secure IoT platform with brokerless and microservice architecture. Int. J. Adv. Comput. Sci. Appl. (2021)
29. Thanh, L.N.T., et al.: Toward a security IoT platform with high rate transmission and low energy consumption. In: Gervasi, O., et al. (eds.) ICCSA 2021. LNCS,

vol. 12949, pp. 647–662. Springer, Cham (2021). https://doi.org/10.1007/978-3-030-86653-2_47

30. Nguyen, T.T.L., et al.: Toward a unique IoT network via single sign-on protocol and message queue. In: Saeed, K., Dvorský, J. (eds.) CISIM 2021. LNCS, vol. 12883, pp. 270–284. Springer, Cham (2021). https://doi.org/10.1007/978-3-030-84340-3_22
31. Thanh, L.N.T., et al.: UIP2SOP: a unique IoT network applying single sign-on and message queue protocol. IJACSA **12**(6) (2021)
32. Thi, Q.N.T., Dang, T.K., Van, H.L., Son, H.X.: Using JSON to specify privacy preserving-enabled attribute-based access control policies. In: Wang, G., Atiquzzaman, M., Yan, Z., Choo, K.-K.R. (eds.) SpaCCS 2017. LNCS, vol. 10656, pp. 561–570. Springer, Cham (2017). https://doi.org/10.1007/978-3-319-72389-1_44
33. Huynh, T.K.N., Dao, T.A., Van Pham, T.N., Nguyen, K.H.V., Tran, N.C., Luong, H.H.: VBlock—blockchain-based traceability in medical products supply chain management: a case study in VietNam. In: Ibrahim, R., Porkumaran, K., Kannan, R., Mohd Nor, N., Prabakar, S. (eds.) International Conference on Artificial Intelligence for Smart Community. LNCS, vol. 758, pp. 429–441. Springer, Singapore (2022). https://doi.org/10.1007/978-981-16-2183-3_42
34. Nguyen, K.T.H., et al.: Domain name system resolution system with hyperledger fabric blockchain. In: Smys, S., Kamel, K.A., Palanisamy, R. (eds.) Inventive Computation and Information Technologies. LNCS, vol. 563, pp. 59–72. Springer, Singapore (2023). https://doi.org/10.1007/978-981-19-7402-1_5
35. Xuan, S.H., et al.: Rew-XAC: an approach to rewriting request for elastic ABAC enforcement with dynamic policies. In: 2016 International Conference on Advanced Computing and Applications (ACOMP), pp. 25–31. IEEE (2016)

EGL: Efficient Graph Learning with Safety Constrains for Heterogeneous Trajectory Prediction

Weijie Lian[1], Yuze Wang[2], Ximu Zeng[2], Shuncheng Liu[2], Yuyang Xia[2], Huiyong Tang[3], and Han Su[1(✉)]

[1] Yangtze Delta Region Institute (Quzhou), University of Electronic Science and Technology of China, Quzhou, China
hansu@uestc.edu.cn

[2] School of Computer Science and Engineering, University of Electronic Science and Technology of China, Chengdu, China
{wangyuze,ximuzeng,liushuncheng,xiayuyang}@std.uestc.edu.cn

[3] Department of Information Centre, Zibo 148 Hospital, Zibo 255300, China

Abstract. Autonomous vehicles follow predicted trajectories to avoid obstacles and to drive safely. Any trajectory prediction algorithms that ignore the importance of safety constraints may lead to collisions in the real world, especially in a complex environment. In this paper, we tackle this problem with a novel trajectory prediction algorithm called Efficient Graph Learning (EGL), which utilizes road network information, self-attention mechanism, and safety constraints. In EGL, in order to obtain the road network information, real-world images are embedded in Convolutional Neural Networks (CNN), and then a Global Graph is constructed to present road network information and vehicle characteristics. The Global Graph is sent to an encoder-decoder model, which contains a Graph Attention Network(GAT) layer and Long Short-Term Memory (LSTM) layer in both the encoder and decoder. When the LSTM in the decoder outputs the results, EGL can generate all the predicted trajectories simultaneously. We evaluate EGL on some datasets. Experiment results show that EGL performs well in heterogeneous datasets, and outperforms other state-of-the-art methods with improving average displacement error (ADE) and final displacement error (FDE) by 7% and 27% respectively.

Keywords: Trajectory Prediction · Heterogeneous Traffic · Graph Attention Mechanism

1 Introduction

Trajectory prediction is an essential part of the autonomous driving issue since it helps to decide where autonomous vehicles should go in the next time step. Compared with sparse roads, vehicles are more likely to drive unsafely on high-density urban roads, for there exists more environment information, such as

A. El Abbadi et al. (Eds.): DASFAA 2023 Workshops, LNCS 13922, pp. 67–78, 2023.
https://doi.org/10.1007/978-3-031-35415-1_5

(a) Crossroad (b) Vehicle Road (c) Urban Road

Fig. 1. Example of Different Road Environments

pedestrians, lane counts, and crossroads. When the road becomes more complex, a trajectory prediction algorithm is required to take more information into consideration, including ego vehicle information and future trajectories of various targets such as surrounding vehicles, pedestrians, and bicycles. Therefore, it is challenging to make safe trajectory predictions on high-density urban roads. In this paper, to address the aforementioned challenge, we focus on the driving safety of trajectory prediction in complex driving environments. In addition, to make sure autonomous vehicle drive appropriately as well as safely, we consider the area of collision as a safety constraint and evaluate it accurately.

Missing road network information is a common limitation in the existing trajectory prediction methods. In other words, some methods [1,11] only consider the information about vehicles around them, but rarely take the information about the road network into consideration. These methods might perform well in a simple environment, like highways in [2] and other low-density situations. However, ignoring road network information can be very dangerous for complex urban roads. For example, if the trajectory prediction method does not consider the road network information, autonomous vehicles may hit the flower beds around the roadside in the next few time steps. Likewise, if the trajectory of pedestrians is not considered, autonomous vehicles might rush into pedestrians later. In addition, many prediction methods [7,12] do not consider safety constraints. In this paper, safety constraints are used to ensure that the predicted results do not lead to a collision between two agents, including vehicles, pedestrians, and bicycles.

To address the limitation of previous works, we propose a method called Efficient Graph Learning (EGL). EGL solves the following problems: (1)lack of road network information. To make a precise and safe prediction, autonomous vehicles should pay attention to road information and notice the differences in the road environment. As shown in Fig. 1(a), autonomous vehicles need to pay attention to pedestrians when passing crossroads, especially roads with a zebra crossing. Figure 1(b) shows a vehicle road where autonomous vehicles need to pay attention to the information of nearby vehicles. Figure 1(c) shows an urban road where autonomous vehicles are required to pay attention to the surrounding complex environment. (2)lack of safety constraints. Adding safety constraints allows agents to reasonably avoid colliding with other agents when making trajectory predictions, and thus ensure the safety of predictions. (3)lack of efficiency. If a method cannot predict the trajectories of all agents simultaneously, we need to use multiple models and run them respectively in multi-agents cases. This may waste a lot of computation resources.

The proposed EGL is a heterogeneous trajectory prediction method with the main contributions as follows: (1)We use Convolutional Neural Networks (CNN) to extract features of all environment images as road network information and correspond agents' features with road network information features to form a Global Graph at each time step. (2)We define security constraints in the loss function, which makes the predicted trajectory safe enough to avoid a collision. (3)We also use a batch mechanism to output the predicted trajectory of all agents in time t simultaneously. (4)Extensive experiments are conducted on a real-world dataset. Experiment results show that EGL achieves relatively good accuracy, and outperforms other state-of-the-art methods by improving average displacement error (ADE) and final displacement error (FDE) by 7% and 27% respectively.

2 Problem Definition

In this section, we briefly define the preliminary description of trajectory prediction and summarize the concepts in EGL.

Trajectory Point. At each time step, the position of an agent is represented by two-dimension coordinates. We assume that we can capture a set of heterogeneous traffic agents denoted as $P = \{p_i\}_{i=1,2,3...n}$, within the capture range of cameras, radars, and other sensors. Fixed frequency based on the time step of videos is used to capture agents. Since there are a few classes of agents, a trajectory sample point p includes not only the coordinate but also the category. For any time step t in a scene, the trajectory point of an agent p_i is represented as $f_i^t = \{x_i, y_i, w_i, l_i, v_{ix}, v_{iy}, t_i\}$, where (x_i, y_i) is the coordination of two-dimension space, w_i and l_i represent the width and length of the agent respectively; v_{ix} and v_{iy} represent the velocity in x-axis and y-axis; t_i denotes the type of agent. To simplify the complexity of the model, we consider 4 types of heterogeneous traffic agents in this paper, that is, $t_i \in \{1, 2, 3, 4\}$, where 1, 2, 3, 4 denote car, truck, pedestrian, and bike, respectively. The above definition can be naturally extended to any type of traffic agent.

Problem Definition. In this part, we introduce the problem definition in detail. For any time step t, EGL preprocesses each scene to obtain the trajectory point of every agent and the type of agent. With the observed trajectory points of all captured agents P in the time interval $[t - r + 1, t]$, EGL predicts the two-dimensional position of agents p_t ($p_t \in P$) during the future time interval $[t + 1, t + k]$. r and k are the histories and future length of time respectively.

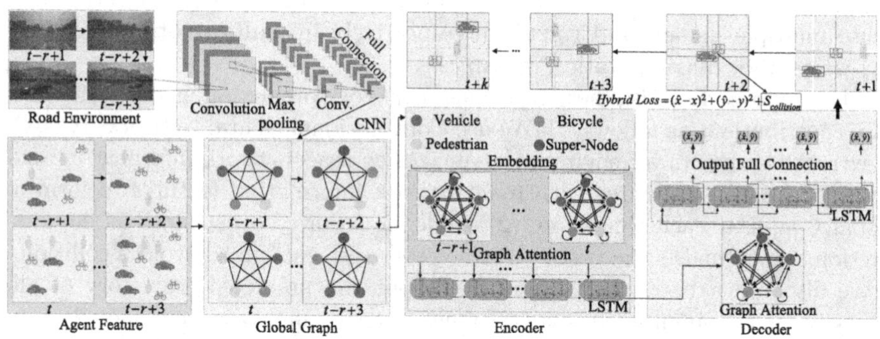

Fig. 2. Model Overview

3 Methodology

3.1 Model Framework

In this section, we introduce the proposed heterogeneous trajectory prediction method, EGL. As Fig. 2 shows, there are three parts in EGL: (1) feature extraction of the road network information; (2) encoder; (3) decoder. Firstly, with the input image which contains road network information, the feature of the road network in each time step is obtained by CNN. The road network feature is presented as a Super-Node in Global Graph, while other nodes in Global Graph present agents' features in vector form. Moreover, all nodes in Global Graph, including the Super-Node, are connected to each other. In this way, Global Graph can present the global connections of all agents well. Secondly, Global Graph is the input of Graph Attention Network(GAT) and LSTM of the encoder. For decoder, The GAT layer in the decoder gets the output of encoder, a fully-connected layer uses the result of LSTM layer as input to get the prediction trajectory of every agent.

3.2 Feature Extraction of Road Network Information

In EGL, CNN is used to extract road network information from images, since it outperforms other methods when we make a few attempts with different methods. Inspired by previous methods and to speed up the training process, the CNN architecture is designed as shown in Fig. 2. The CNN includes two convolutional layers, a max pooling layer, and a fully connected layer to concatenate the agents feature as an n-dimensional output vector. The process of CNN can be represented as:

$$e^t = CNN(image) \tag{1}$$

where $e^t \in R^L$ holds the information about the environment in the input image, existing as a Super-Node in Global Graph. e^t is calculated at each time step since the image and the environment for agents changes during time.

At the time t, we represent all trajectory points of agents as $\mathbf{F} = \{f_1^t, f_2^t, ..., f_N^t\}$, $f_i^t \in R^K$ where $\mathbf{N}$ is the number of agents on the road, $\mathbf{K}$ is the number of features in each node. The category of each agent needs also to be represented. In EGL, since there are four classes of agents, a four-dimension vector encoded by a one-hot encoding method, t_i, is used to illustrate the category of the *i-th* agent. For example, $t_i = [1, 0, 0, 0]$ means that the agent is in the first class, that is, cars.

$$\widetilde{f}_i^t = \theta(f_i^t) \tag{2}$$

where $\widetilde{f}_i^t \in R^{K'}$.

After that, we use a linear fully connected layer to extend the feature vector to the same dimension as the road network information vector.

$$\alpha_i^t = a(\mathbf{W}\widetilde{f}_i^t) \tag{3}$$

where $\alpha_i^t \in R^L$, L is the dimension of the feature vector of both feature nodes and the Super-Node, and $\mathbf{W} \in R^{L \times K'}$.

$\mathcal{A}^t = \{\alpha_1^t, \alpha_2^t, ..., \alpha_N^t, e^t\}$ is all the nodes in the Global Graph at time t, which includes all feature nodes and a Super-Node. In this way, Global Graph can present the features of road networks and agents. At each time step, a Global Graph is built to present the connection of all agents in the image, since the agents that show at different time steps are inconsistent. To simplify the input, we assume that the number of agents in all time steps remains the same. Since we believe that all agents have a certain influence on each other, the Global Graph is a fully connected graph. All $\mathcal{A}^t$ during the time interval of $[t - r + 1, t]$ are combined as the input of the encoder.

3.3 Encoder

In this section, we briefly introduce the encoder. As Fig. 2 shows, encoder consists of two components, a GAT layer, and an LSTM layer. Using the Global-Graph as input, GAT can calculate the attention weight between every two nodes and can transform the Global-Graph into attention vectors. The attention vectors can be later used in LSTM layer.

There are two reasons for the design of the GAT layer in the encoder. First, GAT solves the drawbacks of earlier approaches based on graph convolutions or their approximations. Second, GAT uses a masked self-attentional layer. In detail, the GAT layer can transform Global Graph into higher-level features, which use a shared linear transformation, parametrized by a weight matrix, $\mathbf{W} \in R^{L' \times L}$, which is applied to every node.

For any two nodes α_i^t, α_j^t, which are connected with an edge (α_i^t, α_j^t), the attention mechanism is applied to get the attention weight (i.e., α_{ij}^t) that indicates the importance of the edge. α_{ij}^t is the attention weight of the edge (α_i^t, α_j^t), which is calculated with a weight vector u^T.

$$\alpha_{ij}^{t} = \frac{exp\left(LeakyReLu(\boldsymbol{u}^{T}\left[\mathbf{W}a_{i}^{t}||\mathbf{W}a_{j}^{t}\right]\right))}{\sum_{k\in\mathcal{N}_{i}} exp\left(LeakyReLu(\boldsymbol{u}^{T}\left[\mathbf{W}a_{i}^{t}||\mathbf{W}a_{k}^{t}\right]\right)} \tag{4}$$

where $\cdot^{T}$ represents transposition and $||$ is the concatenation operation. After we get a_{ij}^{t}, the $\alpha_{i}^{t'}$ of the new $\mathcal{A}^{t}$ can be represented by

$$\alpha_{i}^{t'} = \sigma(\sum_{j\in\mathcal{N}_{i}} \alpha_{ij}^{t}\mathbf{W}\alpha_{j}^{t}) \tag{5}$$

Multi-head Attention. With the multi-head attention mechanism, we use K independent attention mechanism to execute the transformation of Eq. 4 and then concatenate the result to get the new α^{t}.

$$\alpha_{i}^{t'} = ||_{k=1}^{K}\sigma(\sum_{j\in\mathcal{N}_{i}} \alpha_{ij}^{t}\mathbf{W}^{K}\alpha_{j}^{t}) \tag{6}$$

or we can output the average of them.

$$\alpha_{i}^{t'} = \sigma(\frac{1}{K}\sum_{k=1}^{K}\sum_{j\in\mathcal{N}_{i}} \alpha_{ij}^{t}\mathbf{W}^{K}\alpha_{j}^{t}) \tag{7}$$

Whatever we chose the concatenate or average operation, we can get a new $\mathcal{A}^{t} = \left\{\alpha_{1}^{t'}, \alpha_{2}^{t'}, ..., \alpha_{N}^{t'}, e^{t'}\right\}$.

The $\mathcal{A}^{t}$ at each time step during the time interval $[t-r+1, t]$ is a unique one, which means $\mathcal{A}^{t}$ needs to be calculated again at every time step. The collection of $\mathcal{A}^{t}$ is represented as $\mathcal{A} = \left\{\mathcal{A}^{t-r+1}, \mathcal{A}^{t-r+2}, ..., \mathcal{A}^{t}\right\}$, which is the output of GAT layer in the encoder.

The next layer in the encoder is an LSTM layer. In this paper, we input the $\mathcal{A} = \left\{\mathcal{A}^{t-r+1}, \mathcal{A}^{t-r+2}, ..., \mathcal{A}^{t}\right\}$into LSTM layer, which outputs a hidden state as follows:

$$h^{t} = LSTM\left(v^{t}, h^{t-1}; W\right) \tag{8}$$

where the v^{t} is the embedding vector of $\mathcal{A}^{t}$, h^{t-1} denotes the hidden state at the $t-1$ time step, and W is the parameters of LSTM layer. The final hidden form h^{t} is calculated by recursively executing LSTM. h^{t} represents the last time step before we start making predictions, which contains the vehicle position information for that time step.

3.4 Decoder

Similar to the encoder, the decoder also includes a GAT layer and an LSTM layer. GAT of the decoder is used to capture the interactions between each predicted agent, while LSTM of the decoder is utilized to acquire the interactions and predict the future trajectories. To simplify the representation, we express the GAT layer as G and denote the parameters of GAT as W.

$$g^t = G\left(h^t; W\right) \tag{9}$$

LSTM can be a sequence generator since it can decode hidden states and predict results one at a time. As shown in Fig. 2, the output of the first hidden layer is served as the next input of the LSTM to complete the self-looping operation of the LSTM. For example, to predict trajectories in the future time interval $[t+1, t+k]$, h^t is regarded as the input of LSTM layer at time step $t+1$. EGL uses the weighted score g^t as the LSTM input and then outputs the position of agents at the next time step through a linear layer. The output is related to the required time interval for prediction. The first output is

$$h^{t+1} = LSTM\left(g^t, h_0; W\right) \tag{10}$$

where h_0 is the initial hidden state. Any output in the time interval $[t+1, t+k]$ can be represented as:

$$h^{t+x} = LSTM\left(h^{t+x}, h^{t+x-1}; W\right)(x > 1) \tag{11}$$

Finally, the output of the decoder is calculated by a linear layer as follows:

$$(\hat{x}, \hat{y})^{t+x} = Linear\left(h^{t+x}; W\right) \tag{12}$$

Hybrid Loss Function. Safety is a crucial constraint in heterogeneous traffic. To evaluate the safety of predicted trajectories, EGL calculates the size of the collision area as area loss. The total loss function combines the L2 distance and area loss as follows:

$$loss = \frac{1}{n}\sum_{i=1}^{n}\left((\hat{x}_\gamma^{t+i} - x_\gamma^{t+i})^2 + (\hat{y}_\gamma^{t+i} - y_\gamma^{t+i})^2 + S_{collision}^{t+i}\right) \tag{13}$$

where x_γ^{t+i} and y_γ^{t+i} are the real values of the spatial coordinates at $t+i$, $S_{collision}^{t+i}$ is the area of collision of two or more than two agents.

4 Experiments

4.1 Experimental Setup

Datasets. We evaluate the proposed method, EGL, on a publicly available dataset, ApolloScope [12]. ApolloScope is collected from the urban scene with high-density traffics. This satisfies the requirement of heterogeneous trajectory prediction in EGL. More specifically, ApolloScope contains real-world trajectories of heterogeneous traffic agents, including vehicles, bicycles, and pedestrians. The trajectory dataset in ApolloScope consists of camera-based images, LiDAR-scanned point clouds, and manually annotated trajectories.

Evaluation Metrics. EGL is compared with baseline methods based on two metrics, average displacement error (ADE) and final displacement error (FDE). ADE is the Euclidean distance between the ground truth trajectories and the predicted trajectories averaged overall predicted time steps. FDE is the Euclidean distance between the ground truth destination and the predicted destination at the last predicted time step (Table 1).

Table 1. ADE and FDE of baselines and EGL

ADE/FDE	Baselines				EGL	
Target Categories	LSTM-ED	S-LSTM	TrafficPredict	TraPHic	Single-head GAT	**Multi-head GAT**
Vehicle	8.49/14.65	8.37/13.75	7.94/10.58	6.85/10.71	6.05/8.25	**5.56/7.45**
Pedestrian	5.44/7.65	5.02/6.23	5.33/7.42	3.17/5.43	4.71/5.73	**4.48/5.23**
Bicycle	8.21/12.96	7.68/11.42	7.26/12.86	5.22/7.12	6.09/7.44	**5.06/5.57**
Average	7.52/12.22	7.23/11.04	7.08/10.07	5.42/8.42	5.53/6.78	**5.13/6.11**

Baseline. We compare EGL with the previous methods.

- LSTM Encoder-Decoder (LSTM-ED): A traditional architecture widely used in trajectory prediction.
- Social LSTM (S-LSTM): An LSTM-based model equipped with grid-based social pooling layers that are used to predict pedestrian trajectories in crowds.
- TrafficPredict: An LSTM-based architecture based on a 4D graph to learn interactions between different agents.
- TraPHic: An LSTM-based network using horizon state-space convolution and neighbor state-space convolution to learn interactions between agents in different ranges.

Implementation Details. We use datasets collected in heterogeneous environments as data in experiments. In addition, we use a multi-head attention mechanism in the GAT of the encoder and use two layers of self-attention to calculate the average result. The LSTM of the encoder is set to 2 layers and has 100 units of the hidden state. In the same way, the GAT of the decoder is set to two-layer self-attention and the LSTM of the decoder is set to 2 layers. EGL uses SGD [15] as an optimizer for every 50 epochs and sets batch size as 100. In this paper, a time step is set as 0.5 s. EGL observes a trajectory for 100 time steps and then predicts the future trajectory in the next 40 time steps.

4.2 Evaluation of Trajectory Prediction

To evaluate the validity of EGL, we compare the experiment results of baselines and EGL and give a reasonable explanation of the superiority of EGL.

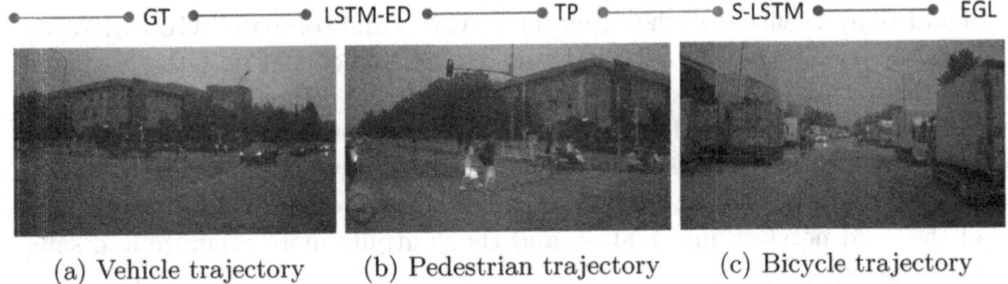

(a) Vehicle trajectory (b) Pedestrian trajectory (c) Bicycle trajectory

Fig. 3. Predicted Trajectory

Effectiveness of Multi-head Attention. In this experiment, we try to analyze the effectiveness of the multi-head attention mechanism in EGL. Specifically, we compare ADE and FDE when the GAT in EGL uses different attention mechanisms, including single-head attention, multi-head attention with concatenation operation, and multi-head attention with the average operation. Experiment results show that ADE and FDE of single-head attention models are lower than those using multi-head attention, while the effect of multi-head attention with concatenation operation is lower than multi-head attention with the average operation. The reason for these results is that the multi-head attention mechanism can pay attention to the subtle cues of agents and can average the results, which can help to save storage space and to achieve better results.

Comparison of EGL and Baselines. ADE and FDE are used to compare the performance of baselines and EGL. LSTM-ED has the worst performance among all the methods. This is caused by the limitation of the number of features encoded in LSTM-ED. Because both S-LSTM and TrafficPredict attempt to jointly anticipate every agent at once, their performances are comparable. The forecasts typically have a bigger variation because they don't have target agents and can't concentrate on every encounter. Additionally, neither of these two models can fully capture the heterogeneous influence. Due to its capacity to learn interactions across many ranges, TraPHic performs better than earlier approaches. It still is unable to discern between interactions with various agents, though. Due to the fact that we account for every traffic agent and assign them adaptive weights to recognize heterogeneous influence, the proposed method is notably superior to other approaches. Compared with others, EGL improves ADE and FDE by 7% and 27% respectively.

4.3 Case Study

A case study about performance in different traffic environments is implemented, using images in the ApolloScope dataset as input. As shown in Fig. 3, there are three different environments. Specifically, Fig. 3(a) shows a car goes straight through the intersection. Figure 3(b) shows pedestrians in the zebra crossing. Figure 3(c) shows a bicycle on the road where parked vans are around.

In addition, we visualize the experiment results and compare EGL with other methods such as LSTM-ED, S-LSTM, and TrafficPredict. In Fig. 3, the predicted trajectories of these methods are illustrated in different colored lines. Experiment results show that compared with the other three methods, EGL achieves the best performance in all three environments. Road network information plays a key role in the accuracy and safety of predicted trajectories. EGL makes full use of the road network information and thus outputs more accurate and safer prediction results.

5 Related Work

In this section, we use the concept of Graph Neural Networks and LSTM models for sequence prediction. In the following work, we will present the relevant research.

5.1 Graph Neural Network

In the research of previous work, Convolutional Neural Networks (CNNs) [13] have been successfully applied to tackle problems such as image classification. Graph Neural Networks (GNN) [5] was introduced by Gori and Scarselli et al. (2009) as a generalization of recursive neural networks that can directly deal with a more general class of graphs, e.g. cyclic, directed, and undirected graphs. Then there is an increasing interest in generalizing convolutions to the graph domain, such as Graph Convolutional Networks [19].

5.2 LSTM for Sequence Prediction

There is a large amount of data in the form of sequences, including sound, text, and video, which can be used in the application of deep learning. Machine translation [3], and speech recognition [18], to mention a few, have all benefited greatly from the use of sequence generation in recurrent neural networks (RNNs), such as LSTM [9] and Gated Recurrent Units (GRU). Although RNNs perform well in these areas, a single RNNs model cannot achieve good results. To address this problem, inspired by Vaswani [16], who proposed an attention mechanism, and Veličković [17], who proposed a graph attention network, we add the attention mechanism before the LSTM layer and get a good result. Some research [1,4,7,8,10,14] propose to use lstm sequence prediction for pedestrian trajectory prediction, and one of which [6] proposes the impact of pedestrian trajectory on vehicles.

6 Conclusion

In this paper, we propose a novel trajectory prediction method, EGL, with road network information and safety constraints. The self-attention mechanism, LSTM, and encoder-decoder framework are used in EGL. Moreover, EGL extracts road network information by making full use of the road network images and combines it with the agent feature to form a graph, called Global Graph, in each time step. The Global Graph is put into GAT to obtain the attention weight between every two agents. The LSTM layer in the encoder is used to observe historical trajectory confidence before feeding it into the next GAT, and the LSTM layer in the decoder gets the prediction sequence and then predicts multiple agents at the same time. Experiment results show that road network information and safety constraints can assign reasonable attention to agents, and EGL can predict accurate and safe trajectories in different scenes.

Acknowledgment. This work is partially supported by NSFC (No. 61972069, 61836007, 61832017, 62272086), Shenzhen Municipal Science and Technology R&D Funding Basic Research Program (JCYJ20210324133607021), and Municipal Government of Quzhou under Grant (No. 2022D037, 2021D022).

References

1. Alahi, A., Goel, K., Ramanathan, V., Robicquet, A., Fei-Fei, L., Savarese, S.: Social LSTM: human trajectory prediction in crowded spaces. In: Proceedings of the IEEE Conference on Computer Vision and Pattern Recognition, pp. 961–971 (2016)
2. Altché, F., de La Fortelle, A.: An LSTM network for highway trajectory prediction. In: 2017 IEEE 20th International Conference on Intelligent Transportation Systems (ITSC), pp. 353–359. IEEE (2017)
3. Bahdanau, D., Cho, K., Bengio, Y.: Neural machine translation by jointly learning to align and translate. arXiv preprint arXiv:1409.0473 (2014)
4. Diao, Y., Su, Y., Zeng, X., Chen, X., Liu, S., Su, H.: Astral: an autoencoder-based model for pedestrian trajectory prediction of variable-length. In: Rage, U.K., Goyal, V., Reddy, P.K. (eds.) DASFAA 2022. LNCS, vol. 13248, pp. 214–228. Springer, Cham (2022). https://doi.org/10.1007/978-3-031-11217-1_16
5. Gori, M., Monfardini, G., Scarselli, F.: A new model for learning in graph domains. In: Proceedings of the 2005 IEEE International Joint Conference on Neural Networks, vol. 2, pp. 729–734 (2005)
6. Gupta, A., Johnson, J., Fei-Fei, L., Savarese, S., Alahi, A.: Social GAN: socially acceptable trajectories with generative adversarial networks. In: Proceedings of the IEEE Conference on Computer Vision and Pattern Recognition, pp. 2255–2264 (2018)
7. Haddad, S., Wu, M., Wei, H., Lam, S.K.: Situation-aware pedestrian trajectory prediction with spatio-temporal attention model. arXiv preprint arXiv:1902.05437 (2019)
8. Helbing, D., Buzna, L., Johansson, A., Werner, T.: Self-organized pedestrian crowd dynamics: Experiments, simulations, and design solutions. Transp. Sci. **39**(1), 1–24 (2005)
9. Hochreiter, S., Schmidhuber, J.: Long short-term memory. Neural Comput. **9**(8), 1735–1780 (1997)
10. Liu, J., Shahroudy, A., Xu, D., Wang, G.: Spatio-temporal LSTM with trust gates for 3D human action recognition. In: Leibe, B., Matas, J., Sebe, N., Welling, M. (eds.) ECCV 2016. LNCS, vol. 9907, pp. 816–833. Springer, Cham (2016). https://doi.org/10.1007/978-3-319-46487-9_50

11. Liu, S., Chen, X., Wu, Z., Deng, L., Su, H., Zheng, K.: HeGA: heterogeneous graph aggregation network for trajectory prediction in high-density traffic. In: Proceedings of the 31st ACM International Conference on Information & Knowledge Management, pp. 1319–1328 (2022)
12. Ma, Y., Zhu, X., Zhang, S., Yang, R., Wang, W., Manocha, D.: TrafficPredict: trajectory prediction for heterogeneous traffic-agents. In: Proceedings of the AAAI Conference on Artificial Intelligence, vol. 33, pp. 6120–6127 (2019)
13. O'Shea, K., Nash, R.: An introduction to convolutional neural networks. arXiv preprint arXiv:1511.08458 (2015)
14. Rudenko, A., Palmieri, L., Herman, M., Kitani, K.M., Gavrila, D.M., Arras, K.O.: Human motion trajectory prediction: a survey. Int. J. Robot. Res. **39**(8), 895–935 (2020)
15. Ruder, S.: An overview of gradient descent optimization algorithms. arXiv preprint arXiv:1609.04747 (2016)
16. Vaswani, A., et al.: Attention is all you need. In: Advances in Neural Information Processing Systems, vol. 30 (2017)
17. Veličković, P., Cucurull, G., Casanova, A., Romero, A., Lio, P., Bengio, Y.: Graph attention networks. arXiv preprint arXiv:1710.10903 (2017)
18. Xiong, W., Wu, L., Alleva, F., Droppo, J., Huang, X., Stolcke, A.: The Microsoft 2017 conversational speech recognition system. In: 2018 IEEE International Conference on Acoustics, Speech and Signal Processing (ICASSP), pp. 5934–5938. IEEE (2018)
19. Zhang, S., Tong, H., Jiejun, X., Maciejewski, R.: Graph convolutional networks: a comprehensive review. Comput. Soc. Netw. **6**(1), 1–23 (2019)

MeFormer: Generating Radiology Reports via Memory Enhanced Pretraining Transformer

Fang Li[1], Pengfei Wang[1(✉)], Kuan Lin[2], and Jiangtao Wang[1]

[1] East China Normal University, Shanghai 200062, China
{fang.li,pfwang}@stu.ecnu.edu.cn, jtwang@sei.ecnu.edu.cn

[2] Aerospace Information Research Institute, Chinese Academy of Sciences, Beijing 100190, China
linkuan@aircas.ac.cn

Abstract. Writing a radiology image report is a very time-consuming and tedious task. Using AI to generate the report is an efficient approach, but there are still two significant challenges. First, the model requires to be fine-tuned regularly with the increasing number of patients; Secondly, the quality of text generation needs to be improved because medical observations are complex. In order to solve above challenges, we propose Memory Enhanced Pretraining Transformer (MeFormer). It uses the pretrained Vision Transformer, which efficiently reduces the number of training parameters and transfers fruitful knowledge for the downstream task. At the same time, memory module was introduced into Transformer. The salient pattern in radiology reports are memorized through this design, and they can serve as cross-references during text generation, moderately enhancing the quality of generated diagnostic reports. Extensive experiments on two datasets show that our method achieves comparable performance to other state-of-the-art methods.

Keywords: Medical report generation · Medical data mining · Transformer model

1 Introduction

Medical images such as radiology and pathology are widely used for clinical diagnosis [8]. When analyzing a radiology image, radiologists must identify both normal and abnormal areas before writing a corresponding report that incorporates their medical knowledge and clinical practice experience. In some large hospitals, many radiology images are generated daily, which is a massive workload for radiologists and can be misdiagnosed or missed for human doctors [3]. In order to free radiologists from the time-consuming and tedious task and at the same time to help inexperienced radiologists, the automatic generation of diagnostic radiology reports has become an essential task in clinical practice. This

A. El Abbadi et al. (Eds.): DASFAA 2023 Workshops, LNCS 13922, pp. 79–94, 2023.
https://doi.org/10.1007/978-3-031-35415-1_6

technology can save time and reduce the tediousness of report writing while improving diagnostic accuracy.

In recent years, computer vision algorithms have achieved great success in the task of generating captions for natural images [25]. Similarly, radiology report generation has become a hot research topic, attracting many researchers to study it. However, this task faces some difficulties. **Long text description problem:** radiology reports are long narrative texts consisting of multiple long sentences, including regular and abnormal features, as well as salient observations. Current image captioning algorithms mainly focus on generating short texts and are unable to produce satisfactory result. **Model update problem:** In the medical field, models often needs to be fine-tuned based on local datasets. This fine-tuning process requires rapid iteration and fast convergence.

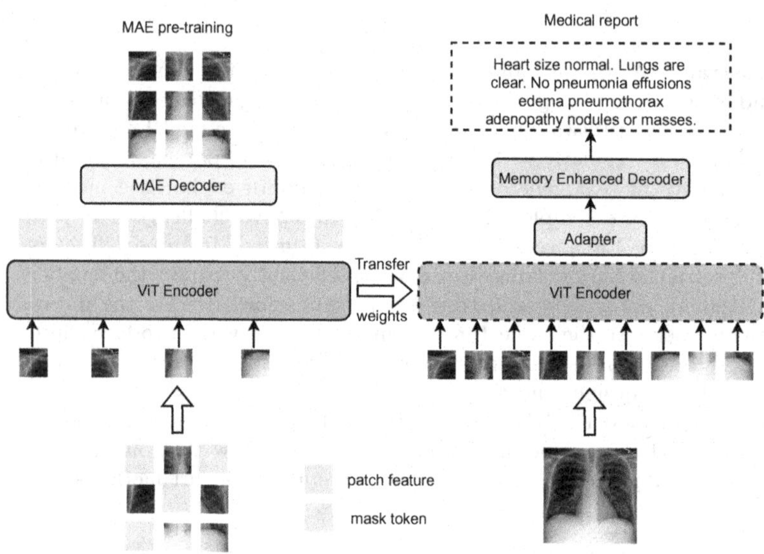

Fig. 1. Pipeline for MeFormer. In the first stage (shown on the left), ViT is pre-trained with MAE as an encoder. In the second stage (shown on the right), the pre-trained ViT weights are transferred to initialize the model encoder and the adapter is added to reduce semantic gap. Then the whole Memory Enhanced Decoder is finetuned to perform radiology report generation.

Many currently existing algorithms for radiology report generation [17,27, 35] are mainly based on the CNN-HRNN [17] architecture, together with the attention mechanism, is a classical encoder-decoder architecture. This type of method is divided into two steps; in the encoding stage, visual features of the input image will be extracted using a visual feature extractor (CNN), and in the decoding stage, the entire radiology report is generated using RNN. Such RNN-based generation methods usually face a series of problems, including the lack of fluency of the generated diagnostic reports, insufficient language diversity, and even the inclusion of repetitive statements, and a series of problems such as poor text generation quality, as well as the inability to parallelize the computation due to the characteristics of the RNN model itself, long computation time and difficulty in handling the dependencies between remote locations.

With the development of the Transformer [29] model, compared to RNN, Transformer can parallelize the computation, computation speed is faster, and at the same time has a strong text generation capability. Chen et al. [6,7] utilized the Transformer as the decoder of the model while cooperating with the memory network can achieve better radiology report generation. Liu et al. [23] proposed the Posterior-and-Prior Knowledge Exploring-and-Distilling approach (PPKED), which is a Transformer model divided into three parts, PoKE incorporates posterior knowledge such as some keywords of radiology, PrKE incorporates relevant reports and medical knowledge graph, and MKD is responsible for the final text generation. However, in these designs, the Transformer-based model needs to consider the need for rapid model updates in real-world scenarios. There are inconsistencies and redundancies in the previously generated sentences.

To solve above problems, this paper proposes the Memory Enhanced Transformer (MeFormer) model, as shown in Fig. 1. This algorithm mainly addresses the high time cost of model updates in the medical field and the problem of unsatisfactory quality in the generation of long text diagnostic reports. The algorithm uses a pre-trained visual model as the visual encoder for the entire model. When we need to fine-tune the model based on new training data, we only need to update the text generation part of the model to accelerate the convergence of the model; for the current unsatisfactory results in long text diagnosis report generation, we introduce external memory to assist in text generation, by introducing external memory, we can model and remember the similar patterns of different medical diagnosis texts during the text generation process, which can help Transformer to decode and generate longer text medical reports with richer content.

We propose an effective and simple approach to radiology report generation named Memory Enhanced Transformer (MeFormer) model. The main contributions of this paper are summarized as follows:

- We utilize MAE [11] pre-training to extract the visual representation of radiology images using a pre-trained ViT [10] model. During training, we do not update the parameters of ViT, which significantly reduces the number of parameters to be updated and accelerates the convergence time of the model. This approach improves the overall performance of the model.
- We introduce an external memory module into the Transformer decoder. Medical reports often have a highly patterned nature, and external memory can help the model to save these patterns and use them as mutual references in the report generation process. This improvement can significantly enhance the quality of the generated reports.

- Our approach has been tested in a large number of experiments, and we can prove that it achieves comparable performance to other state-of-the-art methods.

2 Related Work

2.1 Image Captioning

The task of image caption [5,30] has attracted a large number of researchers to study it. These current approaches mainly employ an encoder-decoder architecture, with most models employing CNNs as encoders to extract visual features of images and RNNs as decoders to generate textual descriptions of images. Among them, Show-Tell [30] is the most classical neural image caption model, which provides an end-to-end framework that uses the image features extracted by the CNN as the input to the LSTM. These frameworks proposed so far [1,22,25,27,34] achieve state-of-the-art results on this task. Nevertheless, these methods generate knowledge in a simple sentence. In contrast, radiology reports need to generate paragraphs containing multiple sentences, each containing observations for a specific medical region.

2.2 Report Generation

Since radiology reports are highly schematic, some early retrieval-based work existed. For example, Liu et al. [24] found that a simple retrieval-based approach could achieve outstanding results on this task. Li et al. [20] combined retrieval-based and template-based generation methods also achieved some results on this task. Although retrieval-based approaches can achieve better results, they still have some drawbacks, requiring larger datasets or a large number of pre-constructed templates to guide the generated sentences of the model. The generation of radiology reports requires visual and textual data to train the model, which naturally requires linking image features with textual features.

As the self-attention mechanism is widely used, there are now some methods [17,36] that use the self-attention mechanism, where Jing et al. [17] proposed a multi-task, combined with the co-attention hierarchical model to automatically predict keywords and paragraphs. Yin et al. [36] proposed a topic-matching mechanism that projects sentence RNNs and corresponding ground truths to the same embedding space so that they have the same semantics. This approach requires a large amount of task-relevant additional annotated data and external medical datasets or knowledge and may still need to be made available. Therefore this type of approach cannot be widely applied and developed. Moreover, labels and disease categories only sparsely cover the reported information, leading to relatively loosely correlated visual and textual features.

With the widespread use of the Transformer model, some new algorithms have used Transformer as the text generator of the model recently. Compared with RNN, Transformer can compute in parallel, with faster computation speed and stronger text generation capability. Chen et al. [6,7] utilized Transformer as the decoder of the model and, together with the memory network, achieved better text generation results. Inspired by R2Gen [7], this paper also uses memory to enhance the Transformer text generation effect; PPKED [23] uses the Transformer, where textual and visual features are aligned in the Transformer encoder, and prior external knowledge, knowledge graphs, and textual information from existing reports are introduced.

2.3 Self-Supervised Methods in Medical Imaging

Recent advancements in self-supervised representation learning show that contrastive learning (CL) [4,12] and masked image modeling (MIM) [11,33] are effective pretraining strategies for the Vision Transformer (ViT) in the public domain where infinite images are available. Contrastive learning encourages augmentations (views) of the same input to have more similar representations than augmentations of different inputs. This concept might not work directly for radiography imaging because the negative pairs usually appear too similar. [14,37] introduce multimodal information into the CL framework, but they all depend on paired medical images and texts, so they encounter the lacking data problem. MedCLIP [31] decouples images and texts for multimodal contrastive learning using a semantic matching loss based on medical knowledge. On the contrary, MIM masks a set of image patches before input into the Transformer and reconstructs these masked patches at the output, which is good at conserving fine-grained textures embedded in the image context. [32,38] investigate this framework for medical Image analysis, such as disease classification and segmentation. Unlike them, we take masked autoencoders (MAE) as a pretraining task for their simplicity and efficiency. We use a slim adapter for the cross-modal downstream task without re-training again. In summary, we use single modality data to pretraining and apply for cross-modality tasks.

3 The Proposed Method

3.1 Problem Definition

Here we give the problem formulation, given a radiology image that is encoded as I, our goal is to go about generating a descriptive radiology report $R = \{y_1, y_2, ..., y_N\}$, where y represents the generated words and N represents the number of generated words.

3.2 Model Architecture

The MeFormer architecture is shown in the Fig. 2, MeFormer consists of components (1) encoder that contains ViT and Adapter encoder, (2) Memory Enhanced decoder, and (3) Memory Updater.

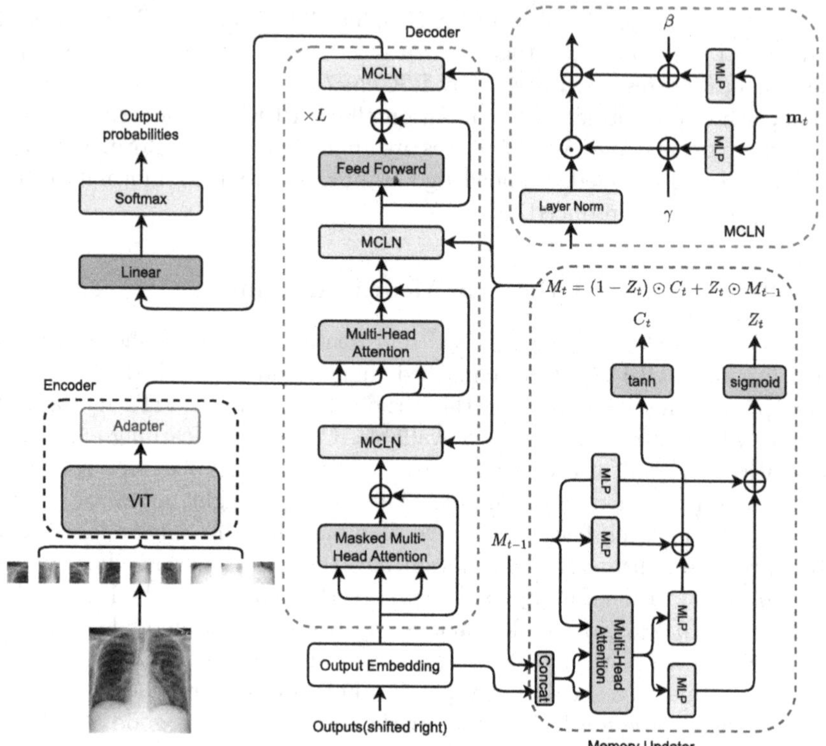

Fig. 2. MeFormer Model Architecture, the module in the red dotted line in the figure is MCLN, is introduced in Sect. 3.3, the module in the blue dotted line in the figure is memory updater, which is introduced in Sect. 3.3, and the module in black dotted line in the figure is an encoder, which contains ViT that is frozen in the training stage and adapter encoder. (Color figure online)

Pre-trained ViT with Masked Autoencoders. We use a pre-trained ViT as our visual feature extractor. This section illustrates the encoder, decoder, and loss function in MAE.

The encoder of MAE is ViT. Encoder input is many fixed size image patches, which contains unmasked patches and masked patches, and ViT will only calculate the unmasked patches.

The input of MAE's decoder is the output of the encoder, and the goal of pre-training is to reconstruct the pixel values of the masked patches. Decoder will not be used in downstream tasks.

The loss of MAE is the reconstruction loss, explicitly using the mean squared error by predicting the pixel values of the masked patches.

In this paper, given a chest X-ray image Img, its visual features $\mathbf{X}$ are extracted by a pre-trained CNN (e.g., VGG [28], ResNet [13], etc.) or ViT [10]. The result of the image having been encoded by the pre-trained ViT can be used as an input sequence for the subsequent modules of the model, and this process can be formalized as

$$\{\mathbf{x}_1, \mathbf{x}_2, ..., \mathbf{x}_N\} = f_{ve}(Img) \tag{1}$$

where $f_{ve}(\cdot)$ represents visual feature extractor.

The specific pre-training method is described in detail in the section experiment later.

Adapter. Since the pre-training task of our ViT is to reconstruct the pixel values of images. The task of this paper is text generation, there is a particular gap between these two tasks, so we introduce a 3-layer standard Transformer encoder as an adapter to reduce this gap. The 3-layer Transformer encoder has very few parameters, which will not significantly increase the training time. The adapter can simultaneously exploit the knowledge stored in the original MAE pre-training and then freshly update knowledge from generating the report. The adapter input is the visual feature $\mathbf{x}_i$ extracted by the pre-trained ViT, and the output is the hidden state $\mathbf{h}_i$, and this process can be formalized as

$$\{\mathbf{h}_1, \mathbf{h}_2, ..., \mathbf{h}_N\} = f_e(\mathbf{x}_1, \mathbf{x}_2, ..., \mathbf{x}_N) \tag{2}$$

where $f_e(\cdot)$ represents Transformer encoder.

Memory Enhanced Decoder. As shown in Fig. 2, the main part of the decoder comes from the Transformer decoder part. For the layer normalization (LayerNorm) in each decoder layer, this paper uses MLCN to improve it and introduces an external memory. The decoding process can be formalized as

$$y_t = f_d(\mathbf{h}_1, \mathbf{h}_2, ..., MLCN(Mem(y_1, y_2, ..., y_{t-1}))) \tag{3}$$

where $f_d(\cdot)$ represents the decoder, the details of Mem and MCLN will be described later.

Training Target. According to the previous description, the entire text generation process can be defined recursively by the following equation as

$$p(Y|Img) = \prod_{t=1}^{T} p(y_t|y_1, ..., y_{t-1}, Img) \tag{4}$$

where $y = \{y_1, y_2, ..., y_T\}$ is the target text sequence, and the optimization objective of the model is to optimize the negative conditional log-likelihood of Y to maximize $p(Y|Img)$ given Img. Where θ is the model parameter.

3.3 Memory Updater

For any related image, specific patterns may be shared between them in the text report. These patterns can be used as cross-references in the text generation process, improving the quality of the text report generation. For example, this pattern, "The heart is normal in size." and "The heart size within normal limits." will often appear in reports of similar images and show consistency. In order to allow the model to take advantage of this feature, we introduce an external memory module to enhance the Transformer's ability to learn patterns and facilitate the interaction between the computational model and the generative process.

We introduce an external memory module that uses a memory matrix to store important pattern information during each iteration to meet our expectations. Each row of the matrix represents some pattern information. The memory matrix is updated iteratively during text generation, and the previous memory matrix is used during each update. At time step t, the generated memory matrix $\mathbf{M}_{t-1}$ of the previous iteration is used as the query of the multi-head attention. Then $\mathbf{M}_{t-1}$ is concatenated with the output of the previous iteration as the keys and values of the multi-head attention. The Transformer module in this paper has H heads of the multi-head attention module, and by changing linearly, there are H query matrices, key matrices, and value matrices. For each head, we get the query matrix, key matrix, and value matrix in the memory module, which can be formalized as

$$\begin{aligned} \mathbf{Q} &= \mathbf{M}_{t-1}\mathbf{W_q} \\ \mathbf{K} &= [\mathbf{M}_{t-1}; \mathbf{y}_{t-1}]\mathbf{W_k} \\ \mathbf{V} &= [\mathbf{M}_{t-1}; \mathbf{y}_{t-1}]\mathbf{W_v} \end{aligned} \tag{5}$$

$[\mathbf{M}_{t-1}; \mathbf{y}_{t-1}]$ represents the matrix $\mathbf{M}_{t-1}$ and the matrix $\mathbf{y}_{t-1}$ concatenated in the row direction. The $\mathbf{W_q}$, $\mathbf{W_k}$, and $\mathbf{W_v}$ are the three linear layer weights of query, key, and value, respectively, after which S_t is obtained by multi-headed attention calculation. Inspired by this work [19], as shown in part contained by the blue dashed line on the right side of Fig. 2. The external memory module in this paper is used to enhance the video caption. Since the images and texts within a short video clip have a powerful correlation, the external memory can be used to remember similar features and thus enhance the text generation capability of the Transformer. Inspired by this paper, we found that medical diagnosis texts are similar and can also use this external memory to enhance text generation. The design of the memory updater in this paper can be formalized as

$$\begin{aligned} \mathbf{S}_t &= \text{MultiHeadAtt}(\mathbf{M}_{t-1}, [\mathbf{M}_{t-1}; \mathbf{y}_{t-1}], [\mathbf{M}_{t-1}; \mathbf{y}_{t-1}]) \\ \mathbf{C}_t &= \tanh(\mathbf{W}_{mc}\mathbf{M}_{t-1} + \mathbf{W}_{sc}\mathbf{S}_t + \mathbf{b_c}) \\ \mathbf{Z}_t &= \text{sigmoid}(\mathbf{W}_{mz}\mathbf{M}_{t-1} + \mathbf{W}_{sz}\mathbf{S}_t + \mathbf{b_z}) \\ \mathbf{M}_t &= (1 - \mathbf{Z}_t) \odot \mathbf{C}_t + \mathbf{Z}_t \odot \mathbf{M}_{t-1} \end{aligned} \tag{6}$$

where $\odot$ represents Hadamard product, $\mathbf{W}_{mc}$, $\mathbf{W}_{sc}$, $\mathbf{W}_{mz}$, $\mathbf{W}_{sz}$ represents learnable weights, $\mathbf{b_c}$, $\mathbf{b_z}$ represents learnable bias term. $\mathbf{C}_t$ is the internal cell state, $\mathbf{Z}_t$ is the update gate that determines which information from the previous memory state is retained, thus reducing redundancy and maintaining consistency of the generated passages.

To introduce memory into the transformer decoder, we use Memory-driven Conditional Layer Normalization (MCLN) [7].

As shown in the module in the red dotted line in Fig. 2, MCLN combines layer normalization and external memory to enhance the Transformer's decoding capabilities. For this purpose, MCLN introduces two critical parameters, γ and β, for scaling and moving the learned features.

$$f_{mcln}(\mathbf{r}) = (\gamma + f_{mlp}(\mathbf{m}_t)) \odot \frac{\mathbf{r} - \mu}{v} + (f_{mlp}(\mathbf{m}_t) + \beta) \tag{7}$$

where $\mathbf{r}$ represents the output of the previous module, μ and v represent the mean and standard deviation of $\mathbf{r}$, $f_{mlp}(\cdot)$ represents the multilayer perceptron and $\mathbf{m}_t$ represents the vector obtained by concatenating the memory matrix $\mathbf{M}_t$ by rows.

4 Experiment

4.1 Datasets, Metrics and Settings

Two publicly available datasets were chosen for this paper's experiments: the IU-Xray [9] and the MIMIC-CXR [18]. The ViT model was pre-trained using images from CheXpert [15].

IU-Xray. [9] is a widely used benchmark dataset frequently used to evaluate the performance of radiology report generation methods. The dataset contains 7,470 chest radiographs and 3,955 radiology reports associated with the images.

MIMIC-CXR. [18] is by far the largest radiology image dataset, and this dataset contains 473,057 chest X-ray images from 63,478 patients and 206,563 corresponding reports.

CheXpert. [15] is a large multitasking dataset of chest X-rays, and the dataset contains 224,316 chest X-ray images from 65,240 patients.

For the IU-Xray and MIMIC-CXR datasets, we follow the way Li et al. [20] preprocessed the data to exclude unreported images from the dataset. After that, we use the traditional way of dataset partitioning. Specifically, the IU-Xray dataset was divided into a training/validation/test set with a ratio of 7:1:2; for the MIMIC-CXR dataset, the official split was used.

Metrics. This paper adopt the widely-used metrics, BLEU [26], METEOR [2] and ROUGE-L [21]. In particular, BLEU and METEOR are originally proposed for machine translation evaluation. ROUGE-L is designed to measure the quality of summaries.

Settings. Our visual feature extractor is ViT-base, pre-trained on the CheXpert dataset using the MAE [11] method, with pre-training hyperparameters referenced to MAE. The parameters of the visual feature extractor are frozen throughout the model training. Eight heads and 512 hidden units are used in MHA, and the optimizer uses Adam with a batch size of 48, a learning rate of 3e−4, a weight decay of 5e−5, and amsgrad.

Table 1. Performance of the proposed MeFormer and other state-of-the-art algorithms on the MIMIC-CXR and IU-Xray datasets. A higher value denotes better performance in all columns.

Dataset	Methods	BLEU-1	BLEU-2	BLEU-3	BLEU-4	METEOR	ROUGE-L
MIMIC-CXR	CNN-RNN	0.299	0.184	0.121	0.084	0.124	0.263
	AdaAtt	0.299	0.185	0.124	0.088	0.118	0.266
	Att2in	0.325	0.203	0.136	0.096	0.134	0.276
	Up-Down	0.317	0.195	0.130	0.092	0.128	0.267
	Transformer	0.314	0.192	0.127	0.090	0.125	0.265
	R2Gen	0.353	0.218	0.145	0.103	0.142	0.277
	CMN	0.353	0.218	0.148	0.106	0.142	0.278
	PPKED	0.360	0.224	0.149	0.106	**0.149**	**0.284**
	MeFormer	**0.393**	**0.235**	**0.155**	**0.110**	0.148	0.272
IU-Xray	HRNN	0.439	0.281	0.190	0.133	–	0.342
	CoAtt	0.455	0.288	0.205	0.154	–	0.369
	HRGR-Agent	0.438	0.298	0.208	0.151	–	0.322
	CMAS-RL	0.464	0.301	0.210	0.154	–	0.362
	Transformer	0.396	0.254	0.179	0.135	–	0.342
	R2Gen	0.470	0.304	0.219	0.165	0.187	0.371
	CMN	0.475	0.309	0.222	0.170	0.191	0.375
	PPKED	0.483	0.315	0.224	0.168	–	**0.376**
	MeFormer	**0.492**	**0.318**	**0.225**	**0.170**	**0.203**	0.371

4.2 Main Results

In order to compare the algorithm proposed in this paper with currently existing methods in the field of image caption and radiology report generation, the following methods were selected in the text as the baseline methods for comparing the implementations. In the comparison experiments, the results of the experiments are taken from the original paper if not specifically stated. i.e., CNN-RNN [30], AdaAtt [25], Att2in [27], Up-Down [1], Transformer [7], HRGR-Agent [20], CMAS-RL [16], R2Gen [7], CMN [6], PPKED [23]. The baseline, as well as the algorithm proposed in this paper, is shown in Table 1. On the MIMIC-CXR dataset, the memory slots size is 2, the BLEU metric is ahead of the other models in all aspects, and the METEOR metric is almost equal to PPKED, with only the

ROUGE-L metric having some differences. On the IU-Xray data set, the memory slot size is 1. Better results on BLEU compared to other algorithms, and the METEOR is also much better than the other algorithms and only slightly lower than the other algorithms in the ROUGE-L. Overall the algorithm MeFormer proposed in this paper is quite effective and has greater advantages.

As shown in Fig. 3, we can find that for the normal images on the left side, we generate descriptions that are consistent with the ground truth, covering the parts that are usually described, whether the lungs are normal, whether the heart size is normal, and other issues. For the abnormal images on the right side, the text we generate can be descriptive for normal areas, such as the marked red words "heart" and "lung", and can also determine symptoms such as pneumothorax or effusion. In particular, our method can detect degenerative changes in the spine.

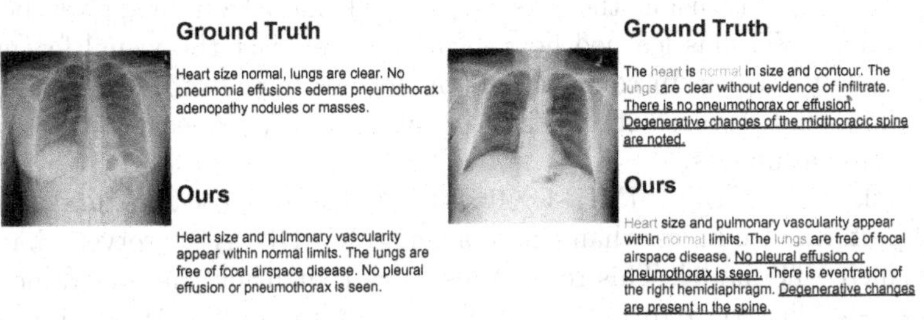

Fig. 3. Sample cases of MeFormer on IU-Xray. The words marked in red indicate some of the organs mentioned in the report, and the orange words are degree descriptions. Underlined sentences indicate the sentences where our results match the ground truth descriptions. (Color figure online)

4.3 Quantitative Analysis

In order to investigate how the various components of the model affect the quality of the model-generated text, this paper conducts some ablation experiments as a way to analyze the effect of pretrained ViT on the quality of the model-generated text and the effect of external memory modules on the model.

To demonstrate the effectiveness of the present method, we have tested it on two benchmark datasets, and the results show in Table 1 that our proposed method outperforms the existing models in most of the metrics. Also, in this paper, an ablation experiment was conducted on the MIMIC-CXR dataset, and the results are shown in Table 2.

4.4 Effect of Pretrained ViT

The BASE model, using vanilla Transform as the text generator of the model, and the pretrained ViT-base model on ImageNet as the visual feature encoder,

Table 2. Ablation experiments of our method. The Base model is the implementation of vanilla Transformer models.

DATA	MODEL	NLG METRICS					
		BLEU-1	BLEU-2	BLEU-3	BLEU-4	METEOR	ROUGE-L
MIMIC-CXR	BASE	0.319	0.198	0.134	0.096	0.129	0.272
	+pretrained ViT	0.340	0.211	0.143	0.103	0.137	**0.278**
	+MeFormer	0.378	0.223	0.145	0.103	0.138	0.265
	+pretrained ViT+MeFormer	**0.393**	**0.235**	**0.155**	**0.110**	**0.148**	0.272

also do not update the parameters of this part during the model training. Compared with the BASE+pretrained ViT model, BLEU, METEOR, and ROUGE-L all show moderate improvement. This experimental result can show that using the pre-trained ViT as the visual feature encoder of the model is very effective compared to directly using the pre-trained ViT model on ImageNet because the pre-trained ViT has learned how to properly extract the visual features of the chest X-ray images. In contrast, the model pretrained on ImageNet cannot extract the visual features of the chest X-ray images well because the training data are natural images.

In particular, on the MIMIC-CXR dataset, the proposed method achieves optimal performance in less than 8 h on a single NVIDIA A100 server, thanks to a pre-trained ViT model. This result illustrates that using the pre-trained ViT model can significantly reduce the training time of the model. When fine-tuning the model with new data, the fine-tuning process can be done faster, significantly reducing the time cost.

4.5 Effect of Memory

In the training stage of the model, the initial memory uses randomly obtained word embeddings from the training set. In the inference stage, several word embeddings are randomly selected from the word list as the initial memory. In addition to this way of initializing the memory matrix, we have also tried using the identity matrix and random matrix initialization. However, the effect could be worse than the way adopted in this paper, probably because the memory matrix needs better initialization to improve the effectiveness of the Transformer. By observing the experimental results, BASE+MeFormer and BASE+pretrained ViT+MeFormer are ahead of the BASE in most NLG (Natural language generation) metrics. This phenomenon demonstrates the effectiveness of introducing external memory modules into the Transformer since the text in radiology reports is highly patterned, and these patterns are reasonably modeled.

4.6 Effect of Memory Size

To show the impacts of memory size, as shown in Table 3, we experiment with different memory sizes on two datasets and found that there is a consistent phenomenon for both datasets, i.e., larger memory size significantly degrades the performance of the model. Smaller memory size benefits the performance of the model. Smaller memory slots can be used as information supplements for the Transformer decoder while saving valuable memory space. At the same time, a larger memory size will have too much redundant and invalid information, thus bringing many adverse effects on the model text generation process. Also, larger memory sizes introduce more parameters and take up more memory during the training or inference process. This observation shows that the proposed model effectively and efficiently uses memory for learning in radiology report generation tasks.

Table 3. The effect of memory size on the model

dataset	memory slots	NLG metrics					
		BLEU-1	BLEU-2	BLEU-3	BLEU-4	METEOR	ROUGE-L
MIMIC-CXR	1	0.384	0.234	**0.157**	**0.113**	0.147	**0.278**
	2	**0.393**	**0.235**	0.155	0.110	**0.148**	0.272
	3	0.377	0.222	0.144	0.100	0.143	0.268
	4	0.349	0.202	0.128	0.087	0.153	0.257
IU-xray	1	**0.492**	**0.318**	0.225	0.170	0.203	0.371
	2	0.470	0.310	**0.228**	**0.174**	0.182	**0.396**
	3	0.484	0.306	0.216	0.156	**0.208**	0.373
	4	0.469	0.310	0.228	0.174	0.181	0.396

5 Conclusion

In this paper, we propose an effective memory-based Transformer model called MeFormer, where external memory can model and remember similar patterns across radiology reports, thus helping the Transformer to decode and generate meaningful and robust radiology reports. The convergence of the model and its effectiveness can be accelerated by using a pre-trained visual feature extractor. The experimental results show that the proposed method achieves comparable performance to other state-of-the-art methods on two benchmark datasets.

Acknowledgements.. We thank editors and reviewers for their suggestions and comments. This work was supported by NSFC grants (No. 62136002), National Key R&D Program of China (2021YFC3340700), Shanghai Trusted Industry Internet Software Collaborative Innovation Center and National Trusted Embedded Software Engineering Technology Research Center (NTESEC).

References

1. Anderson, P., et al.: Bottom-up and top-down attention for image captioning and visual question answering. In: Proceedings of the IEEE Conference on Computer Vision and Pattern Recognition, pp. 6077–6086 (2018)
2. Banerjee, S., Lavie, A.: Meteor: an automatic metric for MT evaluation with improved correlation with human judgments. In: Proceedings of the ACL Workshop on Intrinsic and Extrinsic Evaluation Measures for Machine Translation and/or Summarization, pp. 65–72 (2005)
3. Brady, A., Laoide, R.Ó., McCarthy, P., McDermott, R.: Discrepancy and error in radiology: concepts, causes and consequences. Ulst. Med. J. **81**(1), 3 (2012)
4. Chen, T., Kornblith, S., Norouzi, M., Hinton, G.: A simple framework for contrastive learning of visual representations. In: International Conference on Machine Learning, pp. 1597–1607. PMLR (2020)
5. Chen, X., et al.: Microsoft COCO captions: data collection and evaluation server. arXiv preprint arXiv:1504.00325 (2015)
6. Chen, Z., Shen, Y., Song, Y., Wan, X.: Cross-modal memory networks for radiology report generation. arXiv preprint arXiv:2204.13258 (2022)
7. Chen, Z., Song, Y., Chang, T.H., Wan, X.: Generating radiology reports via memory-driven transformer. arXiv preprint arXiv:2010.16056 (2020)
8. Delrue, L., Gosselin, R., Ilsen, B., Van Landeghem, A., de Mey, J., Duyck, P.: Difficulties in the interpretation of chest radiography. In: Coche, E., Ghaye, B., de Mey, J., Duyck, P. (eds.) Comparative Interpretation of CT and Standard Radiography of the Chest. Medical Radiology, pp. 27–49. Springer, Heidelberg (2011). https://doi.org/10.1007/978-3-540-79942-9_2
9. Demner-Fushman, D., et al.: Preparing a collection of radiology examinations for distribution and retrieval. J. Am. Med. Inform. Assoc. **23**(2), 304–310 (2016)
10. Dosovitskiy, A., et al.: An image is worth 16 $\times$ 16 words: transformers for image recognition at scale. arXiv preprint arXiv:2010.11929 (2020)
11. He, K., Chen, X., Xie, S., Li, Y., Dollár, P., Girshick, R.: Masked autoencoders are scalable vision learners. In: Proceedings of the IEEE/CVF Conference on Computer Vision and Pattern Recognition, pp. 16000–16009 (2022)
12. He, K., Fan, H., Wu, Y., Xie, S., Girshick, R.: Momentum contrast for unsupervised visual representation learning. In: Proceedings of the IEEE/CVF Conference on Computer Vision and Pattern Recognition, pp. 9729–9738 (2020)
13. He, K., Zhang, X., Ren, S., Sun, J.: Deep residual learning for image recognition. In: Proceedings of the IEEE Conference on Computer Vision and Pattern Recognition, pp. 770–778 (2016)
14. Huang, S.C., Shen, L., Lungren, M.P., Yeung, S.: Gloria: a multimodal global-local representation learning framework for label-efficient medical image recognition. In: Proceedings of the IEEE/CVF International Conference on Computer Vision, pp. 3942–3951 (2021)

15. Irvin, J., et al.: CheXpert: a large chest radiograph dataset with uncertainty labels and expert comparison. In: Proceedings of the AAAI Conference on Artificial Intelligence, vol. 33, pp. 590–597 (2019)
16. Jing, B., Wang, Z., Xing, E.: Show, describe and conclude: On exploiting the structure information of chest X-ray reports. arXiv preprint arXiv:2004.12274 (2020)
17. Jing, B., Xie, P., Xing, E.: On the automatic generation of medical imaging reports. arXiv preprint arXiv:1711.08195 (2017)
18. Johnson, A.E., et al.: MIMIC-CXR-JPG, a large publicly available database of labeled chest radiographs. arXiv preprint arXiv:1901.07042 (2019)
19. Lei, J., Wang, L., Shen, Y., Yu, D., Berg, T.L., Bansal, M.: MART: memory-augmented recurrent transformer for coherent video paragraph captioning. arXiv preprint arXiv:2005.05402 (2020)
20. Li, Y., Liang, X., Hu, Z., Xing, E.P.: Hybrid retrieval-generation reinforced agent for medical image report generation. In: Advances in Neural Information Processing Systems, vol. 31 (2018)
21. Lin, C.Y.: Rouge: a package for automatic evaluation of summaries. In: Text Summarization Branches Out, pp. 74–81 (2004)
22. Liu, F., Ren, X., Liu, Y., Wang, H., Sun, X.: simNet: stepwise image-topic merging network for generating detailed and comprehensive image captions. arXiv preprint arXiv:1808.08732 (2018)
23. Liu, F., Wu, X., Ge, S., Fan, W., Zou, Y.: Exploring and distilling posterior and prior knowledge for radiology report generation. In: Proceedings of the IEEE/CVF Conference on Computer Vision and Pattern Recognition, pp. 13753–13762 (2021)
24. Liu, G., et al.: Clinically accurate chest x-ray report generation. In: Machine Learning for Healthcare Conference, pp. 249–269. PMLR (2019)
25. Lu, J., Xiong, C., Parikh, D., Socher, R.: Knowing when to look: adaptive attention via a visual sentinel for image captioning. In: Proceedings of the IEEE Conference on Computer Vision and Pattern Recognition, pp. 375–383 (2017)
26. Papineni, K., Roukos, S., Ward, T., Zhu, W.J.: Bleu: a method for automatic evaluation of machine translation. In: Proceedings of the 40th Annual Meeting of the Association for Computational Linguistics, pp. 311–318 (2002)
27. Rennie, S.J., Marcheret, E., Mroueh, Y., Ross, J., Goel, V.: Self-critical sequence training for image captioning. In: Proceedings of the IEEE Conference on Computer Vision and Pattern Recognition, pp. 7008–7024 (2017)
28. Simonyan, K., Zisserman, A.: Very deep convolutional networks for large-scale image recognition. arXiv preprint arXiv:1409.1556 (2014)
29. Vaswani, A., et al.: Attention is all you need. In: Advances in Neural Information Processing Systems, vol. 30 (2017)
30. Vinyals, O., Toshev, A., Bengio, S., Erhan, D.: Show and tell: a neural image caption generator. In: Proceedings of the IEEE Conference on Computer Vision and Pattern Recognition, pp. 3156–3164 (2015)
31. Wang, Z., Wu, Z., Agarwal, D., Sun, J.: MedCLIP: contrastive learning from unpaired medical images and text. arXiv preprint arXiv:2210.10163 (2022)
32. Xiao, J., Bai, Y., Yuille, A., Zhou, Z.: Delving into masked autoencoders for multi-label thorax disease classification. arXiv preprint arXiv:2210.12843 (2022)
33. Xie, Z., et al.: SimMIM: a simple framework for masked image modeling. In: Proceedings of the IEEE/CVF Conference on Computer Vision and Pattern Recognition, pp. 9653–9663 (2022)
34. Xu, K., et al.: Show, attend and tell: neural image caption generation with visual attention. In: International Conference on Machine Learning, pp. 2048–2057. PMLR (2015)

35. Xue, Y., et al.: Multimodal recurrent model with attention for automated radiology report generation. In: Frangi, A.F., Schnabel, J.A., Davatzikos, C., Alberola-López, C., Fichtinger, G. (eds.) MICCAI 2018. LNCS, vol. 11070, pp. 457–466. Springer, Cham (2018). https://doi.org/10.1007/978-3-030-00928-1_52
36. Yin, C., et al.: Automatic generation of medical imaging diagnostic report with hierarchical recurrent neural network. In: 2019 IEEE International Conference on Data Mining (ICDM), pp. 728–737. IEEE (2019)
37. Zhang, Y., Jiang, H., Miura, Y., Manning, C.D., Langlotz, C.P.: Contrastive learning of medical visual representations from paired images and text. arXiv preprint arXiv:2010.00747 (2020)
38. Zhou, L., Liu, H., Bae, J., He, J., Samaras, D., Prasanna, P.: Self pre-training with masked autoencoders for medical image analysis. arXiv preprint arXiv:2203.05573 (2022)

Cache-Enhanced InBatch Sampling with Difficulty-Based Replacement Strategies for Learning Recommenders

Yucheng Li[1], Defu Lian[1(✉)], and Jin Chen[2]

[1] University of Science and Technology of China, Hebei, China
liycustc@mail.ustc.edu.cn, liandefu@ustc.edu.cn
[2] University of Electronic Science and Technology of China, Chengdu, China
chenjin@std.uestc.edu.cn

Abstract. Negative sampling techniques are prevalent in learning recommenders to reduce the computational cost over the entire corpus, but existing methods still have a significant overhead for re-encoding out-of-batch items. Inbatch sampling is a more practical strategy that regards items in the mini-batch as negatives, although it suffers from exposure bias. Several researchers attempt to alleviate the bias by cache mechanism, which supplements more items for better approximation, but none of them sufficiently evaluate the information level of different items and further exploit them. In this paper, we propose a Cache-Enhanced InBatch Sampling with Difficulty-Based Replacement Strategy for Learning Recommenders that heuristically and adaptively updates the cache depending on the designed training difficulty of negative samples. Specifically, the cache is updated based on the average and standard deviation with respect to the training difficulty, which correspond with the estimated first-order and second-order moments, in which way the items with high averages and high uncertainties have a higher probability of being restored. Thus, the historical informative items in training are more effectively explored and exploited, leading to superior and rapid convergence. The proposed DBRS is evaluated on four real-world datasets and outperforms the existing state-of-the-art approaches.

Keywords: Recommender systems · Information retrieval · Neural Networks · Sampling

1 Introduction

The information overload has been a significant issue due to the rapid development of Internet technologies. In order to overcome this challenge, recommender systems are now widely employed in a variety of situations, including advertising and e-commerce, to enhance the user experience while also bringing considerable commercial benefits. To meet the strict requirements of short latency, current practical recommender systems often use a two-stage technique (retrieval-ranking) [6,7,9], with the retrievers aiming to retrieve a small fraction of relevant items from a huge number of items. The learning of retrievers is

A. El Abbadi et al. (Eds.): DASFAA 2023 Workshops, LNCS 13922, pp. 95–108, 2023.
https://doi.org/10.1007/978-3-031-35415-1_7

often reduced to the extreme multi-class classification problem, which has been proposed in [7,24]. However, computing an extreme classification loss over the entire corpus incurs a huge computational cost [11], making negative sampling a prevalent strategy for training retrievers.

By drawing a few items from the entire corpus as negatives in accordance with a certain distribution, the negative sampling approaches allow us to effectively estimate the softmax partition function. A corpus-wide negative sampling strategy requires reloading and re-encoding items outside of the batch, which, however, imposes a considerable overhead. Inbatch sampling gives an alternative solution without a significant increase in time overhead, however it suffers from exposure bias in training and inaccurate estimation of the softmax partition function. Some current researches [5,23,24] are devoted to correcting this bias and improving the effectiveness of inbatch sampling. Since inbatch sampling strategies also naturally suffer from limitations on the size of the sample pool (batch size), it is not feasible to simply increase the batch size limited by GPU memories. Therefore, to get over the limitation of item variety in mini-batch, a lot of researches introduce a cache to reuse items from the training history.

However, the existing sampling cache replacement methods have the following drawbacks: 1) high-difficulty items in the training history are not sufficiently utilized, i.e., there is no definition about the training difficulty of negative items, and 2) the uncertainty of item difficulty is not taken into account for cache replacing. As the training process continues, the difficulty of the items would vary. Some items may have a relatively lower average difficulty with limited sampled times, resulting in a higher level of uncertainty in their training difficulty. These items should be sampled more frequently to provide more confident information. Thus, we propose an adaptive heuristic cache replacement strategy for inbatch sampling by first designing the training difficulty of negative samples and updating the cache by tracking the average difficulty and variance of the items. This allows for more efficient exploration and utilization of items. In our experiments, we show a significant improvement in recommendation metrics by using a cache with the same size as the mini-batch.

2 Related Work

In this paper, we focus on the negative sampling for retrievers. Sampling based on popularity or uniform [16,18] distribution is the most straightforward among the various corpus-wide sampling methods. According to subsequent researches [4,8,14,21,22,25,27], adaptive sampling of hard samples leads to better convergence, but it comes at a significant cost when applied to the full corpus. Negative samplings will lead to a considerable bias in the estimation of the softmax partition function. By utilizing importance sampling [1,2,12] and the kernel-based method [3], several researches attempt to estimate the softmax more accurately.

Inbatch sampling [10] is a more practical and efficient method for retrievers because negative items do not need to be re-encoded, but it suffers from exposure bias in online systems. Some current researches are devoted to correcting this bias. [24] proposes to correct the sampling bias of inbatch softmax based on popularity, and it is successfully adopted in YouTube recommendations. The MNS [23] is developed by Google Play researchers, and it combines negative samples from both inbatch sampling and global uniform sampling, where the latter draws samples from the whole candidate corpus to reduce bias. [20,26] increases the sample pool by introducing a queue to keep previous mini-batches. [15] proposes to cache the previously computed embedding of all items and estimates the softmax by GumbelMax. Recently, [5] proposes that some items with higher scores have higher information, and thus counts the occurrences of the items in the negative sample set and updates the cache accordingly. However, existing methods can not quantitatively identify the information supplied to the training process by negative samples, and how to efficiently explore and exploit samples outside the batch remains an issue that has to be solved.

3 Cache-Enhanced Inbatch Sampling

3.1 Problem Formulation

The retrieval task is prevalently considered as an extremely large-scale multi-classification in the current researches. Given a set $\mathcal{U} = \{u_m\}_0^M$ of users and a set $\mathcal{I} = \{i_n\}_0^N$ of items, each pair of implicit data point indicates the interacted item and user, e.g., (u, i), where $u \in \mathcal{U}$ and $i \in \mathcal{I}$, M represents the number of users and N represents the number of items. The two-tower model is commonly employed in retrieval tasks due to its capacity to quickly retrieve a subset of the most relevant items for a given user query. In this paradigm, the user-side and the item-side features are typically encoded respectively to generate k-dimensional implicit vectors: $f_{\mathcal{U}} : \mathbb{R}^{d_{\mathcal{U}}} \rightarrow \mathbb{R}^k$, $f_{\mathcal{I}} : \mathbb{R}^{d_{\mathcal{I}}} \rightarrow \mathbb{R}^k$, where, respectively, $d_{\mathcal{I}}$ and $d_{\mathcal{U}}$ stand for the dimensions of the item's and user's original features. The final score of the user u and item i is typically obtained by a simple similarity function such as dot product [11,17], i.e., $s_\theta(u,i) = f_{\mathcal{U}}(u)^T \cdot f_{\mathcal{I}}(i)$. We consider minimizing the log-softmax loss to optimize the two-tower model.

3.2 Inbatch Sampling and Native Caches

The naive log-softmax loss takes the entire item corpus to calculate the partition function, which results in remarkable computational cost caused by numerous items. Thus, the sampled softmax loss with the inbatch sampling is widely used to estimate the native softmax loss:

$$\mathcal{L} = -\frac{1}{|\mathcal{B}|} \sum_{(u,i)\in\mathcal{B}} \log \frac{\exp\left(s_\theta(u,i)\right)}{\sum_{j\in\mathcal{S}} \exp\left(s_\theta(u,j)\right)} \tag{1}$$

where $\mathcal{B} = \{(u_1, i_1), \cdots, (u_{|\mathcal{B}|}, i_{|\mathcal{B}|})\}$ represents the set of user-item pairs in the current mini-batch, θ is the parameter of the model and $\mathcal{S}$ is the set of sampled negative items. For the two-tower model, sampling out-of-batch items as negative samples requires reloading and re-encoding them, which imposes a non-negligible overhead. A more practical strategy is to perform inbatch negative sampling, which takes the rest of the items in the current batch as negative samples, i.e., $\mathcal{S} = \mathcal{I}_{\mathcal{B}}$, where $\mathcal{I}_{\mathcal{B}} = \{i|(u,i) \in \mathcal{B}\}$ is the set of items in the current mini batch. The sampled items are subject to the popularity distribution in this way. This is similar to utilizing the popularity distribution as the proposal distribution [24], and the loss function corrected by importance sampling is shown as follows:

$$\mathcal{L} = -\frac{1}{|\mathcal{B}|} \sum_{(u,i)\in\mathcal{B}} \log \frac{\exp\left(s_\theta\left(u,i\right) - \log pop(i)\right)}{\sum_{j\in\mathcal{S}} \exp\left(s_\theta(u,j) - \log pop(j)\right)} \tag{2}$$

where $pop(i)$ represents the popularity of item i. Specifically, the number of occurrences f_i of each item i in the dataset is counted, and $pop(i)$ is calculated by normalization, i.e., $pop(i) = \frac{f_i}{\sum_j f_j}$. However, this sampling strategy is still constrained by the batch size and cannot adaptively select informative items since the popularity of each item is static during training, which may result in oversampling of easy samples.

The inbatch sampling suffers from the selection sample bias caused by the exposure bias, and thus the valuable long-tailed items have less chance of being sampled. The cache mechanism is proposed to provide more variety of sampled items to alleviate the selection sample bias, whose loss function has two parts as following [5]:

$$\mathcal{L} = -\frac{1}{|\mathcal{B}|} \sum_{(u,i)\in\mathcal{B}} \left((1-\lambda) \log \frac{\exp\left(s_\theta(u,i)\right)}{\sum_{j\in\mathcal{S}_u} \exp\left(s_\theta(u,j)\right)} + \lambda \log \frac{\exp\left(s_\theta(u,i)\right)}{\sum_{j\in\mathcal{S}'_u} \exp\left(s_\theta(u,j)\right)} \right) \tag{3}$$

where $\mathcal{S}_u$ and $\mathcal{S}'_u$ represent sampled items from batch and cache, respectively. λ is a hyperparameter to control the influence of each part. Importance resampling is performed separately within the batch and cache to generate $\mathcal{S}_u$ and $\mathcal{S}'_u$, and the respective sampling weights are $p_{\mathcal{B}}(j|u) = \frac{\exp(s'_\theta(u,j))}{\sum_{k\in\mathcal{B}} \exp(s'_\theta(u,k))}$ and $p_{\mathcal{C}}(j|u) = \frac{\exp(s'_\theta(u,j))}{\sum_{k\in\mathcal{C}} \exp(s'_\theta(u,k))}$, where $s'_\theta(u,j) = s_\theta(u,j) - \log pop(j)$. Specifically, We use $p_{\mathcal{B}}$ to sample items for u from mini-batch $\mathcal{I}_{\mathcal{B}}$ to generate $\mathcal{S}_u$, then $p_{\mathcal{C}}$ to sample items from cache $\mathcal{C}$ to generate $\mathcal{S}'_u$. About how the cache is implemented, MNS [23] utilizes a uniformly sampled set as a cache and XIR [5] caches the more frequently sampled items. However, the informative item is not defined and used explicitly in these existing cache replacement approaches.

3.3 Inbatch Sampling With Difficulty-Based Replacement Strategies

Regarding the BPR [19] loss, which penalizes more the sampled items having higher scores than positive items, we design a difficulty-based replacement weight for negative samples to measure the difficulty of negative items for training. If a negative sample has a higher score, the gradient induced by the negative gets larger and thus contributes more to the model training. By this means, the training difficulty is defined as follows:

$$g(j|u,i;\theta) = \sigma(s_\theta(u,j) - s_\theta(u,i)), j \in \mathcal{S}_u \cup \mathcal{S}'_u \tag{4}$$

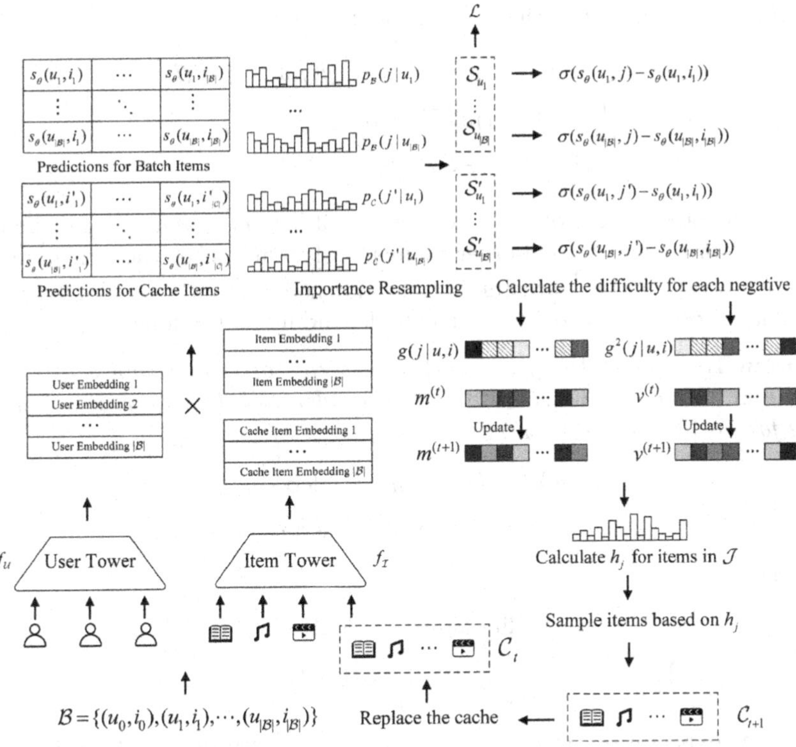

Fig. 1. Inbatch Sampling with Difficulty-Based Replacement Strategies

where $g(j|u,i;\theta)$ represents the training difficulty of negative sample j given the user-item pair (u,i) and the current parameter θ. The value by which the negative sample score exceeds the positive is mapped by a sigmoid function σ between 0 and 1 to indicate the degree of difficulty of the negative. In this way, the more the negative sample score exceeds the positive sample, the higher the confidence that the negative is misclassified, and the more gradients such a negative sample brings in the training process. Therefore, such negative samples are defined as difficult negatives, which brings more information to model.

However, the items within the mini-batch suffer from exposure bias, i.e., the items follow the long-tail distribution. Thus, the long-tailed items have more uncertainty for the training difficulty since they have a much lower chance to be sampled. From this perspective, items would be more explored according to the uncertainty of the difficulty. Here we calculate the average and variance with respect to the difficulty of each sample, where a sliding average approach is employed to adaptively update during model training. Before deriving the average and the variance, the first-order moment and the second-order moment are estimated according to the following equations:

$$\begin{aligned} m_j^{(t)} &= (1-\alpha)m_j^{(t-1)} + \alpha \cdot g(j|u,i), j \in \mathcal{J} \\ v_j^{(t)} &= (1-\alpha)v_j^{(t-1)} + \alpha \cdot g^2(j|u,i), j \in \mathcal{J} \\ \mathcal{J} &= \{j|j \in \mathcal{S}_u \cup \mathcal{S}'_u, \forall u \in \mathcal{B}\} \end{aligned} \tag{5}$$

where $\mathcal{J}$ denotes the set of sampled negative items, $m_j^{(t)}$ and $v_j^{(t)}$ are the estimated first-order and second-order moments of item j after the t-th iteration, respectively. $m_j^{(0)}$ and $v_j^{(0)}$ are first initialized to zero vectors. Accordingly, we can get unbiased estimations of the moments depending on the following theorem.

Theorem 1. *Suppose* $\{g_1, g_2, \cdots, g_t\}$ *is a series of i.i.d. samples of random variable* Δ. *Let* $\delta = \mathbb{E}[\Delta], \zeta = \mathbb{E}[\Delta^2]$. *Consider a sliding average estimation where for* $i \in [t]$ *and* $\alpha \in (0,1)$,

$$\begin{aligned} m^{(t)} &= (1-\alpha)m^{(t-1)} + \alpha \cdot g_t \\ v^{(t)} &= (1-\alpha)v^{(t-1)} + \alpha \cdot g_t^2 \end{aligned} \tag{6}$$

Then we have

$$\lim_{t\to\infty} \mathbb{E}[m^{(t)}] = \mathbb{E}[\Delta] \tag{7}$$

$$\lim_{t\to\infty} \mathbb{E}[v^{(t)}] = \mathbb{E}[\Delta^2] \tag{8}$$

Proof. The expectations of $m^{(i)}$ and $v^{(i)}$ can be derived directly from Eq. 6:

$$\mathbb{E}[m^{(i)}] = (1-\alpha)\mathbb{E}[m^{(i-1)}] + \alpha \cdot \delta \tag{9}$$

$$\mathbb{E}[m^{(i)}] - \delta = (1-\alpha)\left(\mathbb{E}[m^{(i-1)}] - \delta\right) \tag{10}$$

From Eq. (10), we know that $\mathbb{E}[m^{(i)}] - \delta$ is isoperimetric series. Thus we have the expectation bias of $m^{(t)}$:

$$\mathbb{E}[m^{(t)}] - \delta = (1-\alpha)^t\left(m^{(0)} - \delta\right) \tag{11}$$

Equation 11 indicate as $t \to \infty$, the expectation bias tends to 0. Similarly, we can prove that the bias of second-order moment approaches to 0.

We show that the unbiased estimation of the first-order moment and second-order moment are given through a sliding average when t tends to infinity in Theorem 1, therefore, we can implement the cache replacement strategy by leveraging the average and variance of the difficulty. When we update the cache, the items with larger averages and larger variances tend to be selected because larger averages indicate that the items are more informative, while larger variances mean that the items are still uncertain and valuable to explore.

$$\begin{aligned} h_j &= \frac{1}{Z}(m_j + d_j) \\ Z &= \sum_{i\in\mathcal{J}} m_i + d_i \end{aligned} \tag{12}$$

More specifically, we design the replacement weight of items as Eq. 12, where the standard deviation $d_i = \sqrt{v_i - m_i^2}$ is used to capture the uncertainty of the items. The likelihood of items being cached increases with increasing weight.

Algorithm 1: Difficulty-Based Cache Replacement Strategy (DBRS)

Input: Training data $\mathcal{D} = \{(u,i)\}$, Index of iterations t, current Cache $\mathcal{C}_t$, Hyperparameter α, λ

Output: Model parameters Θ, Cache for next iteration $\mathcal{C}_{t+1}$

Initialize the 1st and 2nd moment vectors $\boldsymbol{m} = \{0\}_{i=1}^N$, $\boldsymbol{v} = \{0\}_{i=1}^N$;
if $t = 0$ **then**
    Initialize $\mathcal{C}_0$ with uniformly sampled items
for *mini-batch* $\mathcal{B} \in \mathcal{D}$ **do**
    $\mathcal{U}_\mathcal{B} = \{u|(u,i) \in \mathcal{B}\}$, $\mathcal{I}_\mathcal{B} = \{i|(u,i) \in \mathcal{B}\}$;
    Calculate $s_\theta(u,i), u \in \mathcal{U}_\mathcal{B}, i \in \mathcal{I}_\mathcal{B}$ and $s_\theta(u,i'), u \in \mathcal{U}_\mathcal{B}, i' \in \mathcal{C}_t$;
    for $(u,i) \in \mathcal{B}$ **do**
        Sampling negatives $\mathcal{S}_u$ from batch with $p_\mathcal{B}(j|u) = \frac{\exp s'_\theta(u,j)}{\sum_{k \in \mathcal{I}_\mathcal{B}} \exp s'_\theta(u,k)}, \forall j \in \mathcal{I}_\mathcal{B}$;
        Sampling negatives $\mathcal{S}'_u$ from cache with $p_\mathcal{C}(j|u) = \frac{\exp s'_\theta(u,j)}{\sum_{k \in \mathcal{C}_t} \exp s'_\theta(u,k)}, \forall i \in \mathcal{C}_t$;
        Calculate the difficulty of negative samples $g(j|u,i) = \sigma(s_\theta(u,j) - s_\theta(u,i)), j \in \mathcal{S}_u \cup \mathcal{S}'_u$;
    Update 1st moment and 2nd moment estimations ;
    $m_j = (1-\alpha)m_j + \alpha \cdot g(j|u,i)$;
    $v_j = (1-\alpha)v_j + \alpha \cdot g^2(j|u,i)$;
    `// Replace the cache`
    Calculate the standard deviation based on moments: $d_j = \sqrt{v_j - m_j^2}$;
    Calculate the sample weight $h_j = \frac{1}{Z}(m_j + d_j), Z = \sum_{i \in \mathcal{J}} m_i + d_i$;
    Sample $|\mathcal{C}_t|$ items from $\mathcal{J}$ based on h_j to generate the new cache $\mathcal{C}_{t+1}$;
    Update model parameters based on the loss function, i.e., Eq. (3);

As shown by Fig. 1 and Algorithm 1, our cache replacement strategy entails computing the difficulty for sampled items $\mathcal{S}_u$ and $\mathcal{S}'_u$ by inbatch and cache sampling in each iteration, updating its first-order and second-order moments through sliding average, and refreshing the cache in accordance with the average and uncertainty of the difficulty. It is worth noting that some items might not be present in the negative set $\mathcal{J}$ of the mini-batch, as a result, their m and v will not be updated. On the other hand, some items might show up as negatives of different users in the same batch in which case their m and v will be updated multiple times. We only sample from the negative set $\mathcal{J}$ that has appeared in the mini-batch when we update the cache to prevent wasting to reload the out-of-batch items. We follow the training procedure in XIR [5], hence the loss Eq. 3 is employed for optimizing after negative sampling from the cache and batch individually.

3.4 Complexity Analysis

Compared with existing inbatch sampling methods, our approach additionally increases the overhead of computing the training difficulty and updating the estimations, with a time complexity of $O(|\mathcal{B}| \times |\mathcal{B}|)$, where $|\mathcal{B}|$ is the batch size. The time complexity at updating the cache is $O(N)$, where N is the number of items. Both are significantly smaller compared to the time complexity $O(|\mathcal{B}| \times |\mathcal{B}| \times k)$ for re-encoding the out-of-batch items in the global sampling approach, where k is the embedding dimension.

The space complexity of our cache is $O(2N + |\mathcal{B}|)$ since our strategy necessitates two vectors at length of N to store the first-order and second-order moments for all items in addition to the $|\mathcal{B}|$ items that the cache must keep.

4 Experiments

4.1 Experimental Setup

Baselines. We conduct experiments to compare our method with competitive inbatch sampling strategies proposed in previous researches. All methods treat the retrieval issue as a large-scale multi-classification to optimize the log-sampled-softmax loss, and the following methods vary in their sampling strategies.

- **SSL** performs uniform inbatch negative sampling.
- **SSL-Pop** utilizes the items within the batch and employs item popularity as the proposal distribution, correcting the bias by importance sampling.
- **MNS** [23] mixes an additional portion of uniformly sampled items as negatives besides inbatch sampling. In our implementation, the number of uniformly sampled negatives is the same as the number of inbatch sampled negatives.
- **G-Tower** [24] alleviates the exposure bias by the popularity of streaming estimation. It records the last occurrence step of each item in a fixed-length hash array and reduces hash collisions by using multiple hash functions. In this way, the popularity is estimated as the reciprocal of the average steps. In our implementation, five different hash arrays are used for estimation.
- **BIR** [5] performs importance resampling based on model scores of popularity-based debias to improve the accuracy of the estimation of the softmax partitioning function, and enables personalized negative sampling across users.
- **XIR** [5] further improves BIR by introducing a cache based that combines the negatives sampled from the batch and the cache. It updates the cache by counting the number of occurrences of negatives, and the more occurrences the higher the probability of items being cached.

Datasets. We conducted experiments on four real-world datasets: Gowalla, Amazon, Ta-feng, and Tmall. In the amazon dataset, the items with scores higher than the average are considered as positive samples, while interactions in other datasets are considered as positive samples. These four datasets vary in different recommendation scenarios. The number of specific items, users, interactions, and density are summarized in Table 1. 80% of the interacted items of a user in the datasets are sampled for training and others for testing.

Parameter Configurations. In our implementation, the batch size is set to 2048, the cache size is set by default to the same as the batch size, and the embedding dimension k is set to 32. The l_2 regularization coefficients are tuned from {1e−4, 1e−5, 1e−6}. Except for Sect. 4.4, all experiments employ ID features of users and items. We used the Adam optimizer [13] with a learning rate of 0.001 and set a decay factor of 0.95 per 5 epochs. On each dataset, 100 training epochs were run. Moreover, the hyperparameter lambda is selected from {0, 0.2, 0.5, 0.8, 1.0} and alpha from {1e−1, 1e−2, 1e−3, 1e−4}.

Metrics. The performance of the recommendation retriever is measured by how well positive items are ranked on the test set. We used two widely used metrics: NDCG (Normalized Discounted Cumulative Gain) and RECALL. The default cutoff values for NDCG and RECALL are 10.

Table 1. Statistics of Datasets

Dataset	#User	#Item	#Interaction	Density
Gowalla	29,858	40,988	1,027,464	8.40e−4
Amazon	130,380	128,939	2,415,650	1.44e−4
Ta-feng	19,451	10,480	630,767	3.09e−4
Tmall	125,553	58,058	2,064,290	2.83e−4

4.2 Comparison with Baselines

The baseline and our method are run 5 times with different random seeds and we report the average scores and their standard deviations in Table 2. The results demonstrate that our proposed method outperforms the competitive baselines. In four datasets, our method outperforms XIR by 5.44%, 6.27%, 4.24%, and 7.29% at NDCG@10. As the model is gradually updated during training, our method adaptively selects items with higher difficulty and explores items with higher uncertainty, thereby enhancing the quality of inbatch sampling and making fuller use of inbatch items, leading to better convergence.

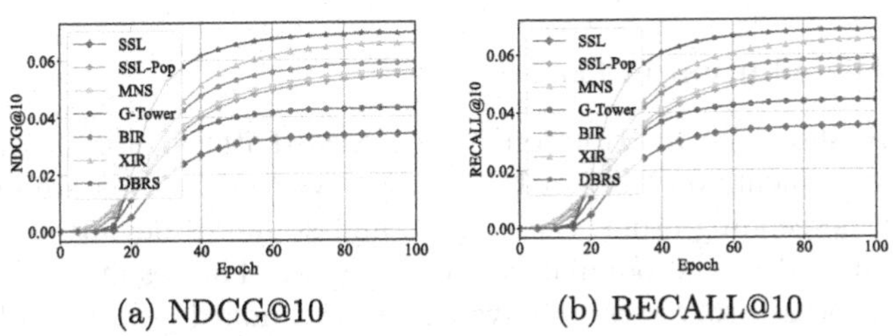

(a) NDCG@10 (b) RECALL@10

Fig. 2. Performances vs. number of epochs on Tmall

Table 2. Comparison with Baselines. $\delta = 1e-4$

	Gowalla	Amazon	Ta-feng	Tmall
Method	NDCG@10	NDCG@10	NDCG@10	NDCG@10
SSL	$0.1381 \pm 5.5\delta$	$0.0718 \pm 3.7\delta$	$0.0437 \pm 9.4\delta$	$0.0340 \pm 3.1\delta$
SSL-Pop	$0.1479 \pm 9.1\delta$	$0.0764 \pm 3.2\delta$	$0.0625 \pm 1.5\delta$	$0.0547 \pm 5.6\delta$
MNS	$0.1486 \pm 8.9\delta$	$0.0781 \pm 4.9\delta$	$0.0634 \pm 1.4\delta$	$0.0561 \pm 4.9\delta$
G-Tower	$0.1500 \pm 8.6\delta$	$0.0764 \pm 4.0\delta$	$0.0496 \pm 8.7\delta$	$0.0429 \pm 2.1\delta$
BIR	$0.1523 \pm 7.9\delta$	$0.0833 \pm 2.7\delta$	$0.0675 \pm 1.3\delta$	$0.0590 \pm 3.2\delta$
XIR	$\underline{0.1543} \pm 1.1\delta$	$\underline{0.0877} \pm 3.4\delta$	$\underline{0.0730} \pm 1.1\delta$	$\underline{0.0658} \pm 2.2\delta$
DBRS	$\mathbf{0.1627} \pm 9.4\delta$	$\mathbf{0.0932} \pm 1.9\delta$	$\mathbf{0.0761} \pm 5.2\delta$	$\mathbf{0.0706} \pm 4.4\delta$
	RECALL@10	RECALL@10	RECALL@10	RECALL@10
SSL	$0.1105 \pm 4.1\delta$	$0.0753 \pm 4.5\delta$	$0.0375 \pm 7.5\delta$	$0.0354 \pm 3.6\delta$
SSL-Pop	$0.1124 \pm 4.4\delta$	$0.0777 \pm 4.3\delta$	$0.0506 \pm 9.2\delta$	$0.0544 \pm 6.1\delta$
MNS	$0.1130 \pm 4.2\delta$	$0.0796 \pm 6.1\delta$	$0.0514 \pm 8.5\delta$	$0.0558 \pm 5.4\delta$
G-Tower	$\underline{0.1176} \pm 6.8\delta$	$0.0798 \pm 5.8\delta$	$0.0415 \pm 7.2\delta$	$0.0440 \pm 1.8\delta$
BIR	$0.1157 \pm 4.6\delta$	$0.0848 \pm 3.5\delta$	$0.0549 \pm 8.6\delta$	$0.0588 \pm 3.1\delta$
XIR	$0.1169 \pm 6.1\delta$	$\underline{0.0895} \pm 4.1\delta$	$\underline{0.0589} \pm 6.1\delta$	$\underline{0.0651} \pm 2.8\delta$
DBRS	$\mathbf{0.1243} \pm 7.5\delta$	$\mathbf{0.0966} \pm 2.7\delta$	$\mathbf{0.0639} \pm 6.8\delta$	$\mathbf{0.0697} \pm 4.3\delta$

Finding 1: According to the results, MNS and XIR which use out-of-batch items for sampling significantly outperform SSL-Pop in terms of recommendation quality, showing that out-of-batch items help to reduce selection bias.
Finding 2: Compared to the state-of-the-art baselines, the recommendation results of DBRS are effectively improved by caching items in accordance with the difficulty principle as shown in Table 2. It indicates that high training difficulty items are critical to the learning process. By making the hard sample gradient dominate the optimization process, the model is continuously updated.
Finding 3: Our method leads to better convergence. Figure 2a, 2b depict the recommendation performance changing curves for each baseline and our method throughout training. These graphs demonstrate how quickly our method could converge to a superior solution.

4.3 Item Distribution in Cache

To intuitively show the distribution of the cached items in the training, we conducted experiments on this. In Fig. 3a, 3b, when the items are ordered by popularity in descending order, the average and standard deviation of training difficulty compared to the popularity are displayed. The results reveal that as the training process progresses, the average and standard deviation of the item difficulty are closer to the popularity distribution, indicating that more popular items typically have more difficulty and uncertainty. This finding supports the validity of using popularity to correct cache sampling.

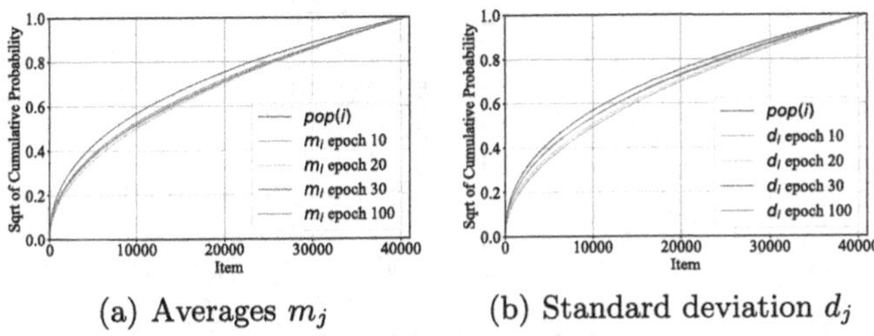

(a) Averages m_j (b) Standard deviation d_j

Fig. 3. Item Distribution vs. number of epochs on Gowalla

Table 3. Performances on Tmall with features

	NDCG@10	RECALL@10	NDCG@20	RECALL@20
SSL	0.0311	0.0329	0.0524	0.0568
SSL-Pop	0.0720	0.0712	0.0999	0.1022
BIR	0.0742	0.0735	0.1032	0.1057
XIR	<u>0.0755</u>	<u>0.0750</u>	<u>0.1067</u>	<u>0.1099</u>
DBRS	**0.0799**	**0.0798**	**0.1118**	**0.1153**

4.4 Effect of Including Item Features

In industrial scenarios, the two-tower model typically contains a number of features, and thus, we intend to show that our method is still valid on the two-tower model with features. Specifically, we use the *SellerId* and *CateId* from the Tmall-buy dataset, which are encoded respectively and joined to the item tower. As shown in Table 3, our method also outperforms the baselines, with a 5.82% relative improvement compared to XIR at NDCG@10.

4.5 Effect of Cache Size

In the previous parts, the cache size is set to batch size by default. Furthermore, we aim to figure out the impact of cache size on the performance of the algorithm, and we conduct experiments on different cache sizes, including {0x, 0.25x, 0.5x, 1x, 2x, 4x} multiples of batch size, where 0x means that the cache is not used, i.e., degraded to BIR. The experimental results are shown in Fig. 4a, which shows that the recommendation performance improves with the growth of the cache size. A larger cache means that more items with high training difficulty are explored and utilized, thus improving the performance of the retriever.

We also explored at how cache size affected training time, as shown in the Fig. 4b. The training time increases with the cache size, which is intuitive because of the additional operations for cache. As can be shown, the training time does not significantly increase when only a tiny cache is adopted, and DBRS outperforms BIR by its cache mechanism.

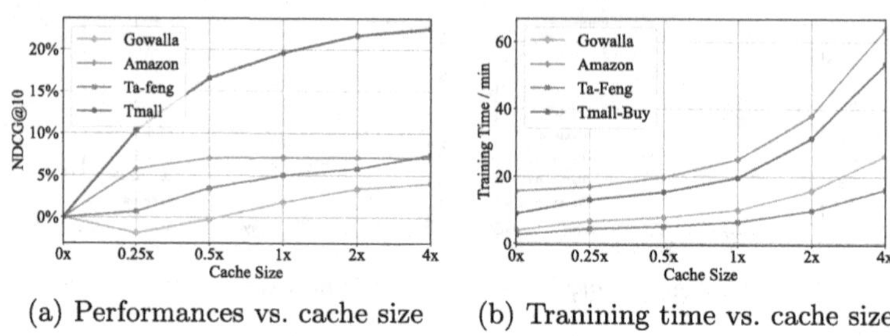

(a) Performances vs. cache size (b) Tranining time vs. cache size

Fig. 4. Effect of Cache Size

4.6 Effect of Hyperparameters

Since the hyperparameter λ in Eq. 3 and the α in the moments estimation algorithm Eq. 5 have a significant effect on the model, we conduct experiments to demonstrate the effect of hyperparameters on the performance. Figure 5a, 5b depict the trend of recommendation performance for each dataset with different values of λ, where λ stands for the proportion of cache sampling in the gradient. In most datasets, a larger λ improves retrieval performance, proving the effectiveness of the cache mechanism. Figure 5c, 5d illustrate how α affects recommendation performance. In contrast to larger α, smaller α converges more slowly but the estimations are more stable.

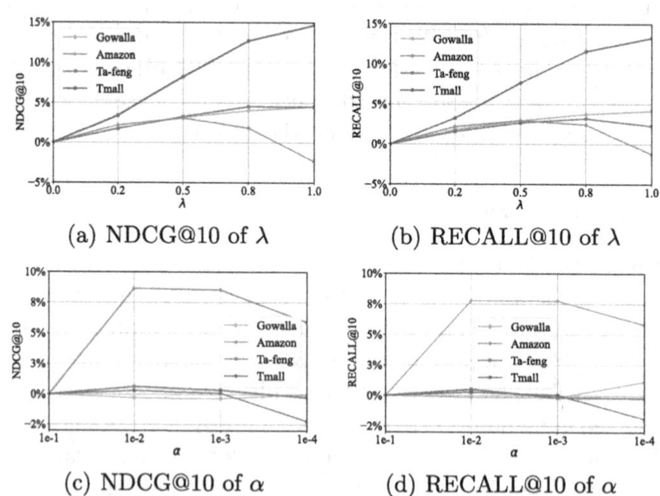

(a) NDCG@10 of λ (b) RECALL@10 of λ

(c) NDCG@10 of α (d) RECALL@10 of α

Fig. 5. Performances vs. hyperparameter

5 Conclusion

In this paper, we propose a novel cache-enhanced inbatch sampling with difficulty-based replacement strategy for learning recommenders by calculating training difficulty and estimating the mean and variance of each item. It allows us to explore and exploit negative items adaptively in the training phase by difficulty-based cache replacement strategy and effectively boosting negative sample difficulty without significantly increasing retriever overhead. The experiment comparisons with existing methods on real-world datasets validate the effectiveness of our method and show that our method is beneficial for inbatch sampling.

References

1. Bai, Y., Goldman, S., Zhang, L.: TAPAS: two-pass approximate adaptive sampling for softmax. arXiv preprint arXiv:1707.03073 (2017)
2. Bengio, Y., Senécal, J.S.: Adaptive importance sampling to accelerate training of a neural probabilistic language model. IEEE Trans. Neural Networks **19**(4), 713–722 (2008)
3. Blanc, G., Rendle, S.: Adaptive sampled softmax with kernel based sampling. In: International Conference on Machine Learning, pp. 589–598 (2018)
4. Burges, C.J.: From RankNet to LambdaRank to LambdaMART: an overview. Learning **11**(23–581), 81 (2010)
5. Chen, J., Lian, D., Li, Y., Wang, B., Zheng, K., Chen, E.: Cache-augmented inbatch importance resampling for training recommender retriever. arXiv preprint arXiv:2205.14859 (2022)
6. Cheng, H.T., et al.: Wide & deep learning for recommender systems. In: Proceedings of the 1st Workshop on Deep Learning for Recommender Systems, pp. 7–10 (2016)
7. Covington, P., Adams, J., Sargin, E.: Deep neural networks for Youtube recommendations. In: Proceedings of the 10th ACM Conference on Recommender Systems, pp. 191–198 (2016)
8. Ding, J., Quan, Y., Yao, Q., Li, Y., Jin, D.: Simplify and robustify negative sampling for implicit collaborative filtering. Adv. Neural. Inf. Process. Syst. **33**, 1094–1105 (2020)
9. Gomez-Uribe, C.A., Hunt, N.: The Netflix recommender system: algorithms, business value, and innovation. ACM Trans. Manage. Inf. Syst. (TMIS) **6**(4), 1–19 (2015)
10. Hidasi, B., Karatzoglou, A., Baltrunas, L., Tikk, D.: Session-based recommendations with recurrent neural networks. arXiv preprint arXiv:1511.06939 (2015)
11. Hu, Y., Koren, Y., Volinsky, C.: Collaborative filtering for implicit feedback datasets. In: 2008 Eighth IEEE International Conference on Data Mining, pp. 263–272. IEEE (2008)
12. Jean, S., Cho, K., Memisevic, R., Bengio, Y.: On using very large target vocabulary for neural machine translation. arXiv preprint arXiv:1412.2007 (2014)
13. Kingma, D.P., Ba, J.: Adam: a method for stochastic optimization. arXiv preprint arXiv:1412.6980 (2014)
14. Lian, D., Liu, Q., Chen, E.: Personalized ranking with importance sampling. In: Proceedings of the Web Conference 2020, pp. 1093–1103 (2020)

15. Lindgren, E., Reddi, S., Guo, R., Kumar, S.: Efficient training of retrieval models using negative cache. Adv. Neural. Inf. Process. Syst. **34**, 4134–4146 (2021)
16. Pan, R., et al.: One-class collaborative filtering. In: Proceedings of ICDM 2008, pp. 502–511. IEEE (2008)
17. Rendle, S.: Factorization machines. In: 2010 IEEE International Conference on Data Mining, pp. 995–1000. IEEE (2010)
18. Rendle, S., Freudenthaler, C.: Improving pairwise learning for item recommendation from implicit feedback. In: Proceedings of the 7th ACM International Conference on Web Search and Data Mining, pp. 273–282 (2014)
19. Rendle, S., Freudenthaler, C., Gantner, Z., Schmidt-Thieme, L.: BPR: Bayesian personalized ranking from implicit feedback. arXiv preprint arXiv:1205.2618 (2012)
20. Wang, J., Zhu, J., He, X.: Cross-batch negative sampling for training two-tower recommenders. In: Proceedings of the 44th International ACM SIGIR Conference on Research and Development in Information Retrieval, pp. 1632–1636 (2021)
21. Wang, J., et al.: IRGAN: a minimax game for unifying generative and discriminative information retrieval models. In: Proceedings of the 40th International ACM SIGIR Conference on Research and Development in Information Retrieval, pp. 515–524 (2017)
22. Weston, J., Bengio, S., Usunier, N.: Large scale image annotation: learning to rank with joint word-image embeddings. Mach. Learn. **81**(1), 21–35 (2010)
23. Yang, J., et al.: Mixed negative sampling for learning two-tower neural networks in recommendations. In: Companion Proceedings of the Web Conference 2020, pp. 441–447 (2020)
24. Yi, X., et al.: Sampling-bias-corrected neural modeling for large corpus item recommendations. In: Proceedings of the 13th ACM Conference on Recommender Systems, pp. 269–277 (2019)
25. Zhang, W., Chen, T., Wang, J., Yu, Y.: Optimizing top-n collaborative filtering via dynamic negative item sampling. In: Proceedings of the 36th International ACM SIGIR Conference on Research and Development in Information Retrieval, pp. 785–788 (2013)
26. Zhou, C., Ma, J., Zhang, J., Zhou, J., Yang, H.: Contrastive learning for debiased candidate generation in large-scale recommender systems. In: Proceedings of the 27th ACM SIGKDD Conference on Knowledge Discovery & Data Mining, pp. 3985–3995 (2021)
27. Zhu, Q., Zhang, H., He, Q., Dou, Z.: A gain-tuning dynamic negative sampler for recommendation. In: Proceedings of the ACM Web Conference 2022, pp. 277–285 (2022)

Syndrome-Aware Herb Recommendation with Heterogeneous Graph Neural Network

Jiayin Huang[1], Wenjing Yue[1,2], Yiqiao Wang[1(✉)], Jiong Zhu[3], and Liqiang Ni[4]

[1] School of Computer Science and Technology, East China Normal University, Shanghai, China
{jyhuang,wjyue,yqwang_01}@stu.ecnu.edu.cn

[2] Shanghai Institute of AI for Education, East China Normal University, Shanghai, China

[3] Shuguang Hospital Affiliated to Shanghai University of Traditional Medicine, Shanghai, China

[4] Shanghai University of Traditional Chinese Medicine, Shanghai, China

Abstract. The herb recommender system usually induces the implicit syndrome representations based on TCM prescriptions to generate related herbs as a treatment to cure a given symptom set. Previous methods primarily focus on modeling the interaction between symptoms (or diseases) and herbs without explicitly considering the syndrome information. As a result, these methods only capture the coarse-grained syndrome information. In this paper, we propose a new method to incorporate the explicit syndrome information for herb recommendation. To model the coarse-grained interaction between diseases and herbs within a specific syndrome class, we employ clustering algorithms to obtain the syndrome class, and apply the graph convolution network (GCN) on multiple disease-herb bipartite subgraphs. Next, we model the fine-grained interaction upon the syndrome-herb graph. Further, we propose a syndrome-aware heterogeneous graph neural network architecture, which integrates the syndrome information into the GCN message propagation process by combining the coarse-grained and fine-grained information of the interactions. The experimental results on the real TCM dataset demonstrate the improvements over state-of-the-art herb recommendation methods, further validate the effectiveness of our model.

Keywords: Herb recommendation · Traditional Chinese medicine · Heterogeneous graph neural network · Representation learning

1 Introduction

Traditional Chinese medicine (TCM) plays an important role in health maintenance for people [5]. The general practice of TCM depends on the treatment based on syndrome differentiation, and TCM syndrome induction can be

---

Jiayin Huang and Wenjing Yue are co-first authors of the article.

A. El Abbadi et al. (Eds.): DASFAA 2023 Workshops, LNCS 13922, pp. 109–124, 2023.
https://doi.org/10.1007/978-3-031-35415-1_8

used for further stratification of the patients' conditions with specific diseases, which could help improve the efficacy of diagnosis and treatment procedures [8]. Figure 1 takes the *Eczema* as an example to demonstrate the central role of syndrome induction in the treatment of specific diseases. In this case, "reddened tongue" and "rapid onset" appear under the "syndrome of dampness-heat in the spleen", therefore the doctor may choose a herb set that can "clear heat, drain dampness and relieve itching" to cure this syndrome. Here the herb set(i.e., medicine) is the "gentian liver-draining decoction", which includes herbs such as "gentian" and "scutellaria". However, If a patient presents with symptoms of a "pale tongue" and a "long course", the doctor may diagnosis this patient with "symptom of blood deficiency and wind-dryness", and chooses herbs such as "caulis Spatholobi" and "radix angelicae sinensis". As we can observe, the herb prescribed can vary for the same disease due to different syndromes. Understanding the underlying syndrome information of the disease can directly influence the selection of herbs.

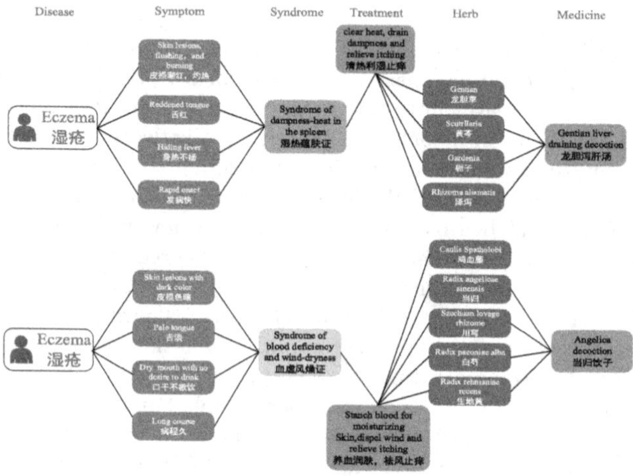

Fig. 1. An example of the therapeutic process in TCM.

In contrast to the traditional item recommendation task, which focuses on a single user, the herb recommendation task is defined as providing a set of herbs to treat a set of symptoms [9] or diseases. Several methods in herb recommendation have explored the relationships between symptoms and herbs. Some researches [7,18,20] models the syndromes as the latent topic for a single symptom, which overlooks the set information of symptoms and the high-order relationship between symptoms and herbs. Recently, GNNs are introduced to tackle the herb recommendation task, SMGCN [10] models the high-order relationship between symptoms and herbs, and incorporates a syndrome induction process into herb recommendation, which can model syndromes based on the set of symptoms. However, it does not explicitly consider the specific information related to each syndrome. In addition, data sparsity poses a significant challenge in recommendation tasks [15]. Actual TCM prescriptions also suffer from data bias in some syndromes. The representation learning for syndromes with sparse data has poor performance, significantly impacting the selection of herbs in such scenarios.

In this paper, we propose to incorporate the explicit syndrome information into herb recommendation, which can leverage additional knowledge to enhance the accuracy of the syndrome induction and treatment. Especially, we consider the herb recommendation task as providing a set of herbs to treat a set of diseases under the specific syndrome. Specifically, we propose a new method named *Syndrome Aware Heterogeneous Graph Neural Network* (SA-HGNN), a heterogeneous neural network model with clustering for the herb recommendation task, which models the higher-order interaction between diseases and herbs under different syndromes. Technically, SA-HGNN consists of two parts: (1) Syndrome clustering module; (2) Syndrome-aware herb recommendation module. In syndrome clustering module, we first learn syndromes embedding and herbs embedding via LightGCN [6] upon the syndrome-herb graph. And then, we apply the K-means algorithm on the syndrome embeddings to obtain the syndrome classes, which can alleviate the data sparsity issue associated with syndromes. In the syndrome-aware herb recommendation module, we use GCN to learn diseases embedding and herbs embedding upon the multiple disease-herb bipartite subgraphs that are classified by the syndrome class. Next, we use the disease induction layer to capture an overall embedding of the diseases set. Finally, we propose a herb prediction layer to predict the possibility of interactions between diseases and herbs for the target syndrome, which considers coarse-grained and fine-grained information about the interactions. In the end, we conduct experiments on a real dataset provided by the hospital, demonstrating the effectiveness of our SA-HGNN method.

The main contributions of this work are as follows.

- We consider the importance of considering the explicit syndrome information when modeling the interactions between diseases and herbs for herb recommendation task.
- We propose an effective herb set recommender, SA-HGNN, which models the interactions between diseases and herbs via syndrome clustering to obtain explicit syndrome information, and uses GCN for data representation learning with syndrome information. We also construct multiple disease-herb subgraphs based on syndrome classes.
- Our model outperforms SOTA baselines on the real dataset, and we demonstrate the effectiveness of SA-HGNN through real-world case studies.

2 Related Work

2.1 Graph Based Herb Recommendation

Nowadays, an increasing number of researchers are using graph representation learning for the herb recommendation task. Graph representation learning-based herb recommendation typically involves encoding TCM entities into a low-dimensional space using the TCM graph, and then recommending herbs based on the embeddings of the entities. Li et al. [11] apply the BRNN [13] to learn the representation of herb words in TCM medical records, with the goal of assisting doctors in writing prescriptions. Ruan et al. [12] utilize the autoencoder model with meta-path analysis to mine the TCM heterogeneous information network.

Furthermore, it is essential to explore multiple symptoms as a whole and identify the underlying syndromes behind them to prescribe appropriate treatments. Jin et al. [10] model the interactions between symptoms and herbs via GCN upon three TCM heterogeneous graphs. The prediction layer uses the MLP to analyze multiple symptoms comprehensively as an implicit syndrome. Jin et al. [9] use the attention mechanism to capture the relationship between symptoms and single herb. Additionally, they introduce entity embeddings extracted from the knowledge graph of TCM, which improves the quality of representation learning. However, few studies explicitly introduce syndrome information and capture the interactions between diseases and herbs based on different syndromes.

2.2 Heterogeneous Graph Neural Networks-Based Recommender Systems

Heterogeneous graphs neural networks have been widely applied in recommender systems due to the diverse types of nodes and edges, which allow for the modeling of complex relationships and interactions between entities. GNNs are applied to different kinds of heterogeneous graphs as follows:

Bipartite Graph. Ying et al. [21] sample a fixed number of neighbors for each node using the random walk strategy and perform GCN to generate node embeddings. Wang et al. [16] use multi-layer GCN on the user-item interaction graph to model higher-order relationships between users and items. LightGCN [6] is optimized from NGCF [16] by removing unnecessary feature conversion weights and nonlinear activation functions. Velickovic et al. [14] propose an attention mechanism based on GCN that enables automatic differentiation of the contribution of neighboring nodes to the central node.

Multi-relation Heterogeneous Graph. Heterogeneous graph neural network models currently extract multiple subgraphs based on the different types of edges present in the graph. The node embeddings obtained through representation learning on each subgraph are then merged to improve the overall quality of the node representations. Wang et al. [17] apply a two-level attention mechanism based on GNN, which can effectively model complex relationships within a graph with multiple subgraphs. Wei et al. [19] treat the interaction under each behavior as a subgraph and utilize contrastive learning to capture the relationship dependencies between multiple behaviors. This approach can effectively capture the complex interactions between behaviors. Chen et al. [3] develop a method, GHCF, to explicitly model high-order relationships among users, items, and behaviors based on GCN. It is achieved through aggregating information of nodes and edges during the GCN message propagation process. This method can lead to improved performance in recommendation systems.

3 Methodologies

In this section, we first define the task of herb recommendation and the present our SA-HGNN, as shown in Fig. 2. Then, we then discuss the working mechanism of SA-HGNN.

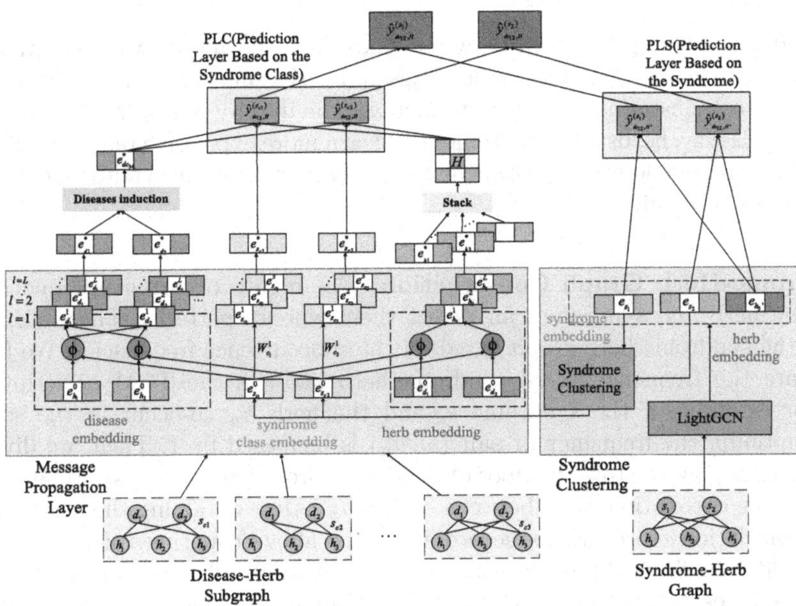

Fig. 2. The overall architecture of SA-HGNN. In all graph, disease nodes are in blue and herb nodes are in green. In syndrome-herb graph, syndrome notes are in yellow. (Color figure online)

3.1 Problem Definition

The herb recommendation task aims to generate a set of herbs to cure a given disease set under a specific syndrome based on the TCM prescriptions. Let $D = \{d_1, d_2, ..., d_M\}$, $H = \{h_1, h_2, ..., h_N\}$, and $S = \{s_1, s_2, ..., s_O\}$ represent all diseases, herbs, and syndromes, respectively. Each prescription $p = \langle\{dc, s, hc\}\rangle$ consists of a symptom set $dc = \{d_1, d_2, ...\}$, a syndrome s, and a herb set $hc = \{h_1, h_2, ...\}$. The disease-herb heterogeneous interactions are denoted as $\{Y^{(1)}, Y^{(2)}, ..., Y^{(O)}\}$, where $Y^{(s)}_{M \times N} = y^{(s)}_{dh} \in \{0, 1\}$ is the heterogeneous disease-herb interaction matrix, which indicates whether disease d has interacted with herb h under the specific syndrome s.

Given a disease set dc and a syndrome s, our model is to obtain an N-dimensional probability vector $\hat{y}_{(dc,H)}^{(s)}$ for all candidate herbs under the syndrome. The input and output are defined as follows:

- Input: Diseases D, Syndromes S, Herbs H, Prescriptions P.
- Output: A probability vector $\hat{y}_{(dc,H)}^{(s)}$, which is provided for all candidate herbs in H given the disease set dc and syndrome s.

3.2 Syndrome Clustering

To model the synergy pattern between syndromes and herbs, we have utilized LightGCN to learn syndromes embeddings and herbs embedding. In addition, we incorporate a clustering algorithm to increase the density of syndrome information. In this way, herbs and syndromes can learn more expressive representations by their own flexible propagation. Next, we will introduce the syndrome clustering module in detail.

Syndrome-Herb Graph Construction. The high cooccurrence frequencies between herb and syndrome can reflect their synergy patterns. Therefore, we build the syndrome-herb graph based on the cooccurrence frequencies. We first calculate the frequency of all syndrome-herb pairs in the TCM prescription P. For example, if the syndrome s_i and the herb h_n co-occur in the same prescription$p$, the frequency of pair (s_i, h_n) is increased by 1. Then, we divide the frequency by the total number of inquiry records that include s_i. Finally, we can get the correlation score between s_i and h_n. After obtaining the syndrome-herb correlation score matrix according to the above counting rules, we set a threshold t to filter the entries. Specifically, we rank all herbs in descending order under each syndrome. If the herb's rank under the syndrome is less than t, we set the corresponding entries to 1; otherwise, we set them to 0. Based on the syndrome-herb correlation score matrix, we can construct the syndrome-herb graph SH.

Syndrome Clustering Based on Nodes Embedding. Upon the syndrome-herb graph, we apply lightGCN to generate the syndrome and herb embeddings as e_s and $e_{h'}$. By referring to the above procedure, the syndromes embedding represent their similarity in relation to the herbs. Finally, based on syndrome embeddings e_s, we can apply K-means clustering to group the syndromes into C syndrome classes e_{s_c}.

3.3 Syndrome-Aware Herb Recommendation

In this section, we design the critical module to model the coarse-grained and fine-grained interaction information between the disease and herbs under the syndrome information, which is shown in Fig. 2.

Message Propagation Layer. In order to learn the disease, herb and syndrome class representations from the coarse-grained interaction, we first construct the multiple disease-herb subgraph, i.e., D-H$^{(s)}$ graph, based on the heterogeneous disease-herb interaction matrix $\{Y^{(s)}\}$, where $s \in C$. Inspired by GHCF [3], we define nodes as the disease d and herb h, the edge is the syndrome class s_c in the disease-herb subgraph. Then the disease and herb node updates its own embedding by propagating and aggregating messages over the subgraph. Specifically, for the disease d, we define the message transferred from its one-hop neighbor herb node h through a specific syndrome class edge s_c as $\phi(e_h^{(0)}, e_{s_c}^{(0)}) = e_h^{(0)} \odot e_{s_c}^{(0)}$, where $e_h^{(0)}$ is the initial herb embedding, $e_{s_c}^{(0)} = e_{s_c}$ is the initial syndrome class embedding that obtained by the syndrome clustering module. $\odot$ indicates the element-wise product of the herb node and syndrome class edge.

Then we conduct the message propagation of disease node in the $l-$th layer:

$$e_d^{(l)} = \sigma\left(\sum_{(h,s_c)\in\mathcal{N}(d)} \frac{1}{\sqrt{|\mathcal{N}_d||\mathcal{N}_h|}} \mathbf{W}^{(l)} \phi(e_h^{(l-1)}, e_{s_c}^{(l-1)})\right) \tag{1}$$

where $\mathcal{N}(d)$ indicate the neighbors of disease d, $\frac{1}{\sqrt{|\mathcal{N}_d||\mathcal{N}_h|}}$ is the symmetrized normalization term, $\mathbf{W}^{(l)}$ denotes the transformation weight matrix of the $l-$th layer, and we use the LeakyReLU as the activation function σ. Likewise, we define the message propagation of herb node in the $l-$th layer as:

$$e_h^{(l)} = \sigma\left(\sum_{(d,s_c)\in\mathcal{N}(h)} \frac{1}{\sqrt{|\mathcal{N}_d||\mathcal{N}_h|}} \mathbf{W}^{(l)} \phi(e_d^{(l-1)}, e_{s_c}^{(l-1)})\right) \tag{2}$$

In this way, we can integrate the syndrome class embedding into the message propagation of the multiple disease-herb subgraph. Especially, the syndrome class embeddings e_{s_c} in the $l-$th layer are transformed as follows:

$$e_{s_c}^{(l)} = \mathbf{W}_{s_c}^{(l)} e_{s_c}^{(l-1)} \tag{3}$$

where $\mathbf{W}_{s_c}^{(l)}$ is the transformation weight matrix in the l-th layer for s_c.

To generate the final embedding of diseases, herbs and syndrome classes, we average their embeddings of L layers, respectively, which enables the representation of information received from neighbors at different depths in the graph,

$$e_d^* = \sum_{l=0}^{L} \frac{1}{L+1} e_d^{(l)};\ e_h^* = \sum_{l=0}^{L} \frac{1}{L+1} e_h^{(l)};\ e_{s_c}^* = \sum_{l=0}^{L} \frac{1}{L+1} e_{s_c}^{(l)} \tag{4}$$

Diseases Induction Layer (DI). The TCM theory emphasizes the integrity of the human body, so it is important to summarize multiple diseases of patients for the herb recommendation. Inspired by SMGCN [10], we use MLP to capture the nonlinear interaction between diseases to induce an overall disease representation, as shown in Fig. 2. Specifically, given the diseases $dc = \{d_1, d_2, ..., d_m\}$

with a multi-hot vector, we first look up the disease embedding e_d^* for each symptom d in d_c. Second, we stack these vectors to build a matrix $e_{dc} \in \mathcal{R}^{m \times z}$, where z is the dimension of e_d^*. Third, we use the average pooling $Mean(\cdot)$ induce an overall disease set representation from e_{dc}. Finally, we apply a single-layer MLP to transform the mean vector as the inducted disease embedding e_{dc}^* for e_{dc}. The above process can be formulated as,

$$e_{dc}^* = ReLU(\mathbf{W}^{mlp} \cdot Mean(e_{dc}) + \mathbf{b}^{mlp}) \tag{5}$$

where $\mathbf{W}^{mlp}$ and $\mathbf{b}^{mlp}$ are the parameters of the MLP layer.

Herb Prediction Layer. In order to generate a herb set to cure the given disease set, we need to calculate the likelihood between the recommend herbs and the disease set across multiple syndrome. Specifically, we propose two prediction layer to capture the coarse-grained and fine-grained interaction information between the disease and herbs under the specific syndrome, respectively.

Prediction Layer Based on the Syndrome Class (PLC) . In order to calculate the likelihood between the recommend herbs and the disease set across multiple syndrome classes, we propose the representation of each syndrome class as a separate prediction layer, which can make predictions for each class independently. Specifically, the likelihood that the all candidate herbs H interact with the disease set dc under the syndrome class s_c is estimated as follows,

$$\hat{y}_{dc,H}^{(s_c)} = {e_{dc}^*}^T \cdot diag(e_{s_c}^*) \cdot e_H \tag{6}$$

where, e_H is the learned embedding matrix stacked by the candidate herbs embedding. $diag(e_{s_c}^*)$ indicates a diagonal matrix that the diagonal elements is $e_{s_c}^*$. As previously stated, we can capture the coarse-grained interaction information between the disease and herbs under the syndrome class.

Prediction Layer Based on the Syndrome (PLS). The PLC can enhance the data and improve the model performance by clustering similar syndromes. However, it could not highlight the unique characteristics of each syndrome. Therefore, we extract the fine-grained syndrome embedding e_s and all herbs embedding matrix $e_{H'}$ from the syndrome clustering module, where the $e_{H'}$ is stacked by the all candidate herbs embedding $e_{h'}$. Then we use the sample inner product to generate the prediction score for the specific syndrome and all candidate herbs,

$$\hat{y}_{dc,H'}^{(s)} = e_s^T e_{H'} \tag{7}$$

Comprehensive Prediction Layer. After obtaining the coarse-grained score $\hat{y}_{dc,H}^{(s_c)}$ and fine-grained score $\hat{y}_{dc,H'}^{(s)}$, we fuse the final prediction score through the hyper parameter λ,

$$\hat{y}_{dc,H}^{(s)} = \hat{y}_{dc,H}^{(s_c)} + \lambda \cdot \hat{y}_{dc,H'}^{(s)} \tag{8}$$

We implement λ in two ways, one is the fixed value, and the other is inversely proportional to the frequency of the syndrome.

Model Optimization. We apply **the efficient non-sampling learning** [3,4] to optimize our SA-HGNN model, the loss function of non-sampling strategy of syndrome s and its corresponding syndrome class s_c is,

$$\begin{aligned}\mathcal{L}_{s_c}(\theta) = & \sum_{(dc,s)\in B} \sum_{h\in H_{dc}^{(s_c)+}} \left(\left(p_h^{(s_c)+} - p_h^{(s_c)-} \right) \hat{y}_{dc,H}^{(s)} 2 - 2 p_h^{(s_c)+} \hat{y}_{dc,H}^{(s)} \right) \\ & + \sum_{i=1}^{z} \sum_{j=1}^{z} \left(\left(e_{s_c,i}^{*} e_{s_c,j}^{*} \right) \left(\sum_{dc\in B} e_{dc,i}^{*} e_{dc,j}^{*} \right) \left(\sum_{h\in H} p_h^{(s_c)-} e_{h,i}^{*} e_{h,j}^{*} \right) \right) \end{aligned} \tag{9}$$

where, B represents a batch of input disease data, $H_{dc}^{(s_c)+}$ is the herb collection whose herb interacts the disease collection dc under the syndrome class s_c. And we define the herb in $H_{dc}^{(s_c)+}$ as positive samples, otherwise, the herb is the negative samples. $p_h^{(s_c)+}$ represents the weight of positive samples and $p_h^{(s_c)-}$ represents the weight of negative samples.

Then, we joint all multiple syndrome classes' loss function [1] to modulate parameters from all heterogeneous data better. The final loss function is proposed as follows,

$$\mathcal{L}(\theta) = \sum_{s_c=1}^{C} \lambda_{s_c} \mathcal{L}_{s_c}(\theta) + \mu \left\|\theta\right\|_2^2 \tag{10}$$

where C is the number of types of syndrome class, λ_{s_c} is a hyper-parameter used to control the influence of the s_c-th syndrome class on joint training. μ and θ are parameters of L_2 regularization, which can prevent model overfitting.

4 Experiments

In this section, we evaluate our proposed SA-HGNN model on the actual data set. In particular, there are several important questions to answer:

RQ1: Can our proposed model outperform the state-of-art herb recommendation approaches?
RQ2: Can our proposed model outperform the state-of-art heterogeneous graph neural network-based models?
RQ3: How effective are our proposed modules(syndrome clustering and prediction layer based on syndrome(PLS))?
RQ4: How does our model performance react to the different hyper-parameter settings?
RQ5: Can our proposed SA-HGNN provide reasonable herb recommendation under different syndromes?

4.1 Dataset

The experimental data set is supported by Shuguang hospital affiliated to Shanghai university of traditional medicine. After data cleaning, we obtained 6535 real medical records of TCM diagnosis and treatment. Specifically, 5573 records include clear syndromes and 962 records are missing syndromes. The dataset has 511 types of diseases, 387 types of herbs, and 252 types of syndromes. The statistics of the dataset is summarized in Table 1.

4.2 Evaluation

Given a disease set, our proposed model generates a herb set to cure the diseases. To evaluate the performance of our approach, we adopt the following measures $Precision@K$, $Recall@K$, and $NDCG@K$, which are commonly used in herb recommendation task [10]. In the experiments, we use $Precision@5$ to decide the optimal parameters. And we truncate the ranked list at 20 for all three measures and report the average metrics for all prescriptions in the test set.

Table 1. Statistics of the evaluation data set.

Dataeset	# Prescriptions	# Diseases	# Syndromes	# Herbs
All	5573	456	252	371
Train	4458	412	252	363
Valid	557	200	158	289
Test	558	204	191	275

4.3 Baselines

Heterogeneous Graph Neural Network-Based Recommendation. In these models, we treat the user as the disease, the herb as the herb. Especially, we consider the relation between users and items as syndromes in the multi-relation heterogeneous graph based model. We select the following SOTA methods:

- GC-MC [2]: It fills in the matrix based on a self-encoder framework, with a GCN used as the encoder and a bilinear model as the decoder. Then, the completed matrix is used as the score for items corresponding to users.
- NGCF [16]: It employs a bipartite user-item graph to explicitly model high-order interactions, thereby obtaining more expressive representations for both users and items.
- LightGCN [6]: It is optimized from NGCF by discarding feature conversion weights and nonlinear activation functions that have no positive influence on collaborative filtering.
- GHCF [3]: It constructs several user-item bipartite graphs based on different behaviors, incorporates the behaviors into GCN message propagation, and merges multiple subgraphs to obtain the final node representation.

Graph Based Herb Recommendation

- SMGCN [10]: It constructs three multiple graphs(i.e., the symptom-herb bipartite graph, symptom-symptom graph, and herb-herb graph), and designs an MLP-based method to induce an overall implicit syndrome representation for each symptom set.

4.4 Parameter Settings

For our proposed SA-HGNN, the embedding size is fixed to 256, the learning rate(lr) is 0.0002, the *drop ratio* is set to 0.5, the μ is set to 1e-5, the number of syndrome clusters is set to 2, the depth of GCN layers is set to 1. The parameter setting of baselines is set as follows: For GCMC, $lr = 0.001$, $drop\ ratio = 0.0$, $\mu = 1e-5$; for LightGCN and NGCF, $lr = 0.001$, $drop\ ratio = 0.5$, $\mu = 1e-5$; for SMGCN, $lr = 0.001$, $drop\ ratio = 0.0$, $\mu = 7e-3$; for GHCF, $lr = 0.0002$, $drop\ ratio = 0.5$, $\mu = 1e-5$. For a fair comparison, we add the DI part to GC-MC, NGCF, LightGCN, and GHCF.

4.5 Performance Comparison

In this section, we demonstrate the overall results compared by baselines, followed by ablation analysis and the influence of critical hyperparameters. In the end, we display the actual cases to demonstrate the effectiveness of our models.

Table 2. Performance of different models. GHCF(*) is the model by adding **Syndrome Clustering** module based on GHCF. r and p are short of Recall and Precision.

Approaches	r@5	r@10	r@20	p@5	p@10	p@20	ndcg@5	ndcg@10	ndcg@20
GC-MC	0.2057	0.3548	0.5519	0.6308	0.5461	0.4256	0.6601	0.6560	0.8209
LightGCN	0.2093	0.3673	0.5728	0.6434	0.5658	0.4435	0.6678	0.6665	0.8213
NGCF	0.2102	0.3682	0.5673	0.6437	0.5656	0.4375	0.6697	0.6686	0.8258
SMGCN	0.2134	0.3720	0.5772	0.6538	0.5726	0.4470	0.6730	0.6696	0.8244
GHCF(*)	0.2201	0.3805	0.5869	0.6774	0.5872	0.4551	0.7007	0.6854	0.8397
SA-HGNN	**0.2238**	**0.3874**	**0.5902**	**0.6875**	**0.5977**	**0.4574**	**0.7116**	**0.6983**	**0.8458**

Overall Comparison (RQ1 & RQ2). The overall comparison is shown in Table 2, the best performing models are bold, while the strongest models are underlined. We can observed that:

(1) The herb recommendation model SMGCN outperforms the bipartite graph-based models(GC-MC, LightGCN, and NGCF), indicating the multiple graph can learn more abundant information of the diseases and herbs.

(2) GHCF(*) consistently outperforms SMGCN, and the improvement is more obvious than other baseline models, the reasons may contain two aspects: 1) GHCF(*) consider the explicit syndromes' information, which can capture more effectiveness information to model the interaction between diseases and herbs; 2) To some degree, syndrome clustering can address the issue of syndrome data sparse and bias. 3) The efficient non-sampling learning module is more effective than traditional negative sampling learning methods.

(3) Our proposed SA-HGNN performs much better than GHCF(*) showing that integrating the fine-grained syndrome information can induce more information, and leverage the characteristics of syndrome information better.

Ablation Analysis (RQ3). To verify the impact of the syndrome clustering module and the prediction layer based on the syndrome(PLS), the ablation study is conducted on the two variants of SA-HGNN. We define SA-HGNN-C to disable the syndrome clustering module in SA-HGNN. Especially, GHCF(*) is a variant of SA-HGNN without PLS. The experimental results are shown in Table 3.

Table 3. Effect of Different Modules on the Dataset.

Approaches	r@5	r@10	r@20	p@5	p@10	p@20	ndcg@5	ndcg@10	ndcg@20
SA-HGNN-C	0.1886	0.3305	0.5143	0.5919	0.5188	0.4051	0.6155	0.6283	0.7738
GHCF(*)	0.2201	0.3805	0.5869	0.6774	0.5872	0.4551	0.7007	0.6854	0.8397
SA-HGNN	**0.2238**	**0.3874**	**0.5902**	**0.6875**	**0.5977**	**0.4574**	**0.7116**	**0.6983**	**0.8458**

For both SA-HGNN-C and SA-HGNN, we can find that when no considering the syndromes clustering, the result significantly declines. It could be because the syndrome data are highly sparse, resulting in sparse disease-herb adjacency matrices for these syndromes. The sparsity can negatively affect the effectiveness of graph representation learning. To address this issue, grouping similar syndromes into clusters can help mitigate the data sparsity problem and enhance performance through data enhancement.

For both GHCF(*) and SA-HGNN, integrating PLS leads the further improvement. This could be because SA-HGNN utilizes PLS to introduce fine-grained syndrome information. This implementation enhances the coarse-grained syndrome class information induced by PLC module. Moreover, it can make better use of the features inherent to syndrome information.

Influence of Hyperparameters (RQ4)

Effect of Cluster Numbers. In clustering algorithms, the number of clusters is an essential factor that affects the effectiveness of clustering. Therefore, in this section, we will analyze how different numbers of clusters impact the model performance. We set the number of syndrome class varied in {2, 3, 4, 5}. From

Table 4. Performance in Different Cluster Number.

Class	Approaches	r@5	r@10	r@20	p@5	p@10	p@20	ndcg@5	ndcg@10	ndcg@20
2	GHCF(*)	0.2201	0.3805	0.5869	0.6774	0.5872	0.4551	0.7007	0.6854	0.8397
	SA-HGNN	**0.2238**	**0.3874**	**0.5902**	**0.6875**	**0.5977**	**0.4574**	**0.7116**	**0.6983**	**0.8458**
3	GHCF(*)	0.2184	0.3808	0.5778	0.6724	0.5882	0.4476	0.7025	0.6927	0.8446
	SA-HGNN	0.2219	0.3872	0.5891	0.6824	0.5964	0.4560	0.7093	0.6966	0.8467
4	GHCF(*)	0.2154	0.3712	0.5678	0.6627	0.5735	0.4399	0.6860	0.6757	0.8306
	SA-HGNN	0.2168	0.3754	0.5728	0.6663	0.5790	0.4437	0.6923	0.6816	0.8357
5	GHCF(*)	0.2111	0.3684	0.5619	0.6498	0.5688	0.4354	0.6772	0.6771	0.8287
	SA-HGNN	0.2149	0.3753	0.5708	0.6602	0.5780	0.4420	0.6866	0.6791	0.8319

Table 4, we can observed that the two-clusters model achieves the best performance. It is likely due to the relatively small dataset, which results in the subgraphs of the two syndrome classes being denser than those of other number classes. This denseness can better help GCN learn embeddings of diseases, syndromes, and herbs. Additionally, when the syndrome data is sparse, the data enhancement performance of clustering becomes more apparent. Moreover, whether the cluster number is 2, 3, 4 or 5, our model consistently outperforms GHCF(*). This verifies the effectiveness of *PLS* again.

Effect of GCN Layer Numbers in Message Propagation Layer Module. To investigate whether GHHGF can benefit from message propagation in deep layers of graph neural networks, we adjust the number of GCN layers in the message propagation layer module, and the experimental results are presented in Table 5. From Table 5, we can find that our model is insensitive to changes in the depth of the message propagation layer. Furthermore, we observe that the effectiveness of the convolution operation is limited when using a two-layer architecture compared to a one-layer architecture. When we increase the depth of the GCN to three layers, we observe a slight reduction in model performance compared to the one-layer architecture. It could be due to the occurrence of overfitting of GCN. Moreover, whether the layer number is 1, 2, or 3, our model consistently outperforms other baselines. This demonstrates the effectiveness of our proposed model SA-HGNN.

Table 5. Effect of Layer Number on SA-HGNN.

Depths	r@5	r@10	r@20	p@5	p@10	p@20	ndcg@5	ndcg@10	ndcg@20
1	**0.2238**	**0.3874**	**0.5902**	**0.6875**	**0.5977**	**0.4574**	**0.7116**	0.6983	0.8458
2	0.2216	0.3871	0.5866	0.6814	0.5961	0.4540	0.7096	**0.6994**	**0.8478**
3	0.2205	0.3867	0.5895	0.6781	0.5955	0.4565	0.7049	0.6951	0.8449

Case Study (RQ5). In this section, we conduct two case studies to verify the rationality of our proposed herb recommendation approach.

Syndrome Cluster Display. We display the distribution of syndromes under two classes in Fig. 3. It can be observe that most syndromes in class one are associated with vital energy and blood, while most syndromes in class two are related to rheumatic fever. This suggests that the syndrome clustering module has performed well, and the final results confirm its accuracy.

syndrome class 1		syndrome class 2	
血虚证(Blood deficiency)	肝肾阴虚证(Yin deficiency of liver and kidney)	风湿热证(Rheumatic fever)	暑热证(summer-heat syndrome)
肺气亏虚证(Lung qi deficiency syndrome)	血瘀证(Blood stasis syndrome)	风热上扰证(Wind-heat disturbing upward)	湿热内蕴证(Damp-heat syndrome)

Fig. 3. The case of syndrome cluster display.

Cases of Herb Recommendation. Figure 4 shows two real examples in the herb recommendation scenario. Given the disease and syndrome, our proposed SA-HGNN generates different herb sets in the same disease under different syndromes. The common herbs between the herb set recommended by SA-HGNN and the ground truth are indicated in bold red font. According to the TCM knowledge, the missing herbs usually have similar effects to the remaining ground-truth herbs. For example, chinese angelica and caulis spatholobi have the same effect of enriching the blood. The above comparative analysis demonstrates that our proposed SA-HGNN can provide reasonable herb recommendations in different syndromes.

Disease	Syndrome	Herb Set	
		SA-HGNN	Ground Truth
湿疮病 (Wet sore disease)	湿热证 (Damp heat syndrome)	牡丹皮(peony bark) 苦参(sophora flavescens) 土茯苓(smilax glabra) 白鲜皮(cortex dictamni) 黄芩(baikal Skullcap) 知母(rhizoma anemarrhenae) **蒲公英(dandelion)** 白花蛇舌草(oldenlandia) 地肤子(broom cypress fruit) 玄参(radix scrophulariae)	苦参(sophora flavescens) 牡丹皮(peony bark) **黄连(coptis chinensis)** 地肤子(broom cypress fruit) **大青叶(isatis leaf)** 白花蛇舌草(oldenlandia) 黄芩(baikal Skullcap) 玄参(radix scrophulariae) 土茯苓(smilax glabra) 白鲜皮(cortex dictamni) 知母(rhizoma anemarrhenae) **牡蒿(artemisia japonica)**
湿疮病 (Wet sore disease)	气虚血瘀证证 (Qi deficiency and blood stasis syndrome)	丹参(salvia miltiorrhiza) **当归(chinses angelica)** 土茯苓(smilax glabra) 白术(atractylodes macrocephala) **地黄(rehmannia glutinosa)** **牡丹皮(peony bark)** **党参(codonopsis pilosula)** 玄参(radix scrophulariae) 牛膝(common Achyranthes) 杜仲(eucommia)	**伸筋草(common clubmoss Herb)** 杜仲(eucommia) **延胡索(corydalis)** **石韦(pyrrosia lingua)** 白术(atractylodes macrocephala) 土茯苓(smilax glabra) **鹿衔草(pyrola)** **鸡血藤(caulis spatholobi)** 丹参(salvia miltiorrhiza) **木瓜(papaya)** 牛膝(common Achyranthes) **防己(caulis spatholobi)** 玄参(radix scrophulariae) **黄芪(astragalus mongholicus)**

Fig. 4. The case of syndrome cluster display.

5 Conclusions

In this paper, we propose a herb recommendation method that explicitly incorporates syndrome information. We apply the syndrome-herb graph and multiple disease-herb graph based on the syndrome class to learn the embedding of symptoms, herbs, syndromes, and syndrome classes. To alleviate the problem of data sparsity, we utilize the syndrome clustering to capture more effectiveness interaction between diseases and herbs. In addition, we induce both coarse-grained and fine-grained syndrome information to utilize the characteristics of syndromes better. In future work, we will perform contrastive learning from different syndromes to improve the embedding quality of TCM entities. At the same time, we intend to incorporate attention mechanisms and TCM knowledge graphs to enable interpretable analyses based on aggregated weights for herb recommendation.

Acknowledgements. We thank editors and reviewers for their suggestions and comments. This work was supported by NSFC grants (No. 62136002), National Key R&D Program of China (2021YFC3340700) and Shanghai Trusted Industry Internet Software Collaborative Innovation Center.

References

1. Argyriou, A., Evgeniou, T., Pontil, M.: Multi-task feature learning. In: Advances in Neural Information Processing Systems, vol. 19 (2006)
2. van den Berg, R., Kipf, T.N., Welling, M.: Graph convolutional matrix completion. arXiv preprint arXiv:1706.02263 (2017)
3. Chen, C., et al.: Graph heterogeneous multi-relational recommendation. In: Proceedings of the AAAI Conference on Artificial Intelligence, vol. 35, pp. 3958–3966 (2021)
4. Chen, C., Zhang, M., Zhang, Y., Liu, Y., Ma, S.: Efficient neural matrix factorization without sampling for recommendation. ACM Trans. Inf. Syst. (TOIS) **38**(2), 1–28 (2020)
5. Cheung, F.: TCM: made in China. Nature **480**(7378), S82–S83 (2011)
6. He, X., Deng, K., Wang, X., Li, Y., Zhang, Y., Wang, M.: LightGCN: simplifying and powering graph convolution network for recommendation. In: Proceedings of the 43rd International ACM SIGIR Conference on Research and Development in Information Retrieval, pp. 639–648 (2020)
7. Ji, W., Zhang, Y., Wang, X., Zhou, Y.: Latent semantic diagnosis in traditional Chinese medicine. World Wide Web **20**, 1071–1087 (2017)
8. Jiang, M., et al.: Syndrome differentiation in modern research of traditional Chinese medicine. J. Ethnopharmacol. **140**(3), 634–642 (2012)
9. Jin, Y., Ji, W., Zhang, W., He, X., Wang, X., Wang, X.: A KG-enhanced multi-graph neural network for attentive herb recommendation. IEEE/ACM Trans. Comput. Biol. Bioinform. **19**, 2560–2571 (2021)
10. Jin, Y., Zhang, W., He, X., Wang, X., Wang, X.: Syndrome-aware herb recommendation with multi-graph convolution network. In: 2020 IEEE 36th International Conference on Data Engineering (ICDE), pp. 145–156. IEEE (2020)

11. Li, W., Yang, Z.: Distributed representation for traditional Chinese medicine herb via deep learning models. arXiv preprint arXiv:1711.01701 (2017)
12. Ruan, C., Ma, J., Wang, Y., Zhang, Y., Yang, Y., Kraus, S.: Discovering regularities from traditional Chinese medicine prescriptions via bipartite embedding model. In: IJCAI, pp. 3346–3352 (2019)
13. Schuster, M., Paliwal, K.K.: Bidirectional recurrent neural networks. IEEE Trans. Signal Process. **45**(11), 2673–2681 (1997)
14. Velickovic, P., et al.: Graph attention networks. Stat **1050**(20), 10–48550 (2017)
15. Volkovs, M., Yu, G., Poutanen, T.: DropoutNet: addressing cold start in recommender systems. In: Advances in Neural Information Processing Systems, vol. 30 (2017)
16. Wang, X., He, X., Wang, M., Feng, F., Chua, T.: Neural graph collaborative filtering. In: Piwowarski, B., Chevalier, M., Gaussier, É., Maarek, Y., Nie, J., Scholer, F. (eds.) Proceedings of the 42nd International ACM SIGIR Conference on Research and Development in Information Retrieval, SIGIR 2019, Paris, France, 21–25 July 2019, pp. 165–174. ACM (2019). https://doi.org/10.1145/3331184.3331267
17. Wang, X., et al.: Heterogeneous graph attention network. In: The World Wide Web Conference, pp. 2022–2032 (2019)
18. Wang, X., Zhang, Y., Wang, X., Chen, J.: A knowledge graph enhanced topic modeling approach for herb recommendation. In: Li, G., Yang, J., Gama, J., Natwichai, J., Tong, Y. (eds.) DASFAA 2019. LNCS, vol. 11446, pp. 709–724. Springer, Cham (2019). https://doi.org/10.1007/978-3-030-18576-3_42
19. Wei, W., Huang, C., Xia, L., Xu, Y., Zhao, J., Yin, D.: Contrastive meta learning with behavior multiplicity for recommendation. In: Proceedings of the Fifteenth ACM International Conference on Web Search and Data Mining, pp. 1120–1128 (2022)
20. Yao, L., Zhang, Y., Wei, B., Zhang, W., Jin, Z.: A topic modeling approach for traditional Chinese medicine prescriptions. IEEE Trans. Knowl. Data Eng. **30**(6), 1007–1021 (2018)
21. Ying, R., He, R., Chen, K., Eksombatchai, P., Hamilton, W.L., Leskovec, J.: Graph convolutional neural networks for web-scale recommender systems. In: Proceedings of the 24th ACM SIGKDD International Conference on Knowledge Discovery & Data Mining, pp. 974–983 (2018)

BDQM

Fault Prediction Based on Traffic Light Data Cleaning

Weijia Feng[1,2], Siyao Qi[1], Wenwen Liu[1], Yunhe Chen[1], Zhangzhen Nie[1], and Jia Guo[1(✉)]

[1] Tianjin Normal University, Binshui Xi Road 393, XiQing District, Tianjin, China
c04s316@bupt.cn

[2] Postdoctoral Innovation Practice Base, Huafa Industrial Share Co., Ltd., Changsheng Road 155, ZhuHai, China

Abstract. Traffic lights play a crucial role in our daily lives. Predicting traffic light failures through collecting traffic light data can avoid safety incidents and economic losses caused by traffic light failures. Due to reasons such as electrical equipment and magnetic fields, the collected data may be inaccurate. Predicting traffic light failures based on inaccurate data is a challenge. This study collects real traffic light data and proposes an AO-LSTM model for traffic light failure prediction, which includes three processing functions to identify inaccurate data and clean it. The processed data is normalized and features are extracted. A mixed health value system is introduced to evaluate the health status of traffic lights, including various metrics such as Euclidean distance, Mahalanobis distance, and current and voltage variance. The Adam optimization algorithm is used to optimize the model parameters. The proposed model is validated based on real data, and the results show that the accuracy of traffic light mode recognition is 99.18% and the deviation of traffic light failure prediction is 1.73% with an accuracy of 96.55%.

Keywords: Traffic lights · Data quality · Pattern recognition · Fault prediction · AO-LSTM

1 Introduction

Traffic lights are widely used in daily life. If the traffic lights fail, the traffic situation will fall into chaos, which will bring safety hazards and economic losses. Therefore, the pattern recognition and fault prediction of traffic lights are becoming more and more important [1]. The pattern recognition of the traffic light can detect the state of the traffic light in real time, and assist the work of the traffic light. Traffic jams and even serious traffic accidents may occur when traffic lights fail [2, 3]. Fault prediction can effectively avoid some unnecessary huge losses of manpower and material resources [4, 5]. How to predict the traffic light failure is particularly important for traffic travel [6, 7]. At present, the traffic light failure data set is missing, and no research has been published on traffic light failure prediction.

Traditional fault prediction is not suitable for traffic lights [8, 9]. Data-driven fault pre-diction technology and statistical reliability-based fault prediction technology are

A. El Abbadi et al. (Eds.): DASFAA 2023 Workshops, LNCS 13922, pp. 127–137, 2023.
https://doi.org/10.1007/978-3-031-35415-1_9

currently commonly used fault prediction methods [10, 11]. Zhang et al. have proposed log-driven fault detection based on LSTM model [12], and Gao et al. have proposed a Bi-LSTM model based on job and task fault prediction [13]. The LSTM model also has great potential in fault prediction [14]. However, there is a paucity of research in the area of traffic light fault prediction and no relevant results have been published. Research on traffic light failures faces many challenges, including the lack of public datasets, the lack of publicly published algorithms related to traffic lights, and the self-collected data contain dirty data.

The data we collected has data quality problems, which will affect the prediction accuracy of the model. The main reasons for data quality problems include: The environment will affect the stability of the internal components of the traffic light, resulting in data collection; data loss will occur during data transmission and uploading.

To improve the data quality, we propose three processing functions, namely the classification function, the detection function, and the threshold function. The classification function classifies the data based on the type of event, the detection function detects any abnormal data, and the threshold function calculates the thresh-old for abnormal data based on the event type. By implementing these functions, we can effectively eliminate dirty data from the dataset to improve the data quality.

This paper proposes an AO-LSTM model for pattern recognition and fault prediction, which improves data quality by identifying and removing abnormal data, and completes the pattern recognition of traffic lights through normalization processing and feature extraction. When the traffic light is working, it can judge the color of the currently displayed traffic light in real time, which is convenient for maintenance. Constructed a mixed health value system, using Euclidean distance, Mahalanobis distance, and the variance of current and voltage to estimate the data health value, and used Adam optimization algorithm to train model parameters, predict when the traffic light has not failed, and carry out timely Intervention can effectively reduce the hidden dangers and losses caused by the failure of traffic lights. Good results have been obtained through experimental verification.

The main contributions of this paper are as follows:

- Analyze and model the problem of traffic light failure prediction;
- This paper proposes an AO-LSTM model to improve data quality and solve the problem of traffic light pattern recognition and fault prediction;
- A real experimental environment is built to verify the effectiveness of the proposed method.

In this paper, the following sections are presented. Section 1 presents the background of fault prediction; Section 2 presents the problem statement and the principle formulation; Section 3 presents the structural model for traffic light fault prediction; Section 4 presents the scenario construction, data processing methods, experimental design and result analysis for traffic light data acquisition; Section 5 concludes.

2 Problem Formulation

This section mainly introduces the problem solved by the model. The pattern recognition and fault prediction of traffic lights can be stated as:

The input signal denotes as q^t, $1 \leq t \leq T$, q^t includes current and voltages are represented by v^t and a^t. The t is the time period, and T symbolizes the total time period. The output current signal after the normalization is n_t.

$$n_t = N(q^t) \tag{1}$$

In (1) the $N(q^t)$ is the Fast Fourier Transform normalization function. For ease of calculation, need to normalize the input signal. The estimated health degree can be converted to calculate the distance between SHP and input signal. The distance calculates using the mahalanobis distance, Euclidean distance, Variance of current and voltage, and the estimated health status of the equipment at time t marked as H_t.

$$H_t = d(n_t, SHP) \tag{2}$$

In (2), $d(n_t, SHP)$ is computed based on mahalanobis distance, Euclidean distance, and Variance of current and voltage.

$$d(n_t, SHP) = \frac{1}{4}[M(n_t, SHP) + E(n_t, SHP) + V(a_t, v_t)] \tag{3}$$

In the Eq. (3) $M(n_t, SHP)$ is the mahalanobis distance function. And the $E(n_t, SHP)$ is the Euclidean distance function.

The $V(a_t, v_t)$ is the Variance of current and voltage function, as follows:

$$V(a_t, v_t) = \sqrt{\frac{1}{t}\sum_{t=1}^{T}(a_t - \overline{a})} + \sqrt{\frac{1}{t}\sum_{t=1}^{T}(v_t - \overline{v})} \tag{4}$$

The health status of the equipment can be represented by the health degree, the estimated health degree and load characteristics of electrical equipment are used as training data and sent to the LSTM model for training.

LSTM is used to the time series load characteristics. The main concept in the LSTMs is the forget gates and the memory gates. They control the information passed along the sequence so that the long-term correlation of the data is accurately extracted at every time instance. In LSTM, H_t as the input vector to the LSTM unit, i_t is the input gate activation vector, f_t is the forget gate activate vector, O_t is the output gate activation vector, c_t is the cell state vector, h_t is the hidden state vector.

$$i_t = \sigma(w_i H_t + u_i h_{t-1} + b_i) \tag{5}$$

$$f_t = \sigma(w_f H_t + u_f h_{t-1} + b_f) \tag{6}$$

$$O_t = \sigma(w_o H_t + u_o h_{t-1} + b_o) \tag{7}$$

$$c_t = f_t \odot c_{t-1} + i_t \odot \tanh(w_c H_t + u_c h_{t-1} + b_c) \tag{8}$$

$$h_t = o_t \odot \tanh(c_t) \tag{9}$$

3 Methodology

This section mainly introduces the framework of the AO-LSTM model.

In the fault prediction part of the traffic light, firstly, the data quality is improved by identifying and removing abnormal data, the input data is normalized by fast Fourier transform, and then feature extraction is performed by three-level convolution, thereby completing the pattern recognition of traffic light data. The hybrid distance value of the data is calculated to estimate the health level. The data with health values are then sent to the Deep LSTM network and trained using the Adam algorithm to predict subsequent changes in health levels and complete traffic light fault prediction. The specific process framework is shown in Fig. 1.

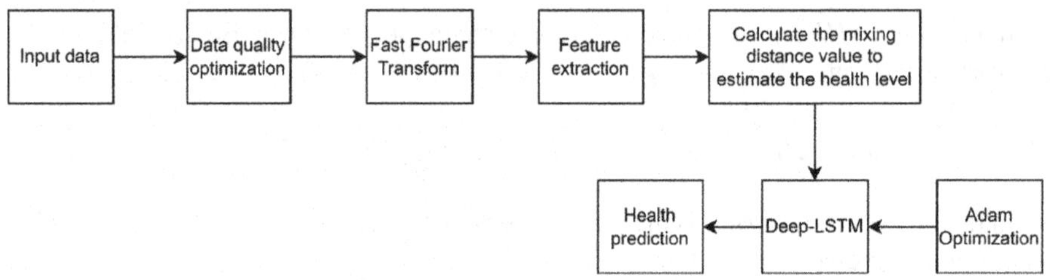

Fig. 1. Fault prediction framework

In order to improve the accuracy of the dataset and enhance data quality, we identify and remove outliers. We design three processing functions, namely the classification function, the detection function, and the threshold function. The classification function classifies the data based on the type of event, the detection function detects any abnormal data, and the threshold function calculates the threshold for abnormal data based on the event type. We use these functions to optimize data quality and improve data accuracy.

The input data is first normalized by the Fast Fourier Transform (FFT) for subsequent feature extraction. "Fast Fourier Transform" is a general term for efficient and fast computational methods for computing the discrete Fourier transform (DFT) using a computer. Using this algorithm allows the computer to greatly reduce the number of multiplications when computing the discrete Fourier transform, and in particular, the more sampling points being transformed, the more significant the computational savings of the FFT algorithm. The normalized data is convolved in three layers and pooled in one layer, which completes the feature extraction of the data and thus completes the pattern recognition of the traffic light data.

We use the data of the normal operation of the traffic light to cluster and get the cluster center under each mode. A mixed health value system is constructed by calculating the Euclidean space distance between the data and the cluster center, the Markov distance, and the variance of current and voltage to estimate the health value of the data.

The data labeled with health classes and time series are fed into the Deep-LSTM model for prediction. Long Short Term Memory networks (LSTM) is a special type of RNN model that can learn long-term dependencies [15]. The Deep-LSTM is divided into four layers, i.e., input layer, LSTM layer, fully connected layer and regression output layer, and the LSTM layer consists of five basic units, i.e., memory block, memory unit, input gate, output gate and forgetting gate. Among them, the memory cell has a recurrent self-connected linear cell, and the input gate, output gate and forgetting gate are used to decide what data to save and what data to forget and when to read them [16]. In Deep-LSTM, the model will make predictions for the uncoming time intervals and add the predicted values from the previous time interval to the data predicted in the next time interval to continuously learn updates. The specific framework architecture of LSTM is shown in Fig. 2.

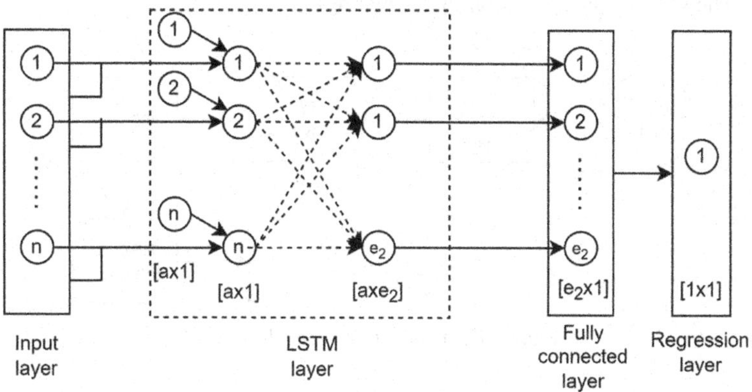

Fig. 2. LSTM framework

We use the Adam optimization algorithm for training the model parameters of Deep-LSTM. in the Adam algorithm, each network weight (parameter) maintains a learning rate and is adjusted individually as the learning unfolds, and the adaptive learning rate for different parameters is calculated by the first and second order moments of the gradient. The advantage of Adam is mainly that after bias correction, each iteration The learning rate has a definite range, making the parameters relatively smooth.

4 Experiments

This section demonstrates the analysis and discussion of the results of the proposed pattern recognition and fault prediction technology, and presents the accuracy of pattern recognition and fault prediction through comparison and analysis with the real data collected.

4.1 Data Collection Environment

We collected real data of traffic lights at 108 intersections in Tianjin Haihe Education Park. Traffic lights include red, green and yellow lights. The data we collect is the data of the traffic light as a whole, not the data of specific color lights. This data collection method does not need to install equipment inside the traffic light, but only needs to install a set of data collection equipment to collect the overall data of the traffic light, which effectively reduces the cost. It is worth noting that the data collection is in terms of current confluences at opposite intersections, not circuit confluences at the entire intersection.

In this paper, we use a real traffic light device to build a data collection environment and collect real data for experiments. Use the self-developed intelligent circuit breaker for data collection, convert the signal into a digital signal by the A/D converter, send it to the single-chip microcomputer to process the signal, and select the RS-232 bus and MAX3232 chip to upload the data to the PC for subsequent follow-up experiment and analysis. The specific scene diagram is shown in Fig. 3.

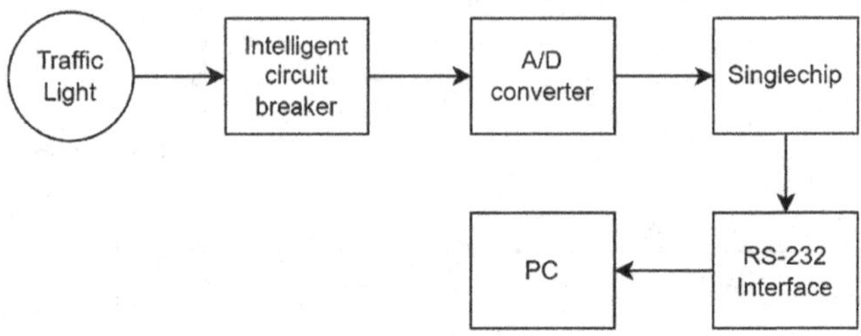

Fig. 3. Data collection scene diagram

In order to collect data on traffic light failures, we set up a real experimental environment and artificially control traffic light failures. The experimental environment is shown in Fig. 4.

Fig. 4. Experimental environment

Traffic lights failures mainly include abnormal lighting color and abnormal lighting quantity. The traffic light failure is shown in Fig. 5.

The real light data include six types. Total events in data is 54672, the health traffic light event count is 54375 accounting for 99.4% of the total event and the unhealth traffic light event count is 297 accounting for 0.6% of the total event. The health red-light signal

Fig. 5. Fault traffic light

event count is 29373 accounting for 53.7% of total event. The health yellow-light signal event count is 11232 accounting for 20.5% of total event. The health green-light signal event count is 13770 accounting for 25.2% of total event. The unhealth red-light signal event count is 141 accounting for 0.3% of total event. The unhealth yellow-light signal event count is 96 accounting for 0.2% of total event. The unhealth green-light signal event count is 60 accounting for 0.1% of total event. The date source includes health and unhealth current and voltages data, show in Table 1.

Table 1. Data Source Details

Number of total data	54672	100.0%
Number of red-light signal data	29373	53.7%
Number of green-light signal data	13770	25.2%
Number of yellow-light signal data	11232	20.5%
Number of unhealth red-light signal data	141	0.3%
Number of unhealth green-light signal data	60	0.1%
Number of unhealth yellow-light signal data	96	0.2%
Number of total data	54672	100.0%
Number of red-light signal data	29373	53.7%
Number of green-light signal data	13770	25.2%

4.2 Experimental Environment

The pattern recognition and fault prediction experiments are all carried out in the real building scenario, using the Wimdows11 operating system Core i5 processor, with 2.2GHz and 4GB RAM, using real collected data, and using Python and Tensorflow1.7.

4.3 Performance Indicators

In this section, we define performance indicators to analyze the experimental results. The prediction deviation uses the following formula:

$$\varphi(H_n, R_n) = \frac{\sum_{n=1}^{N} \sqrt{(H_n - R_n)^2}}{\sum_{n=1}^{N} R_n} \tag{10}$$

Equation (10) calculates the deviation rate of predicted health value, H_n represents the health value of nth data, R_n represents the health value of nth real data.

The prediction accuracy rate μ uses the following formula:

$$\mu(n, N) = \frac{n}{N} \tag{11}$$

In prediction accuracy rate Eq. (11), n represents the number of accurate prediction events, N represents the number of all events.

4.4 Pattern Recognition Results

We use health data to conduct experiments. The model recognizes the color of the traffic light at the sampling time by calculating the Euclidean distance, compares and judges the color of the traffic light at the sampling time, and writes the recognition result into the data in the form of a label. Finally, the result data set is output to the graph with time as the abscissa, current as the ordinate, and color as the distinction. The output results are shown in Fig. 6.

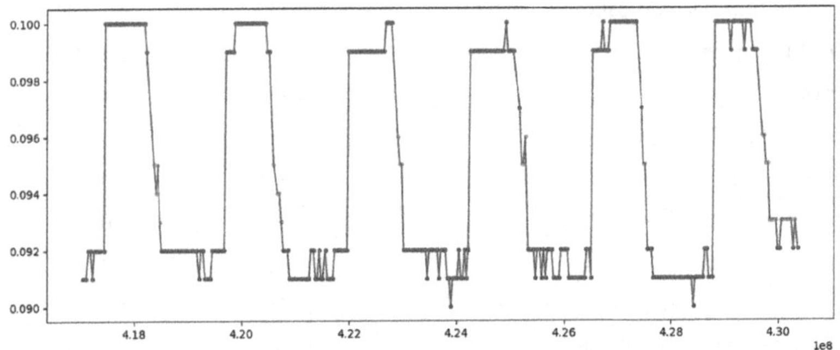

Fig. 6. Pattern recognition results

Figure 6 is a partial interception of the output results, the abscissa is the acquisition time value, and the ordinate is the current. In Fig. 6, the results of pattern recognition are distinguished by color, red corresponds to the traffic light recognition result is green, yellow corresponds to the traffic light is yellow, and green corresponds to the traffic light is red. From the figure, we can clearly distinguish the result of traffic light pattern recognition. The color of the traffic light changes periodically, the order is green-yellow-red, and the recognition effect is very good.

According to the experiment, the number of wrongly identified sampling points is 148, and the total number of sampling points in the data set is 18,000. The calculated pattern recognition accuracy rate is 99.18%.

We conduct pattern recognition on 10 data sets from A1 to A10, and Table 2 shows the obtained recognition accuracy. The collection scenarios, sampling intervals, and collection load characteristic parameters of the 10 data sets are the same, but the collection time and number of sampling points are different. It can be seen from Table 2 that the recognition accuracy is between [99.10%, 99.43%] interval, indicating that the pattern recognition technology proposed in this study has a higher recognition accuracy.

Table 2. Pattern recognition results of multiple datasets

Dataset number	A1	A2	A3	A4	A5	A6	A7	A8	A9	A10
Sampling duration	10 min	40 min	70 min	100 min	130 min	15 min	45 min	75 min	105 min	145 min
Recognition accuracy	99.25%	99.43%	99.17%	99.08%	99.35%	99.16%	99.21%	99.10%	99.14%	99.19%

4.5 Fault Prediction Results

In actual use, the failure rate of traffic lights is not high. In order to make the experimental results more accurate, in the experimental scene, we added the data collected from the faulty traffic lights. In the fault prediction technology experiment, we put the data set into the fault prediction model. The model can obtain the predicted value of health at each time period, and output the result to a graph with time as the abscissa and health value as the ordinate. The output result is shown in Fig. 7.

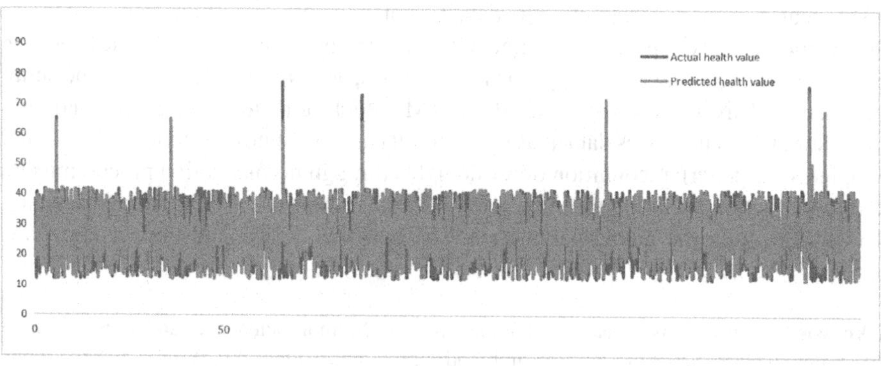

Fig. 7. Fault Prediction results

Figure 7 is a partial interception of the output results. The abscissa is the time conversion value, the ordinate is the health value, the yellow curve is the predicted health value, and the blue curve is the real health value of the data set after calculation. It can be clearly seen from the figure that the coincidence degree of the two curves is high, and the health value indicates the health status of the traffic light, and most of them are stable in the [10, 45] interval; the two lines have peaks in many places, indicating that there is a fault. The peak time mostly coincides with the real fault time.

In this experiment, the prediction deviation between the health value predicted by the fault prediction model and the actual health value is 1.73%. The model predicts a total of 87 failures, and 84 failures actually occur. The prediction accuracy rate is 96.55%, and the prediction accuracy is relatively high.

Fault prediction is carried out on five data sets from B1 to B5. Table 3 shows the prediction accuracy and the deviation between the predicted health value and the actual health value. The collection scenarios and collection load characteristic parameters of these five data sets are the same, but the collection time and the number of sampling points are different. It can be seen from Table 3 that the recognition accuracy rate is between [96.39%, 96.64%] interval, indicating that the fault prediction technology proposed in this study has a higher recognition accuracy rate.

Table 3. Fault prediction results of multiple datasets

Dataset number	B1	B2	B3	B4	B5
Sampling duration	30 min	60 min	90 min	180 min	300 min
Prediction deviation	1.76%	1.68%	1.81%	1.74%	1.70%
Prediction accuracy	96.47%	96.64%	96.39%	96.55%	96.58%

5 Conclusion

In this paper, we use a real traffic light device to build a data collection environment and collect real data for experiments. The data we collect is the data of the traffic light as a whole, not the data of specific color lights. It is worth noting that the data collection is in terms of current confluences at opposite intersections, not circuit confluences at the entire intersection. This paper proposes an AO-LSTM model for pattern recognition and fault prediction, which improves data quality by identifying and removing abnormal data, and completes the pattern recognition of traffic lights through normalization processing and feature extraction, with an accuracy rate of 99.18%. And use the Adam optimization algorithm to train the model parameters. The bias was 1.73% and the accuracy was 96.55%.

Acknowledgement. This research is funded by the Natural Science Foundation of China (NSFC):61602345; National Key Research and Development Plan: 2019YFB2101900; Application Foundation and Advanced Technology Research Project of Tianjin (15JCQNJC01400).

References

1. Gomez, A., Alencar, F., Prado, P., Osorio, F., Wolf, D.: Traffic lights detection and state estimation using hidden markov models. In: 2014 IEEE Intelligent Vehicles Symposium Proceedings, June 2014, pp. 750–755 (2014)
2. Diaz-Cabrera, M., Cerri, P.: Traffic light recognition during the night based on fuzzy logic clustering. In: Moreno-Díaz, R., Pichler, F., Quesada-Arencibia, A. (eds.) EUROCAST 2013, Part II. Lecture Notes in Computer Science, vol. 8112, pp. 93–100. Springer, Berlin Heidelberg (2013). https://doi.org/10.1007/978-3-642-53862-9_13
3. Diaz-Cabrera, M., Cerri, P., Medici, P.: Robust real-time traffic light detection and distance estimation using a single camera. Expert Syst. Appl. **42**(8), 3911–3923 (2015)
4. Cai, Z., Li, Y., Gu, M.: Real-time recognition system of traffic light in urban environment. In: 2012 IEEE Symposium on Computational Intelligence for Security and Defence Applications (CISDA), July 2012, pp. 1–6 (2012)
5. Agarwal, M., Paul, B.C., Zhang, M., Mitra, S.: Circuit failure prediction and its application to transistor aging. In: 25th IEEE VLSI Test Symposium (VTS 2007) (2007)
6. Baldoni, L.M., Rizzuto, M.: On-line failure prediction in safety-critical systems. Future Gener. Comput. Syst. **45**, 123–132 (2015)
7. Wang, S., Wang, X., Wang, S., Wang, D.: Bi-directional long short-term memory method based on attention mechanism and rolling update for short-term load forecasting. Int. J. Electr. Power Energy Syst. **109**, 470–479 (2019)
8. Cheng, Y., Zhu, H., Wu, J., Shao, X.: Machine health monitoring using adaptive kernel spectral clustering and deep long short-term memory recurrent neural networks. IEEE Trans. Ind. Inf. **15**, 987–997 (2019)
9. Zhang, S., et al.: Prefix: switch failure prediction in datacenter networks. In: Proceedings of the ACM on Measurement and Analysis of Computing Systems (2018)
10. Islam, T., Manivannan, D.: Predicting application failure in cloud: a machine learning approach. In: 2017 IEEE International Conference on Cognitive Computing (ICCC) (2017)
11. Yu, J., Chaomurilige, C., Yang, M.-S.: 'On convergence and parameter selection of the EM and DA-EM algorithms for Gaussian mixtures.' Pattern Recognit. **77**, 188–203 (2018)
12. Zhang, K., Xu, J., Min, M.R., Jiang, G., Pelechrinis, K., Zhang, H.: Automated IT system failure prediction: a deep learning approach. In: 2016 IEEE International Conference on Big Data (Big Data), Washington, DC, USA, 2016, pp. 1291–1300 (2016). https://doi.org/10.1109/BigData.2016.7840733
13. Gao, J., Wang, H., Shen, H.: Task failure prediction in cloud data centers using deep learning. IEEE Trans. Serv. Comput. **15**(3), 1411–1422 (2022). https://doi.org/10.1109/TSC.2020.2993728
14. Kumar, J., Goomer, R., Singh, A.: Long short term memory recurrent neural network (LSTM-RNN) based workload forecasting model for cloud datacenters. Procedia Comput. Sci. **125**, 676–682 (2018)
15. Fatih, D.: DeepCoroNet: a deep LSTM approach for automated detection of COVID-19 cases from chest X-ray images. Appl. Soft Comput. J. **103**, 107160 (2021)
16. Sagheer, A., Kotb, M.: Time series forecasting of petroleum production using deep LSTM recurrent networks. Neurocomputing **323**, 203–213 (2018)

Auto-TSA: An Automatic Time Series Analysis System Based on Meta-learning

Tianyu Mu[1], Zhenli Sheng[2], Lekui Zhou[2], and Hongzhi Wang[1(✉)]

[1] Harbin Institute of Technology, Harbin, China
{mutianyu,wangzh}@hit.edu.cn
[2] Hangzhou Huawei Cloud Computing Technologies Co., Ltd., Hangzhou, China
{shengzhenli,zhoulekui}@huawei.com

Abstract. Time series is a necessary data type in both industrial scenarios and data analysis. In this era of explosive data growth, the significant development of sensors has made it possible to obtain massive amounts of time series data. However, the performance of different algorithms for different types of time series data varies greatly. So how to automatically choose an optimal algorithm for different data becomes the key to improving efficiency and saving resources. However, existing cloud services or open-source frameworks are not easy to use for users with little experience in time series analysis, and the user needs to rely on continuous experimentation to choose an algorithm that best suits his scenario. This significantly reduces the efficiency of data analysis. Thus to address this phenomenon, in this paper we propose an automated time series analysis system Auto-TSA. The biggest advantage over existing methods is that we have designed "Automation mode" so that even first-time users can easily use it. Users can automatically obtain the best-performing algorithm and hyperparameter configuration by entering only their own data. The whole process of automatic algorithm selection and hyperparameter optimization is efficient by introducing historical experience. In addition, we also design "Customization mode" for time series analysis experts, which makes it easier to use and more functional through the excellent and simple interface design. We will describe the framework and workflow of the system in detail, and show some samples to guide users to use it quickly.

Keywords: Meta-learning · Time series analysis · Anomaly detection and Forecasting · AutoML

1 Introduction

Time series is a form of data frequently used in industrial production and scientific research. In this era of explosive data growth, a significant amount of time series data is analyzed in various industries (e.g. audio analysis, noise cancellation, sensor signal anomalies monitor, dev ops in public cloud service, etc.) to

The original version of this chapter was revised: Funding program numbers in the Acknowledgement section is added. The correction to this chapter is available at https://doi.org/10.1007/978-3-031-35415-1_24

A. El Abbadi et al. (Eds.): DASFAA 2023 Workshops, LNCS 13922, pp. 138–147, 2023.
https://doi.org/10.1007/978-3-031-35415-1_10

help with trend prediction, anomaly detection, etc. In recent years, the field of time series analysis has gradually become a hot topic among researchers. With a significant number of algorithms proposed [6,7], the current time series analysis algorithms can be classified into the following categories: *time series classification* [9,22], *time series forecasting* [12] and *time series anomaly (outlier) detection* [2]. Along with the proposed algorithms and the progressive research in the time series analysis field, some famous public benchmark data sets have been proposed, the most famous of which is the UCR data set [4]. A significant number of time series analysis algorithms (or models) including statistical algorithms, machine learning (ML) models, and more popular deep learning approaches based on the neural network have continued to be proposed among the three subfields [5,10,11,13,15,21]. Nevertheless, for different data, the performance of these algorithms varies considerably. One learning algorithm cannot outperform another one in all aspects and tasks [17,20].

For such a research field that has been studied for decades, a number of open-source or commercial time series analysis systems (services) have been developed for public use in recent years. We will next analyze the existing popular open-source software libraries and commercial cloud services.

Open Source Library. Sk-time [14] is an open-source time series framework in Python, dedicated to providing an easy-to-use, flexible, and modular ML tool covering the areas of time series prediction, time series classification, time series regression, and time series clustering. However, Sk-time does not currently provide an end-to-end process. Users can only manually perform pipeline building, algorithm selection, and parameter tuning, which have certain domain experience requirements. Tsfresh [3] is also a relatively well-known open-source library in the field of time series, providing automatic extraction and automatic feature filtering in time series feature engineering. Merlion [1] is a time series library that focuses on machine learning, but also AutoML support is not comprehensive enough, only a few algorithms support automatic hyperparameter tuning. ADTK [23] requires the user to specify the specific design of the pipeline, and requires a long script description file, which is very inconvenient for the user to use.

Commercial Cloud Service. AWS Forecast is a commercial forecasting cloud service developed by Amazon, also available as a hosted service. The advantage of AWS Forecast is that it is easy to get started, even for users with no experience in time series forecasting, and Auto-Predictor can be used to create predictive models to analyze and forecast data with a single click. Nevertheless, it has some significant drawbacks. First, AWS Forecast is not flexible enough to use and when a forecaster is created, basic configurations are not allowed to be changed, such as the forecast step size. Second, in non-predefined scenarios, the traditional regression and statistics-based models support fewer features and do not support customized hyperparameters, making it difficult to guarantee prediction results when the data volume is insufficient to train a neural network model. Azure Anomaly Detector [19] is a commercial anomaly detection cloud service developed by Microsoft that provides univariate and multivariate anomaly detection services. It supports univariate and multivariate anomaly detection, including anomalies at the latest time point, outliers at all time points, cloud-based data storage, and multiple model management with synchronous and asynchronous detection. Despite the ease of use, the drawbacks are not negligible. Azure Anomaly Detector cannot upload tag data and cannot directly evaluate the results of the detection. It is also a black-box service, and the model for anomaly detection is not publicly available. Algorithm developers with extensive domain experience can customize very little content, making it suitable for entry-level developers only.

From the above analysis, the current time series analysis system has the following two main drawbacks, which are also the main challenges of our goal. We summarize them as follows.

I. The current system is not fully functional and does not support the entire time series data analysis process. The lack of data preprocessing will reduce the function and performance of the whole system.
II. The existing cloud services or open source libraries are not user-friendly to use, especially for users in other fields who have no experience in time series analysis, a set of convenient interfaces should be designed to enable new users to operate end-to-end with zero or few codes.

We propose Auto-TSA, an automated time series analysis system based on meta-learning. For Challenge I, we designed a data pre-processing module for data anomaly detection and repair. Besides, the feature extraction module is also embedded, which currently supports more than 200 common time series domain features. We integrate the time series classification, prediction, and anomaly detection algorithms commonly used today to meet the types of time series analysis tasks in the industry today. For such a well-functioning system, it is usually required to be operated by experts with extensive domain experience. Therefore, we design an "automated" mode of operation specifically for users in other domains. Users can use it with zero configuration, just upload their own data sets and specify the task types, and Auto-TSA can automatically perform an optimal (for the user's data) algorithm selection based on the user's data and automatically optimize the key hyperparameters in it, achieving the goal of zero code usage and solving Challenge II. Existing work on automatic algorithm selection and hyperparameter optimization mostly use iterative and incremental search, which is not only time-consuming but also usually does not achieve the best results. In Auto-TSA, we extract experience from historical tasks and train a meta-learner with it offline. The experience-based pre-trained learner can quickly select an optimal algorithm for the user's data set online, greatly improving the efficiency of time series analysis. This makes our system not only easy to get started but also historical experience provides acceleration and performance guarantee of the final result from a data-driven perspective. The contributions of this paper are summarized as follows.

1. We first propose an automated system based on meta-learning covering the complete process of time series analysis named Auto-TSA. The whole system includes pre-processing module, feature extraction module, time series

analysis module, and AutoML module (automatic algorithm selection and hyperparameter optimization). Auto-TSA is not only suitable for experts with rich experience in time series analysis, but also for users in other fields who can easily get started with the AutoML module, which automatically selects an optimal algorithm and its hyperparameters that can be used directly based on the user's data.
2. We designed the first knowledge base for meta-learning in the field of time series analysis and proposed an automatic algorithm selection method based on historical experience. With an empirically pre-trained meta-learner and task similarity co-analysis, Auto-TSA can efficiently select an optimal algorithm for user data online and perform hyperparameter optimization. It enables the user to obtain results with zero configuration.
3. We provide use cases to show more clearly how to use our system, and experimental results show that Auto-TSA is more efficient and accurate.

2 Auto-TSA System

2.1 System Architecture and Workflow

The goal of our system is to select an appropriate algorithm with a tuned hyperparameter configuration to gain maximal performance for a given data analysis task. To achieve this goal, we develop an automatic algorithm selection strategy based on historical experiences. The whole workflow of Auto-TSA is shown in Fig. 1. The system design is divided into 4 modules, which are the *data preprocessing* module for data cleaning or repairing; *feature extraction* module to deal with the data features; *AutoML* module for automatic algorithm selection and hyperparameter optimization; *time series analysis* module to integrate common analysis algorithms in three subfields. In the next section, we will introduce the workflow of each module of the system for different usage modes.

Table 1. All algorithms supported in Auto-TSA currently. We have implemented each algorithm with our defined interfaces and classes, which can be called by users themselves or used with the AutoML module. We will gradually expand the library of algorithms in subsequent updates.

Task	Algorithm	Total
Forecasting	Holt-Winters, ETS, ARIMA, SARIMA, Theta, Prophet, Croston, VARIMA, LightGBM, LR, RNN, TCN, N-BEATS, TFT	14
Anomaly Detection	n-sigma, Differentiate, R-Space, SVM, LevelShiftAD, PCAAD, SigmaAD, PeriodAD, IncrementalAD, MARINAD, InterQuartile	11
Classification	Shape DTW, ROCKET, ResNet, BOSS, cBOSS, LSTM, Catch22, ResNet, Inception, MUSE, WEASL, RISE, CNN, FCN	14

Customization Mode. This mode is designed for users with adequate time series analysis experience. After users upload their own data, they can use the

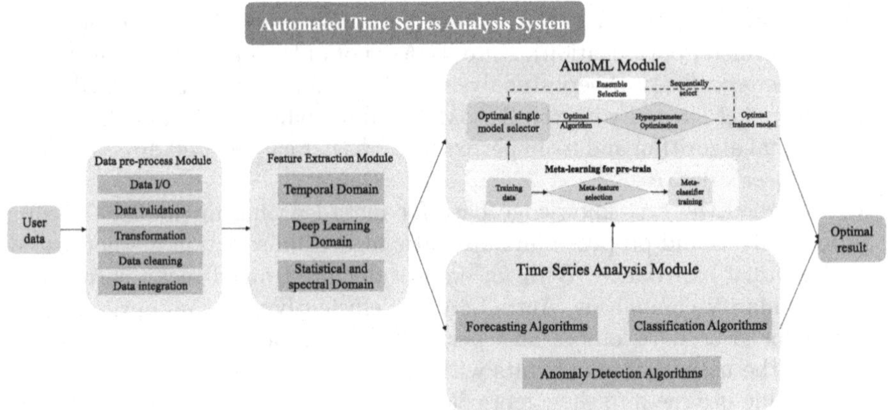

Fig. 1. Workflow of Auto-TSA.

pre-processing module to clean and transform the data, such as Auto-encoder for spatial mapping. Then the feature extraction module analyzes and extracts the features of the data, such as seasonal cycle detection and judgment, and the user can also specify the extracted features by himself. For time series domain experts, they can directly use the algorithms in the time series analysis module for data analysis, and all the currently supported algorithms are shown in Table 1. The workflow of this mode is not complicated and is a standard time series analysis task processing process, but it requires high experience for users and is easy to obtain less than ideal results. Therefore, we recommend another mode for users who do not have rich cross-domain experience.

Automation Mode. This mode is designed for users who are not familiar with the algorithms and performance of time series analysis. Uploading datasets and feature extraction is the same process as Customization mode. Then, users can call the AutoML module to automatically select an optimal processing model based on user data and perform hyperparameter tuning. It uses an empirically pre-trained meta-learner to improve the efficiency of the algorithm selection from a data-driven perspective. The range of algorithm selection is the algorithm integrated into the time series analysis module, and the hyperparameter space is predefined. In other words, our system supports not only customized use by domain experts but also rapidly automated use by users in other domains.

2.2 Automated Model Selection

Definition 1. *Given a task T_{new} uploaded by users, the goal is to find an algorithm with a tuned hyperparameter configuration which can help the algorithm achieve the optimal performance. The objective function is formulated as follows.*

$$(\mathcal{A}, \lambda)^* = \arg\max_{\lambda \in \Lambda} \mathbb{E}_{(T_{old}, T_{new}) \sim \mathcal{T}} \mathcal{V}(\mathcal{L}, \mathcal{A}_\lambda, T_{old}, T_{new}) \tag{1}$$

where the $\mathcal{A}_\lambda$ denotes the model generated by algorithm $\mathcal{A}$ with the hyperparameter configuration λ, and $\mathcal{V}(\mathcal{L}, \mathcal{A}_\lambda, T_{old}, T_{new})$ measures the validation performance of the model generated by $\mathcal{A}_\lambda$ leveraging experience observed from previous tasks T_{old} and evaluated on T_{new}, respectively.

Model selection in the *Automation mode* is based on the meta-learning theory [8], a subfield of AutoML, aims to learn from prior task experience to reduce computation for similar task [18]. In the design of Auto-TSA, experience is extracted, transformed, and filtered for similar historical tasks, which is represented as a one-to-one mapping between historical tasks and their corresponding optimal algorithms that have been validated by a series of experimental results. Each task is represented as a meta-feature vector, which is selected by a reinforcement learning network. In other words, we perform automatic model selection through a pre-trained model based on historical experience, which is very efficient and accurate. We have demonstrated the efficiency of such "train offline, search online" strategy for automatic model selection in our historical work [16]. In addition, our sequential selection of the algorithm and its corresponding hyperparameters will significantly reduce the search space and improve the search efficiency. Suppose that there are x algorithms to be selected, and each algorithm has y hyperparameters with z values. Existing work has a $x \cdot y \cdot z$ search space while it is reduced to $x + y \cdot z$ in ours. Accordingly, the complexity of the approach is reduced from $O(x \cdot y \cdot z)$ to $O(x + y \cdot z)$.

2.3 Sample Demonstrations

In this section, we will introduce how to use Auto-TSA for Time Series analysis with *Customization* mode and *Automation* mode, respectively.

Customization Mode. Here we show two scenarios using Auto-TSA for time series anomaly detection and forecasting respectively. We have highly wrapped the entire sequence interface, making it easier to use than existing python libraries.

```
from autotsa.data.datasets import NAB
from autotsa.visualization import FigureAD
from autotsa.models.detection import InterQuartileAD

# Load the NAB data set
data_loader = NAB()
train_datasets, test_datasets =\
    data_loader.load_train_test(train_ratio=0.15)

# Using the algorithm InterQuartile for AD
model = InterQuartileAD({'c': 6})
model.fit(train_datasets)

# Plot the results
FigureAD(test_datasets,
```

```
    model.detect(test_datasets)).plot()
```

The above code shows how to use the built-in algorithm and data set for anomaly detection, and the result of the anomaly detection is shown in Fig. 2. It is clear that our system is significantly easier to use and does not require the user to implement complex logic and set algorithm parameters.

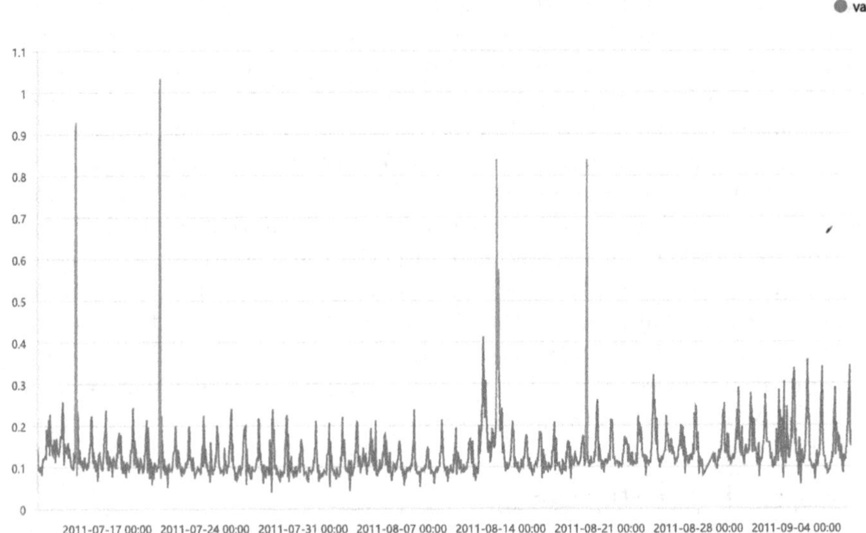

Fig. 2. Result of anomaly detection.

```
from autotsa.data.datasets import AirPassenger
from autotsa.visualization import FigureForecasting
from autotsa.models.forecasting import ARIMA

# Load the AirPassenger data set
data_loader = AirPassenger()
train_datasets, test_datasets =\
    data_loader.load_train_test(train_ratio=0.8)

# Using the algorithm ARIMA for Forecasting
model = ARIMA({'p': 10, 'd': 2, 'q': 7})
model.fit(train_datasets)

# Plot the results
FigureForecasting(test_datasets,
    model.predict(test_datasets)).plot()
```

The above code shows how to use the built-in algorithm and data set for time series forecasting, and the result of the forecasting is shown in Fig. 3. So

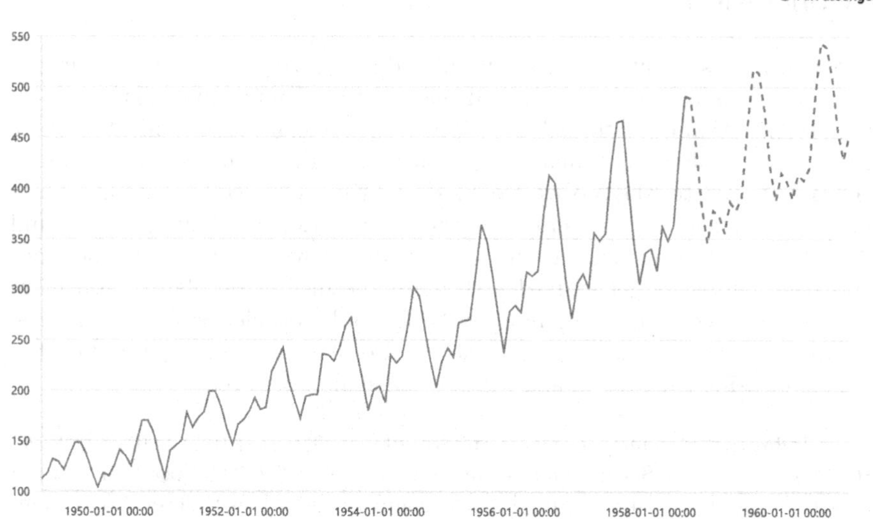

Fig. 3. Result of time series forecasting.

far, we have demonstrated how experienced experts use our interface for time series analysis in two scenarios: forecasting and anomaly detection, respectively.

Automation Mode. We use the time series forecasting scenario mentioned above as an example to show how to use Automation mode for automatic algorithm selection and hyperparameter optimization.

```
from autotsa.data.datasets import AirPassenger
from autotsa.forecasting import alg_selection, hpo

# Load the AirPassenger data set
data_loader = AirPassenger()

# Getting the optimal algorithm
# and its hyperparameter configuration
optimal_alg = alg_selection(data_loader)
parameters = hpo(optimal_alg, data_loader)
print(optimal_alg)
print(parameters)
```

With just two lines of code, the user can know the algorithm that performs optimally for the AirPassenger dataset, and the final output of Auto-TSA is as follows.

```
out:
[Theta]
[seasonal_periods=12, seasonal='auto', theta=2]
```

3 Conclusion

This paper proposes Auto-TSA, an automated time series analysis system covering forecasting, anomaly detection, and classification. By introducing the extraction and utilization of historical experience, a significant improvement is made in the efficiency of the automatic selection of the algorithm. Auto-TSA automates and accelerates the process from a data-driven perspective for three types of tasks in the field of time series analysis. In addition, we have designed and packaged the interface in such a way that even users with little or no experience in this field can easily get started with it. In the future, we plan to add more algorithms to make Auto-TSA more powerful and suitable for more scenarios. Secondly, the design of the GUI interface is also planned, which can reduce the threshold of using the system and achieve the goal of zero code.

Acknowledgement. This paper was supported by NSFC grant (62232005, 62202126, U1866602) and Sichuan Science and Technology Program (2020YFSY0069). This work was done by the first author during his internship at Hangzhou Huawei Cloud Computing Technologies Co., Ltd. We thank the anonymous reviewers for their valuable review comments.

References

1. Bhatnagar, A., et al.: Merlion: a machine learning library for time series. arXiv preprint arXiv:2109.09265 (2021)
2. Blázquez-García, A., Conde, A., Mori, U., Lozano, J.A.: A review on outlier/anomaly detection in time series data. ACM Comput. Surv. (CSUR) **54**(3), 1–33 (2021)
3. Christ, M., Braun, N., Neuffer, J., Kempa-Liehr, A.W.: Time series feature extraction on basis of scalable hypothesis tests (tsfresh-a python package). Neurocomputing **307**, 72–77 (2018)
4. Dau, H.A., et al.: The UCR time series archive. IEEE/CAA J. Automatica Sinica **6**(6), 1293–1305 (2019)
5. Dempster, A., Petitjean, F., Webb, G.I.: Rocket: exceptionally fast and accurate time series classification using random convolutional kernels. Data Min. Knowl. Disc. **34**(5), 1454–1495 (2020)
6. Esling, P., Agon, C.: Time-series data mining. ACM Comput. Surv. (CSUR) **45**(1), 1–34 (2012)
7. Hamilton, J.D.: Time Series Analysis. Princeton University Press, Princeton (2020)
8. Hutter, F., Kotthoff, L., Vanschoren, J.: Automated Machine Learning. Springer, Cham (2019). https://doi.org/10.1007/978-3-030-05318-5
9. Ismail Fawaz, H., Forestier, G., Weber, J., Idoumghar, L., Muller, P.A.: Deep learning for time series classification: a review. Data Min. Knowl. Disc. **33**(4), 917–963 (2019)
10. Ismail Fawaz, H., et al.: Inceptiontime: finding alexnet for time series classification. Data Min. Knowl. Disc. **34**(6), 1936–1962 (2020)
11. Kieu, T., Yang, B., Jensen, C.S.: Outlier detection for multidimensional time series using deep neural networks. In: 2018 19th IEEE International Conference on Mobile Data Management (MDM), pp. 125–134. IEEE (2018)

12. Lim, B., Zohren, S.: Time-series forecasting with deep learning: a survey. Phil. Trans. R. Soc. A **379**(2194), 20200209 (2021)
13. Lines, J., Bagnall, A.: Time series classification with ensembles of elastic distance measures. Data Min. Knowl. Disc. **29**(3), 565–592 (2015)
14. Löning, M., Bagnall, A., Ganesh, S., Kazakov, V., Lines, J., Király, F.J.: sktime: a unified interface for machine learning with time series. arXiv preprint arXiv:1909.07872 (2019)
15. Lucas, B., et al.: Proximity forest: an effective and scalable distance-based classifier for time series. Data Min. Knowl. Disc. **33**(3), 607–635 (2019)
16. Mu, T., Wang, H., Zheng, S., Zhang, S., Liang, C., Tang, H.: Assassin: an automatic classification system based on algorithm selection. Proc. VLDB Endowment **14**(12), 2751–2754 (2021)
17. Peng, Y., Flach, P.A., Soares, C., Brazdil, P.: Improved dataset characterisation for meta-learning. In: Lange, S., Satoh, K., Smith, C.H. (eds.) DS 2002. LNCS, vol. 2534, pp. 141–152. Springer, Heidelberg (2002). https://doi.org/10.1007/3-540-36182-0_14
18. Pinto, F., Soares, C., Mendes-Moreira, J.: Towards automatic generation of metafeatures. In: Bailey, J., Khan, L., Washio, T., Dobbie, G., Huang, J.Z., Wang, R. (eds.) PAKDD 2016. LNCS (LNAI), vol. 9651, pp. 215–226. Springer, Cham (2016). https://doi.org/10.1007/978-3-319-31753-3_18
19. Ren, H., et al.: Time-series anomaly detection service at Microsoft. In: Proceedings of the 25th ACM SIGKDD International Conference on Knowledge Discovery & Data Mining, pp. 3009–3017 (2019)
20. Schaffer, C.: Cross-validation, stacking and bi-level stacking: Meta-methods for classification learning. In: Cheeseman, P., Oldford, R.W. (eds.) Selecting Models from Data, pp. 51–59. Springer, New York (1994). https://doi.org/10.1007/978-1-4612-2660-4_6
21. Shen, Z., Zhang, Y., Lu, J., Xu, J., Xiao, G.: A novel time series forecasting model with deep learning. Neurocomputing **396**, 302–313 (2020)
22. Susto, G.A., Cenedese, A., Terzi, M.: Time-series classification methods: review and applications to power systems data. Big Data Appl. Power Syst. **2018**, 179–220 (2018)
23. Ye, L., Keogh, E.: Time series shapelets: a new primitive for data mining. In: Proceedings of the 15th ACM SIGKDD International Conference on Knowledge Discovery and Data Mining, pp. 947–956 (2009)

BundleRS

Mobile Application Ranking with Transductive Transfer Learning

Xichang Li, Surya Putra Santoso, and Rui Zhang(✉)

Wuhan University of Technology, No. 122 Luoshi Road, Wuhan 430070, Hubei, China
zhangrui@whut.edu.cn

Abstract. The drastic increasing of mobile apps make users feel tough when finding Apps they need. An effective ranking service can filter many unrelated apps and provide users with a high quality result list that users need. Many existed work formed ranking service by considering semantics and built topic models and achieved good results, whereas many of them manually assigned weights when cogitating multiple features, which makes ranking result has biases. On the other hand, some of these work merely aimed at specific store of a country, therefore the algorithm cost and effectiveness when these methods are used in other countries or regions is uncertain. In this paper, we put forward an app ranking framework based on transfer learning (ARFT) which is used to rank apps for a given query and ponder relevance and quality simultaneously. ARFT as well avoided the shortage that classical ranking methods need to assign weights manually when facing multiple features and can pore over all features automatically so that ranking results become more natural. Furthermore, leveraging transfer learning, our framework fit app stores of different countries quickly and initially reflected that there exists inner attribute consistency between apps. Such consistency can largely simplify the complexity of future work in researching app stores. Experiments in real App Store dataset show that our framework can have 50% of precision and 0.96 of NDCG in top 20 Apps reflecting to the query, better than other comparison methods.

Keywords: App ranking · transfer learning · relevance · quality

1 Introduction

App stores like Google Play and App Store provides convenience to developers who would like to promote their apps and users who need to search and download them respectively. With the wide use of app stores, apps increase drastically. However, markets are facing a big problem which is called "information overload", and users may feel tough to find apps they need from the crowded market [1–3]. Therefore, it is essential to build a service for users which can help them find apps in it.

Given a query, apps that rank in the top should have some excellent characteristics than others. Summarized from these literature [4–7], we would like

A. El Abbadi et al. (Eds.): DASFAA 2023 Workshops, LNCS 13922, pp. 151–165, 2023.
https://doi.org/10.1007/978-3-031-35415-1_11

to divide those characteristics into two parts: relevance and quality. Relevance means how much the app does related to that query, and quality refers to whether the app has creditable attributes, i.e. "how good the app does look like". An accurate ranking result should consider both relevance and quality. That is, we should show apps which reflect the query mostly and have best inner properties at the same time to users on the top.

App stores also have traits like multiple features and country differences. Multiple features mean that there may have many different attributes have impact on relevance and quality (e.g. ratings, rating counts, price, etc.). This makes us need to assign weights manually when considering final ranking result, but this makes the ranking service relatively subjective. Currently, many of existed work still manually assigned weights, even though they may considered both relevance and quality [4,8]. Therefore, an objective enough weight assignment process is necessary for app ranking which can render ranking results more natural.

Country differences mean that due to the difference of interests, behaviors and category use of users in different countries, a conclusion in an app store may not suitable for stores of other countries [9,10]. In this situation, traditional methods need to do processing and training for every store of specific country, which makes patterns they found not general.

To ponder all of them, we put forward an app ranking service based on transfer learning (ARFT), which can rank apps corresponding to user query in the store and show them in the top which fit the query best. It can keep both relevance and quality as usual while consider all traits of stores. In the implementation of it, we firstly used category tree [11], Word2Vec [12,13] and LDA [14] to build hidden topics between features and the query and decrease the size of app candidate set, which kept the relevance and quality. Then, we put forward stochastic weighted arithmetic mean (SWAM) to automatically assign weights, which can eliminate subjectivity of ranking service. Finally, we successfully utilized inner attribute consistency to apply framework in different stores of countries through transfer learning and solved country differences. Here, transfer learning is reached by a simple multi-layer perceptron (MLP).

Our core contributions are as follows:

- We put forward a simple improvement of weighted arithmetic mean, SWAM, which can completely eliminate subjectivity of weights assignment in ranking service and automatically consider multiple features.
- We find there has inner attribute consistency between apps by transfer learning. Such consistency can be used to solve country difference problem of app stores.
- We do experiments on the real dataset on App Store. Experiments show that our framework achieved 50% of precision and 0.96 of NDCG in top 20 apps corresponding to a query and keep relevance and quality.

2 Preliminaries

In this section, We first describe a few key concepts, and then introduce our specific goals.

Definition 1 (Feature representation of app). An app can be uniquely represented by a vector of $a = \{f(T), g(D), h(O)\}$, where $T = \{w_1^T, ..., w_{|T|}^T\}$ is word embedding matrix of app title, $D = \{w_1^D, ..., w_{|D|}^D\}$ is word embedding matrix of app description, O refers to the concatenation of other app features, f, g and h are some kinds of mapping relationship.

Definition 2 (Representation of query). A query can be uniquely represented by a word embedding matrix of $q = \{w_1^q, ..., w_{|q|}^q\}$, where $w_i^q \in \mathbb{R}^d$ is embedding vector of word i of the query.

With these two definitions, we can more accurately describe characteristics of app ranking and traits of app store. For app ranking characteristics, relevance refers to the extent to which a can reflect q, denoted as $\phi(a, q)$. Quality refers to the goodness of feature comparison between any two apps, denoted as $\psi(a, q)$. Our framework is dedicated to achieve the following goals:

Problem 1 (Relevance of app ranking). We would like to show apps on the top, while these apps have highest relevance, given a query. In other words, apps on the top should have the feature representation that best satisfies q, i.e. $\arg\max_a \phi(a, q)$.

Problem 2 (Quality of app ranking). These apps should also have the highest quality, i.e. their feature representation will be better than the feature representation of other apps, denoted as $\arg\max_a \psi(a, q)$.

Problem 3 (Weight assignment in multiple features). The importance of $f(T)$, $g(D)$, and $h(O)$ should be automatically considered through a functional relationship $H(a) = H(f(T), g(D), h(O))$, then the problem the subjectivity of assigning weights is solved here.

Problem 4 (Country difference). The intrinsic correlation of the app feature representation should be used to enable the framework to quickly act in app stores in different countries, different queries, and different categories. In other words, the key to solving the country differences problem is that there is no need to train specialized model for specific situations. The intrinsic correlation of the app feature representation is defined as inner attribute consistency.

3 ARFT Design

In this section, we will introduce the details of our framework. Specifically, the whole framework has three parts: generate candidate set, create source/target domain and transfer to target. The workflow of our framework is shown in Fig. 1.

3.1 Generate Candidate Set

The first step of the framework is generating candidate set, while the first step of which is constructing category tree. We implement this easily by checking the

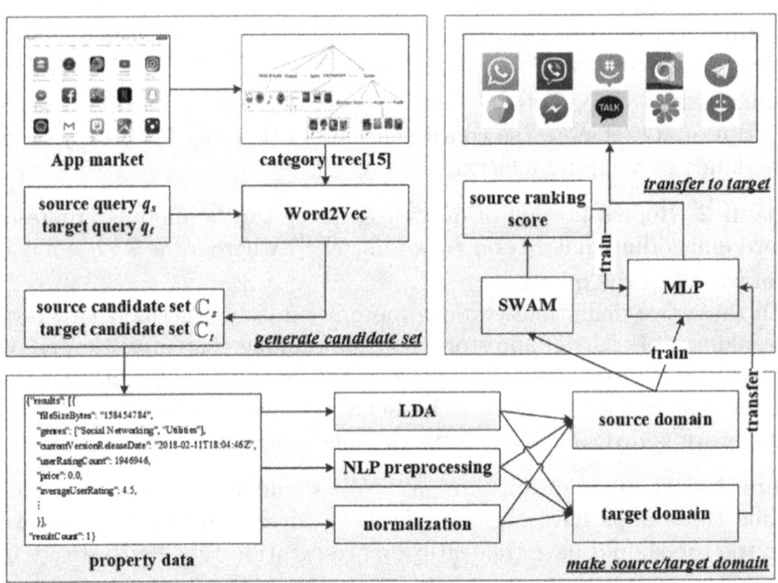

Fig. 1. The workflow of ARFT. Generally it has three parts: generate candidate set, make source/target domain and transfer to target.

existing categories and subcategories in the store. Next, we find a non-leaf node which mostly closed to a given query. However, it has a large probability that not all apps in this node have close relations to the query, even coarse-grained. For example, assume node "social networking" is mostly related to query "chat", we can find Facebook is exactly its child, whereas it is not a chatting app. Furthermore, we found that in the description of Facebook there does not show word "chat". However, in those most popular chatting apps such as Whatsapp and Messenger it shows frequently. Therefore, we prune all leaf nodes if the query is not in the description of it.

Now we only need to find this node. This is completed by comparing the similarity between each non-leaf node and the query. Similarity is calculated by Word2Vec, in which it is more usually known as semantic similarity. Then, we traverse all non-leaf nodes and pick the node which has the maximum similarity as is most related to the query. We use "text8" corpus to train the model[1]. After this and pruning as mentioned above, we generated candidate set, which has coarse-grained relevance.

3.2 Create Source/Target Domain

The next step is to extract features which can represent fine-grained relevance and quality and make them become domain. We directly use the property data

[1] http://mattmahoney.net/dc/text8.zip.

of each app in the candidate set to finish this goal. We can infer that some of the property data are more likely to related to relevance while some others are more related to quality. Attributes related to relevance mean they have to record some functionality behavior which can make us build connection between them and the query. Therefore, these attributes are usually refer to natural words, e.g. title, app description, updating information, etc. Based on these papers [5,8,15–17], we pick title and app description as features in relevance. Similarly, attributes related to quality should record some information so that we can finally infer whether is good based on them. Hence, these attributes can point to ratings, rating counts and so on. Based on these papers [4,8,18–25], in the framework we choose ratings, rating counts, price, file size and latest release date as features in quality. It is noted that what features we should select from property data may need considering. A better advice is using some way to evaluate features before become domain.

After picking features, we need to process them and make them become real domain. We use LDA here to compute topic probability of each app in the candidate set while the topic has highest probability to the query. This is for processing relevance features from property data. Specifically, we at first construct a bag-of-words for all features in relevance for all apps in the set and train LDA with this bag for a given number of topics. The topic which the query mostly like to be is used to represent the query. Then, we concatenate all features in relevance for each app and use trained LDA to gather the probability that the app belongs to the query topic. In the end, this probability as a relevance feature goes into the domain. There are as well some basic NLP pre-processing in this process, including tokenization, lemmatization, etc.

The processing of quality features is much easier. Since they are general data format, we can directly fit them into the domain after normalization, that is, to distribute different features in a similar range so that the model can converge faster. Now we created the domain which can be trained and transferred. We choose a query and its domain as source domain (select which specific query as source query is not important). All others are target domain.

3.3 Transfer to Target

The last step is to train the source domain and transfer the model to target domain. We use SWAM to gather ranking scores of source domain. The basic idea of SWAM is clear: allocating the weights stochastically. We know that stochastically allocate weights for two apps one time may make the former rank higher than the latter, and another stochastic allocation may reverse the ranking. However, if we try many times enough, and count the count each app ranks higher than another, then the one counted more can be seen better than another. Specifically, the ranking score s_a for an app a in the source candidate set is calculated by SWAM:

$$s_a = \frac{\sum_{i=1}^{N} \sum_{j=1}^{|a|} a_j P(a_j)}{N}, \forall_{a_j} \in a, \quad (1)$$

where $P(a_j)$ and N represent the probability (weight) of that feature and iterations respectively. $P(a_j)$ is generated stochasitcally. Since the source domain is a matrix, we could as well use matrix form of SWAM to get ranking score vector S for all apps in the source candidate set $\mathbb{C}_s$ simultaneously:

$$S = \frac{\sum_{j=1}^{N}(AP^T)_j}{N} \in \mathbb{R}^{|\mathbb{C}_s|}, \tag{2}$$

where $A \in \mathbb{R}^{|\mathbb{C}_s| \times |a|}$ is feature matrix, i.e. source domain and $P \in \mathbb{R}^{N \times |a|}$ is probability matrix which is generated stochastically. This score vector is exactly the labels fit into MLP.

Analysis of SWAM. Given two feature representation a^1 and a^2, the difference of their SWAM score is

$$ds = \frac{\sum_{i=1}^{N}[(a_j^1 - a_j^2)\sum_{j=1}^{N}(P_{ij}^T)]}{N}, \tag{3}$$

where $\forall P_{ij} \in P$, $P_{ij} \sim U(0,1)$. The sum of N uniform distribution follows Irwin-Hall distribution [26,27]. The mean of an Irwin-Hall distribution still follows Irwin-Hall distribution. Therefore, the probability distribution function of $p = \frac{\sum_{j=1}^{N}(P_{ij}^T)}{N}$ is

$$\frac{1}{2(N-1)!}\sum_{k=0}^{N}(-1)^k\binom{N}{k}(p-k)^{N-1}\mathrm{sgn}(p-k), \tag{4}$$

According to central limit theorem, when N is large enough, the distribution of p changes to a Gaussian distribution. This indicates that the selection of N has an effect on the final score, and it shows that the score may lose objectivity when N is very large, because most of the p has covered most of the interval of Gaussian distribution, resulting in repetition of p. Therefore, N should be chosen carefully.

Next, we use an MLP to learn patterns of features in $\mathbb{C}_s$ and S as labels. Now this becomes a typical regression task. Therefore, we can use traditional loss function mean squared error (MSE) as a metric which measures the difference between the output of MLP and S.

Finally, we transfer MLP to target domain. Based on previous steps, we will have a candidate set given a target query, which generate target domain. Applying MLP directly in target domain, we can gather the ranking score vector for the target query. Sorting the vector we in the end pick top k apps which have highest scores as the final result of that query. We focus on three different transferring situations: (1) transfer to another target query; (2) transfer to same query but different countries in the store; (3) transfer to both different query and countries in the store. We will show results for all situations in Sect. 4.3. Figure 2 explains the whole transferring process in detail.

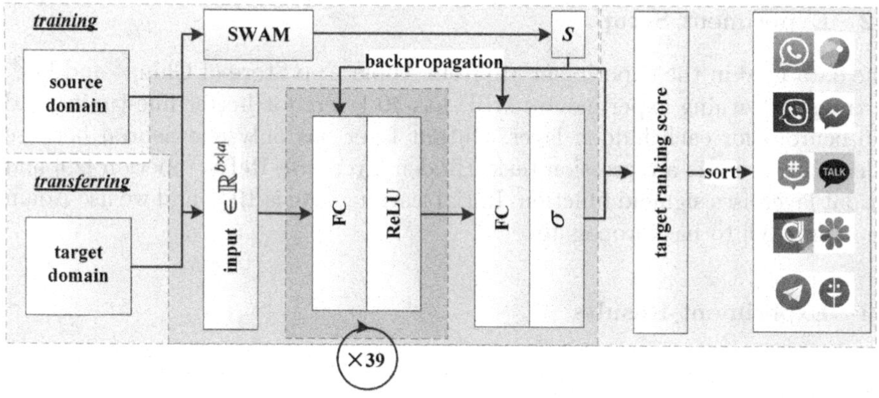

Fig. 2. Process of transfer learning. We only need to train one specific source domain, then the framework can transfer to any target domain.

4 Experiments and Results

4.1 Evaluation Metrics

We use two metrics to evaluate the effectiveness of all methods. The first is the normalized discounted cumulative gain (NDCG) [28]. The general idea is that users have subjective judgment on the relevance of each app to the query results. If the ground-truth ranking is closer to user expectation, then NDCG will be closer to 1, and vice versa. In order to get accurate user ratings, we have manually labeled and classified some apps according to their functionality. Later, we invite dozens of volunteers and ask them to simulate a query and rate these apps. Based on their ratings, we list the NDCG levels and ratings for different expectations in Table 1.

Table 1. NDCG Levels and Ratings

LEVEL	RATING
Excellent	15
Good	7
Fair	3
Poor	0

In addition to NDCG, we also use another metric, Precision@k. It is defined as the proportion of apps related to the first k queries to k. Since we have used NDCG for detailed rating, in this paper we can directly simplify the calculation of Precision@k as

$$\text{Precision@}k = \frac{n_{Excellent} + n_{Good}}{k} \tag{5}$$

4.2 Experiment Setup

The data used in the experiment are from Apple App Store of China[2] and US[3]. In transfer learning experiments, MLP has 40 layers for better fine-tuning and 256 neurons for each hidden layer. Output layer has only one neuron because calculating score is a regression task. Hidden layers use ReLU function [29] and output layer uses sigmoid function. Initial learning rate is 10^{-4} and we use Adam algorithm [30] to backpropagate.

4.3 Experiment Results

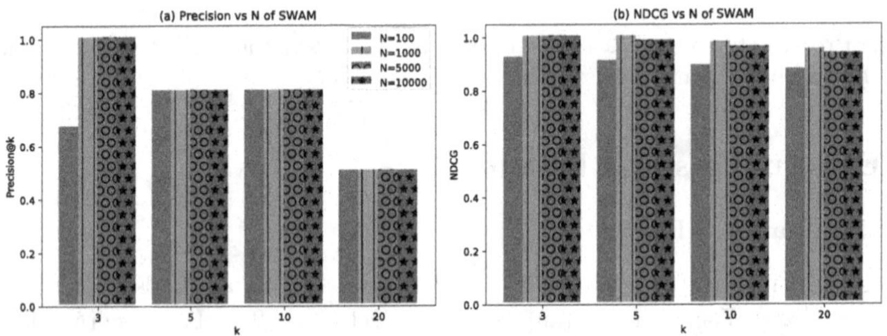

Fig. 3. The N parameter of SWAM is a trade-off. Small N will generated biased weights while large N will make plenty of weights focus on a neighborhood of μ.

Effect of *N* Varies of SWAM. Firstly, we explore the effect of SWAM with different N. We performed multiple experiments on the same candidate set by changing N. Candidate set has a total of 8,213 apps and is of query "chat" where 90% are training set. Figure 3 shows the precision and NDCG result respectively. For precision, $N = 100$ differs with others, worse than others. For NDCG it is further worse and has a drastic decreasing. This indicates that ranking accuracy is related to N and fits our analysis before. Note that when $N = 1000$ we have highest NDCG. When N is too small, Irwin-Hall distribution is biased and may not cover all distribution of p. When N is too large, Irwin-Hall distribution would generated a complete Gaussian distribution so that plenty of P_{ij} fall into a range very closed to μ. Therefore, only a suitable N can make SWAM fully consider all distribution of p at the same time not duplicated. The optimal N has no relation to candidate set size; it only relates to features of apps.

[2] https://itunes.apple.com/cn/genre/ios/id36?mt=8.
[3] https://itunes.apple.com/us/genre/ios/id36?mt=8.

Three Different Transfer Learning Situations. Next, we conduct transfer learning experiments. As mentioned in Sect. 3.3, experiments include three situations: (1) same country, different queries; (2) same query, different countries; and (3) different countries, different queries. (1) uses Chinese app dataset as source and target domain while (2) and (3) use Chinese as source and US as target. Figure 4 shows precision and NDCG result of all situations respectively.

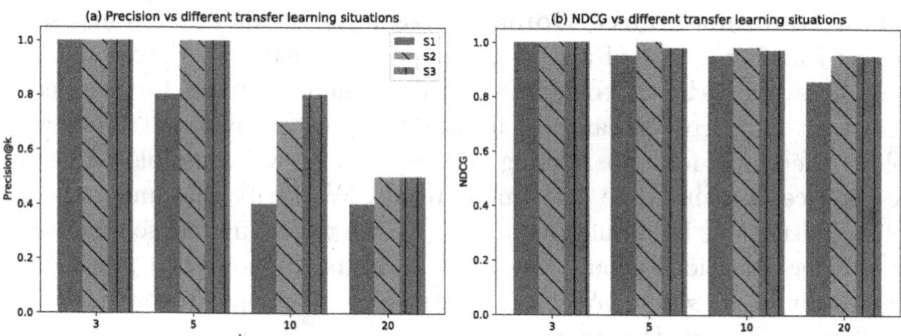

Fig. 4. Precision and NDCG performance of 3 different situations. S1: same country, different queries. S2: same query, different countries. S3: different countries, different queries. The significant lower performance of S1 to others indicates country differences.

For precision, all situations perform similar and good result and even better than direct supervised learning (Fig. 3 left). This shows its effectiveness. Transfer learning can learn the inner attribute consistency, in the meanwhile such consistency can never be represented in supervised learning. For NDCG, generally all of them still perform similar, while situation (1) obviously lower than other situations. We further verified this phenomenon by using different queries of situation (1) and they are still lower. Since only (1) refers to no country, this shows that app stores have country differences and utilizing inner attribute consistency of apps can improve them.

Fine-Tuning of Transfer Learning. In transfer learning, it is possible to utilize pre-trained model to discover app ranking so that model can be more simple and less time-consuming. However, this needs that fine-tuning can remain inner attribute consistency. To verify it, we did fine-tuning experiments for all situations with freezing pre- and post-L layers from the above trained model. For simplicity, L ranges from 1 to 10 for both directions. Situation (1) is denoted as S1, and so on.

Figure 6 shows results of fine-tuning. Among them, (a)–(c) are of precision with freezing pre-L layers from S1 to S3; (d)–(f) are of NDCG; (g)–(i) are precision results for post-L layers and (j)–(l) are NDCG results. For precision, all of them remain same level if freezing not too many layers, no matter what direction it is. This means that precision may not be affected by fine-tuning, which supports that transfer learning can keep inner attribute consistency for app ranking's precision. On the other hand, NDCG performs more stable for all situations. Only few of them dropped off on where freezing too many layers they focus (after freezing ≥ 7 layers). Therefore, their stability further indicates that apps do exist inner attribute consistency in different countries and transfer learning can leverage it for app ranking service.

Effect of Different Source Domain Sizes. As we known, training set size related to learning extent of a model in supervised learning. However, transfer learning gives us a chance to reduce source domain size and further decrease wall-clock training cost because of its ability to catch inner attribute among samples. Therefore, for app ranking service it is as well possible to use only limited source domain while get similar result. We verify this under the most diversified situation, i.e. situation (3). If (3) works for limited source domain, other simpler situations should work either. Figure 5 shows the precision and NDCG result of it, where $p\%$ means extracting $p\%$ apps stochastically from source domain. Target domain remains same.

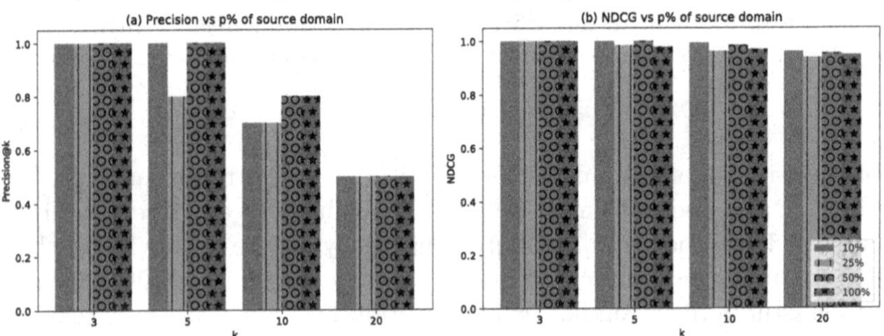

Fig. 5. Presion and NDCG by using different sizes of source domain. A few source domain can reach same result, which verifies inner attribute consistency among apps and the effectiveness of transfer learning.

For precision, changing source domain size have no effect to precision at all. An astonishing fact is when source domain size reduced to 10%, it still remain 100% precision for top 10 apps. This strongly reveals the inner attribute consistency among apps and the suitability of transfer learning in app ranking service. For NDCG it is even more unbelievable: 10% gets highest. It may related to overfit for not too big target domain of (3), nevertheless this show the potential and possibility that we may need very few of apps to transfer to all target ranking. Apparently, for app ranking service transfer learning has its unique advantage.

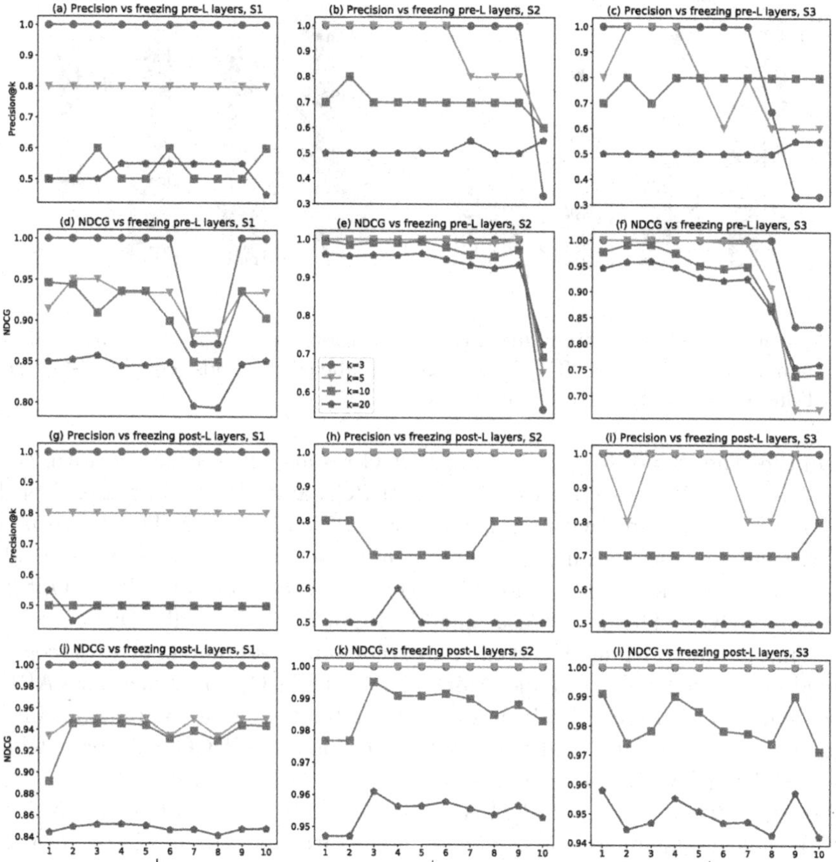

Fig. 6. Fine-tuning results of pre- and post- L layers. (a)–(f): pre-L layers. (g)–(l): post-L layers. Results show that transfer learning can leverage inner attribute consistency of apps to solve country differences because freezing many layers doesn't affect too much performance.

Comparison of Different Methods. This part evaluates the performance of different app ranking methods. Our result is of situation (3) (S3, same below) with 10 layers. Comparison methods include SUCM [11], LMF-Opt [31], PCA and LFM-Auto, where for SUCM and LMF-Opt we pre-labelled some apps. Figure 7(a) shows the result of precision of different methods. LFM-Opt and PCA are significantly lower than ARFT, while SUCM has more similar result. However, we found that top 10 results of SUCM are all pre-labelled apps, which indicates that such method works good only in well-labelled dataset. Moreover, ARFT can transfer directly to new candidate sets without any other training. This verifies the effectiveness of ARFT and shows some advantages when applying transfer learning in app ranking service.

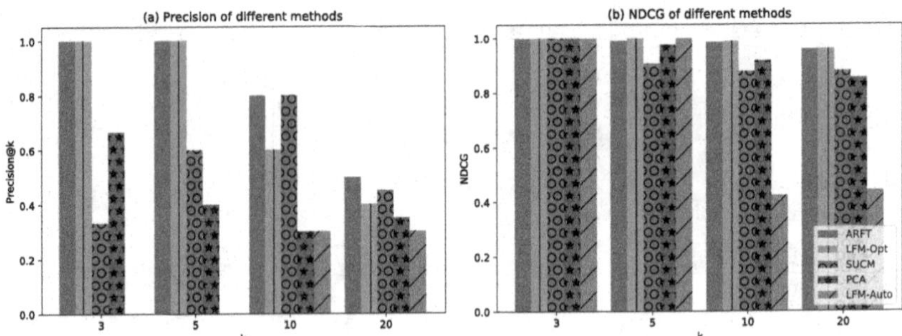

Fig. 7. Comparison of different methods of precision and NDCG. Our method can reach similar precision and NDCG to those pre-labeled methods, e.g. LFM-Opt and SUCM and significantly better than others.

On the other hand, Fig. 7(b) shows NDCG result. All methods perform good when k is small. With k increases, NDCG of PCA and SUCM decreases. Top 10 apps of SUCM are same with ours, while our NDCG is higher. This may due to SUCM lacks considering inner attribute consistency, where it is a recommendation algorithm rather than ranking algorithm. On the other hand, LFM-Opt has very good NDCG. This is because that LFM-Opt can utilize all NDCG levels as labels (fine-grained) while SUCM can only record either-or matrix of apps (coarse-grained). Note that even with fine-grained labels, LFM-Opt only surpasses ARFT a few, which verifies the effectiveness of ARFT further. As a comparison, LFM-Auto is without any label and both of its precision and NDCG are worse, which indicates that LFM cannot gender very good apps ranking without labels.

Note that when k is small LFM-Auto has very low precision but high NDCG. If all apps have same rating, no matter what the rating is, NDCG will be always 1. Therefore, NDCG in app ranking may has some limitation. A simple alternative is using perfect-DCG instead of iDCG as the denominator of NDCG. perfect-DCG assumes that all apps have best rating ideally, i.e. rating $= 15$ in this paper. Figure 8 shows the result using perfect-DCG. Now LFM-Auto has very low NDCG while ARFT still reaches good. However, iDCG has its own meaning since it represents true ranking effect for these apps, while perfect-DCG is just in ideal situation. Hence, a better metric for app ranking may combine both the advantage of iDCG and perfect-DCG so that it can evaluate ideal and practical performance simultaneously. This needs future investigation.

5 Conclusion

In this paper, we proposed an app ranking framework based on transfer learning (ARFT) which can rank apps for a given query. Specifically, ARFT considers relevance, quality, auto-weights and country differences at the same time. The key finding of this paper is inner attribute consistency of apps and transfer learning is a natural and good way to utilize it for ranking service. Experiments on App Store show that ARFT has higher precision and NDCG than other

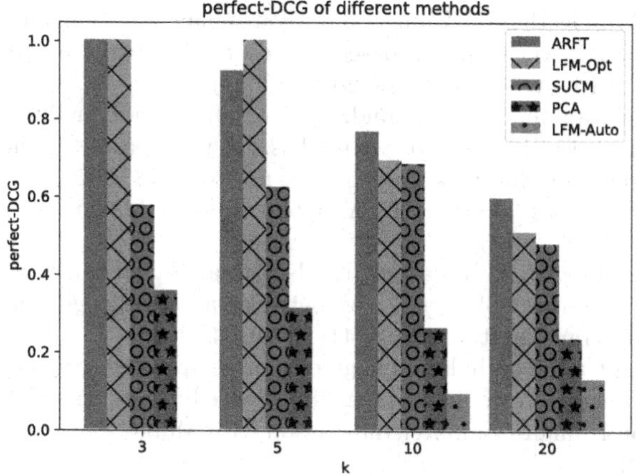

Fig. 8. Comparison under improved metric perfect-DCG. LFM-Opt shows true result with precision while ARFT still reaches good perfect-DCG. This shows its effectiveness in app ranking.

methods. In the future, a good app ranking method should include these aspects: (1) good enough text processing method for query words; this related to NLP; (2) app dataset preprocessing because it is predictable that app markets will have more and more apps; this related to DM; (3) a good ranking algorithm that can effectively and efficiently rank apps and show most related on top; this related to ML; and (4) graphic interactive that makes users have good experience; this related to SE. Therefore, app ranking is not belong to a specific area but a synthesis of many areas.

References

1. Xu, X., Dutta, K., Datta, A., Ge, C.: Identifying functional aspects from user reviews for functionality-based mobile app recommendation. J. Am. Soc. Inf. Sci. **69**(2), 242–255 (2018)
2. Liu, B., Kong, D., Cen, L., Gong, N.Z., Jin, H., Xiong, H.: Personalized mobile app recommendation: reconciling app functionality and user privacy preference. In: Proceedings of the Eighth ACM International Conference on Web Search and Data Mining (WSDM 2015), pp. 315–324. ACM (2015)
3. Xu, Y., et al.: Machine learning-driven apps recommendation for energy optimization in green communication and networking for connected and autonomous vehicles. IEEE Trans. Green Commun. Networking **6**(3), 1543–1552 (2022)
4. Ma, S.P., Lee, S.J., Lee, W.T., Lin, J.H., Lin, J.H.: Mobile application search: a QoS-aware and tag-based approach. EAI Endorsed Trans. Indust. Netw. Intellig. Syst. **2**(4), e6 (2015)
5. Rodrigues, P., Silva, I.S., Barbosa, G.A.R., Coutinho, F.R.D.S., Mourão, F.: Beyond the stars: towards a novel sentiment rating to evaluate applications in web stores of mobile apps. In: Proceedings of the 26th International Conference on World Wide Web Companion, pp. 109–117. International World Wide Web Conferences Steering Committee (2017)

6. Maheswari, M., Geetha, S., Kumar, S.S., Karuppiah, M., Samanta, D., Park, Y.: PEVRM: probabilistic evolution based version recommendation model for mobile applications. IEEE Access **9**, 20819–20827 (2021)
7. Tushev, M., Ebrahimi, F., Mahmoud, A.: Domain-specific analysis of mobile app reviews using keyword-assisted topic models. In: Proceedings of the 44th International Conference on Software Engineering, pp. 762–773 (2022)
8. Datta, A., Kajanan, S., Pervin, N.: A mobile app search engine. Mobile Networks Appl. **18**(1), 42–59 (2013)
9. Lim, S.L., Bentley, P.J., Kanakam, N., Ishikawa, F., Honiden, S.: Investigating country differences in mobile app user behavior and challenges for software engineering. IEEE Trans. Software Eng. **41**(1), 40–64 (2015)
10. Peltonen, E., et al.: The hidden image of mobile apps: geographic, demographic, and cultural factors in mobile usage. In: Proceedings of the 20th International Conference on Human-Computer Interaction with Mobile Devices and Services, p. 10. ACM (2018)
11. Liu, B., Wu, Y., Gong, N.Z., Wu, J., Xiong, H., Ester, M.: Structural analysis of user choices for mobile app recommendation. ACM Trans. Knowl. Discov. Data (TKDD) **11**(2), 17 (2016)
12. Mikolov, T., Sutskever, I., Chen, K., Corrado, G.S., Dean, J.: Distributed representations of words and phrases and their compositionality. In: Advances in Neural Information Processing Systems, pp. 3111–3119 (2013)
13. Le, Q., Mikolov, T.: Distributed representations of sentences and documents. In: International Conference on Machine Learning (ICML 2014), pp. 1188–1196 (2014)
14. Blei, D.M., Ng, A.Y., Jordan, M.I.: Latent dirichlet allocation. J. Mach. Learn. Res. **3**, 993–1022 (2003)
15. Liu, Y., Liu, L., Liu, H., Li, S.: Information recommendation based on domain knowledge in app descriptions for improving the quality of requirements. IEEE Access **7**, 9501–9514 (2019)
16. Ferrari, A., Spoletini, P., Debnath, S.: How do requirements evolve during elicitation? An empirical study combining interviews and app store analysis. Requir. Eng. 1–31 (2022)
17. Auch, M., Weber, M., Mandl, P., Wolff, C.: Similarity-based analyses on software applications: a systematic literature review. J. Syst. Softw. **168**, 110669 (2020)
18. Shen, S., Lu, X., Hu, Z., Liu, X.: Towards release strategy optimization for apps in google play. In: Proceedings of the 9th Asia-Pacific Symposium on Internetware, p. 1. ACM (2017)
19. Noei, E., Lyons, K.: A survey of utilizing user-reviews posted on google play store. In: Proceedings of the 29th Annual International Conference on Computer Science and Software Engineering, pp. 54–63 (2019)
20. Datta, D., Sangaralingam, K.: Do app launch times impact their subsequent commercial success? Int. J. Big Data Intell. **3**(4), 279–287 (2016)
21. Genc-Nayebi, N., Abran, A.: A systematic literature review: opinion mining studies from mobile app store user reviews. J. Syst. Softw. **125**, 207–219 (2017)
22. Hassan, S., Bezemer, C.P., Hassan, A.E.: Studying bad updates of top free-to-download apps in the google play store. IEEE Trans. Software Eng. **46**(7), 773–793 (2018)
23. Tafesse, W.: The effect of app store strategy on app rating: the moderating role of hedonic and utilitarian mobile apps. Int. J. Inf. Manage. **57**, 102299 (2021)
24. Dąbrowski, J., Letier, E., Perini, A., Susi, A.: Analysing app reviews for software engineering: a systematic literature review. Empirical Software Eng. **27**(2), 43 (2022)

25. Assi, M., Hassan, S., Tian, Y., Zou, Y.: Featcompare: feature comparison for competing mobile apps leveraging user reviews. Empir. Softw. Eng. **26**, 1–38 (2021)
26. Hall, P.: The distribution of means for samples of size n drawn from a population in which the variate takes values between 0 and 1, all such values being equally probable. Biometrika **19**(3/4), 240–245 (1927)
27. Nxx, P.: On the frequency distribution of the means of samples from a population having any law of frequency with finite moments, with special reference to pearson's type ii. Biometrika **19**(3/4), 225–239 (1927)
28. Herlocker, J.L., Konstan, J.A., Terveen, L.G., Riedl, J.T.: Evaluating collaborative filtering recommender systems. ACM Trans. Inf. Syst. (TOIS) **22**(1), 5–53 (2004)
29. Nair, V., Hinton, G.E.: Rectified linear units improve restricted Boltzmann machines. In: Proceedings of the 27th International Conference on Machine Learning (ICML 2010), pp. 807–814 (2010)
30. Kingma, D.P., Ba, J.: Adam: a method for stochastic optimization. arXiv preprint arXiv:1412.6980 (2014)
31. Mnih, A., Salakhutdinov, R.R.: Probabilistic matrix factorization. In: Advances in Neural Information Processing Systems (NIPS 2008), pp. 1257–1264 (2008)

Vessel Trajectory Segmentation: A Survey

Zhipei Yu[1], Hanyue Wu[1], Zhenzhong Yin[1], Kezhong Liu[2], and Rui Zhang[1(✉)]

[1] School of Computer Science and Artificial Intelligence, Wuhan University of Technology, Wuhan 430070, China
{260164,276817,yzzero,zhangrui}@whut.edu.cn

[2] School of Navigation, Wuhan University of Technology, Wuhan 430070, China
kzliu@whut.edu.cn

Abstract. Vessel trajectory data, usually derived from AIS data, serves as a robust foundation for extensive research on vessel movements and behaviors. The task of vessel trajectory segmentation identifies the typical sub-trajectory segments in the vessel trajectories, which are recognized as natural, interpretable, and meaningful basic vessel behaviors. Consequently, vessel trajectory segmentation is essential for machine learning and recommendation in the shipping field. However, a systematic literature review on vessel trajectory segmentation is still absent. In this survey, we provide an overview of vessel trajectory segmentation, covering data description, fundamental concepts, typical methods (both supervised and unsupervised), as well as applications. Furthermore, we discuss the challenges and future directions of vessel trajectory segmentation.

Keywords: Vessel trajectory · Vessel behavior · Trajectory segmentation

1 Introduction

As the Automatic Identification System (AIS) has become more prevalent recently, an increasing amount of vessel trajectory data is being collected. Based on such data, various analysis platforms and applications, including hiFleet[1], cjbeidou[2], shipxy[3], and others, have been developed. These platforms offer a diverse range of services for vessel data research and applications, such as data provisioning, vessel monitoring, track playback, vessel intent forecast, and so on.

Most machine learning tasks in the shipping field perform computing and analysis on vessel trajectory data, which typically involves vessel trajectory segmentation. By segmenting the vessel trajectories, the trajectories can be decomposed into several sub-trajectory segments representing meaningful, continuous basic behaviors [12,41]. Each sub-trajectory segment possesses similar inner data

[1] https://www.hifleet.com/.
[2] http://www1.cjbeidou.com/.
[3] https://www.shipxy.com/.

A. El Abbadi et al. (Eds.): DASFAA 2023 Workshops, LNCS 13922, pp. 166–180, 2023.
https://doi.org/10.1007/978-3-031-35415-1_12

characteristics, allowing the identification of vessel behaviors. As a result, the vessels' subsequent actions can be predicted [25] and recommended, vessel operations can be suggested, and navigation risk management can be implemented.

Despite much individual research on vessel trajectory data, we lack a systematic review of vessel trajectory segmentation to figure out this task and classify existing methods. In this study, we conduct a comprehensive review of vessel trajectory segmentation in terms of its process and methods. Firstly, AIS trajectory data is introduced, and the vessel trajectory segmentation task is defined. Subsequently, the methods of vessel trajectory segmentation are categorized, and the strengths and weaknesses of existing methods are examined. Finally, the limitations of existing methods are highlighted, and suggestions for future research are provided.

2 Vessel Trajectory

2.1 AIS Data

Vessel trajectory data is typically generated by the on-board AIS equipment, which records the vessel's position, speed and other relevant information. Similar to general trajectory data, vessel trajectory data is characterized as being massive, time-varying, space-varying, and dynamic. However, vessel trajectory data possesses two unique features. Firstly, as maritime vehicles, vessels typically exhibit smoother trajectories than land vehicles, as they lack the same level of maneuverability. Secondly, unlike land vehicles that are constrained by road networks, vessel movements are relatively unconstrained by waterways in many water areas. Instead, they are subject to environmental factors, such as ocean currents, tides, and typhoons. Consequently, vessel trajectories exhibit greater uncertainty when confronted with environmental changes or obstacles. Though vessel trajectory data is derived from AIS data, raw AIS data cannot be used directly. Due to the complex water environment and the influence of equipment failure, raw AIS data often contains errors and missing values during transmission and decoding. In addition, the sampling frequencies of different AIS devices may vary. Moreover, vessel trajectory data possesses more attributes when compared to land-based trajectory data. Therefore, AIS data can be characterized by multi-dimensionality, high noise, discontinuity, and non-uniform scale, rendering it challenging to analyze vessel behaviors effectively.

AIS data typically consists of both static and dynamic information, as shown in Table 1. The static information mainly includes the IMO code, the unique Maritime Mobile Service Identifier (MMSI), the call sign used for vessel wireless communication, the vessel name, etc. The dynamic information mainly includes the timestamp, latitude and longitude, speed over ground (SOG), course over ground (COG), and heading of the vessel, etc.

2.2 Vessel Trajectory Segmentation

The process of a vessel's voyage in a given time period can be viewed as composed of several meaningful sequential behaviors. Through supervised or unsupervised

Table 1. Static and dynamic information in AIS data. Static information is unchangeable and is set once. In contrast, dynamic information is constantly updated through various sensors connected to the AIS system and changes continuously over time.

Static Information			Dynamic Information		
Attribute	Type	Example	Attribute	Type	Example
mmsi	string	211311970	timestamp	string	1425175434
imo	string	IMO9628899	longitude	double	122.45476
callsign	string	VRME7	latitude	double	31.021538
length	float	73.0	sog	float	76.0
width	float	14.0	cog	float	197.0
draught	float	3.5	heading	float	46.5
name	string	Hong Xing	status	integer	3
type	integer	70			

methods, the vessel trajectory is divided into continuous sub-trajectories, each of which represents a natural, meaningful basic vessel behavior that can maintain inner trajectory characteristics consistently in semantics. Vessel trajectory segmentation enables a deeper comprehension of the underlying movement patterns and the evolution mechanism within the vessel trajectory. The following are definitions for vessel trajectory segmentation:

Definition 1 (Vessel Trajectory). *A vessel trajectory* $\boldsymbol{T} = (\boldsymbol{p}_1, \cdots, \boldsymbol{p}_n)$ *contains* n *trajectory sampling points, where the sampling point* $\mathbf{p} \in \mathbb{R}^d$ *contains the vessel's spatial coordinates and motion parameters, including longitude, latitude, sog, cog and heading.*

Definition 2 (Sub-Trajectory). *Given a vessel trajectory* $\boldsymbol{T}$, *a sub-trajectory consists of the continuous points of* $\boldsymbol{T}$ *from* $\boldsymbol{p}_a$ *to* $\boldsymbol{p}_b$, *i.e.,* $\boldsymbol{T}_{a,b} = (\boldsymbol{p}_a, \cdots, \boldsymbol{p}_b)$, *where* $1 \leq a \leq b \leq n$, $\|\boldsymbol{T}_{a,b}\| = b - a + 1$.

Definition 3 (Vessel Trajectory Segmentation). *Given a vessel trajectory* $\boldsymbol{T}$, *the objective is to find* $m - 1$ *change points in it, and divide* $\boldsymbol{T}$ *into* m *meaningful sub-trajectories, so that each sub-trajectory segment reflects the basic behavior pattern of the vessel. As shown in Fig. 1, each sub-trajectory before and after change point represents a behavior of the vessel.*

3 Supervised Vessel Trajectory Segmentation Method

Supervised trajectory segmentation typically relies on domain knowledge, wherein thresholds are set for different trajectory features based on experiences, such as time and velocity thresholds. Subsequently, the trajectories are segmented according to these thresholds. Additionally, there is another class of

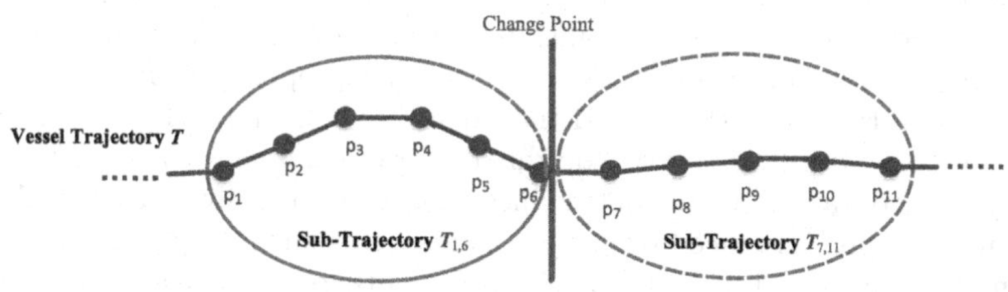

Fig. 1. Vessel trajectory segmentation

supervised methods by predefining some basic behavior patterns, followed by the detection or matching of sub-trajectories.

Supervised trajectory segmentation methods may perform well in specific application scenarios, but have poor generalizability and portability. In addition, these methods rely on domain knowledge or pre-defined patterns and are strongly demand-driven. Consequently, they cannot autonomously discover the behavior patterns of mobile objects based solely on data characteristics.

3.1 Threshold-Based Segmentation

For trajectory segmentation, a natural method is to set thresholds for speed, course, or time, which can detect abrupt changes in the trajectory. These changes can be used to identify sub-trajectory segments that correspond to the object's turning, accelerating, or decelerating. Alternatively, the trajectory can be divided into different behavioral segments by detecting stopping points.

Palma et al. [36] proposed a velocity-based spatial-temporal clustering method to deal with single trajectories. The trajectories are classified into stops and movements based on the velocity of the trajectory points. Xiang et al. [50] proposed a method for automatically extracting stopping points from individual trajectory sequences using SOC (sequence-oriented clustering). Li et al. [27] divided sub-trajectory segments according to velocity threshold and time threshold, respectively. The vessel velocity threshold is set, and the sequence of continuous points with a velocity less than the threshold is regarded as a segment, and the continuous points with a velocity greater than the threshold is regarded as another segment. Similarly some researchers [15] extract the stopping points by time and velocity thresholds and perform trajectory segmentation based on probabilistic logic. However, the change of vessel trajectory position point also includes other features such as the vessel's turning, acceleration and deceleration. So it is difficult to mine other potential behavioral evolution information by simply considering speed change. Buchin et al. [6] consider a variety of features in the trajectory, including position, heading, velocity, speed, curvature, curvature, curvature, and shape. Segmentation is performed based on any one or any combination of these features. The results of segmentation help to understand the implied behavior patterns of the target from its trajectory.

In addition to considering multiple features of trajectory points, some researchers start from the abrupt change points of trajectories to find potential segmentation points. For example, Etemad et al. [10] proposed a supervised trajectory segmentation algorithm WS-II that identifies behavioral changes in time and space by processing trajectory coordinates, trains a binary classifier to segment trajectory data, and finally determines segmentation points based on those predictions. The problem faced by change point detection is its susceptibility to outlier interference, which is highly likely to use outliers as change points, leading to incorrect trajectory segmentation. Yoon et al. [56] takes this problem into account by dividing a given trajectory into uniform segments, each approximating a linear movement at a constant velocity. Then, the spatio-temporal properties of the object movement are used for the detection of outliers, thus weakening the interference generated by outliers on the detection of change points.

Threshold-based segmentation has the advantage of quickly dividing the trajectory into necessary segments. For example, to analyze the speed and heading changes during the vessel's movement, it is easy to obtain the required trajectory segments by using speed and heading thresholds for filtering. However, the problem is that the determination of thresholds often requires experience or domain knowledge, and the analysis of vessel behavior at different scales requires the careful setting of different thresholds. The limitation of such methods is that they are susceptible to outlier interference, and setting thresholds to classify sub-trajectory segments by experience alone requires pre-processing of the trajectory data to eliminate the outliers in them.

3.2 Trajectory Semantic-Based Segmentation

The trajectory semantics includes not only the behavior patterns of vessel movement but also the information jointly extracted from the surrounding environment and domain knowledge. Starting from trajectory semantics, many researchers assign much different semantic information to the changes of position points in the trajectory based on certain domain knowledge, and perform trajectory segmentation by matching trajectory features with semantics. Such methodes are not only concerned with finding segmentation points, but are also interested in assigning appropriate labels to sub-trajectory segments.

Some researchers interpret trajectory semantics as the behavior generated by the interaction between the study object and its surrounding environment. For example, Yan et al. [54] consider the spatial-temporal characteristics of the target trajectory and the surrounding geographical environment information, and convert the original trajectory into labeled sub-trajectories. Similarly, the literatures [17,57] considers the surrounding environment information and matches the semantic information with the trajectory to achieve the trajectory segmentation. Bermingham et al. [4] find stopping points from trajectories, match the stopping points with real locations, and convert the original trajectory into a sequence of visited locations to achieve trajectory segmentation. IB-SMoT [3] (cross-point based detection of stops and moves), CB-SMoT [36] (cluster based detection of stops consider), and DB-SMoT [37] (direction based detection of stops and movement) are three similar methods for matching stopping points with geographical objects. Some researchers [33] have also combined CB-SMoT and DB-SMoT. However, these methods have limitation. The study objects do not necessarily interact with their surroundings at these points.

Vessels usually travel between ports, either for transporting passengers and cargo or for fishing. They have less interaction with their environment, resulting in less semantics to mark. Many researchers have focused on the division of vessels into fishing and non-fishing phases, or have used berth ports as trajectory segmentation points. The aim of DB-SMoT [37] is to automatically find the real locations where vessels carry out fishing activities, but ignores the fact that a complete fishing activity may also include points with small changes in direction. A window-based segmentation algorithm, WBS-RLE, has been proposed by Wu et al./cite[wu2022semantic] in order to segment a trajectory into fishing and non-fishing segments. First, a pre-trained classifier labels windows in the trajectory as fishing or non-fishing, and then run-length encoding combines the labeled windows into the total fishing activity to achieve trajectory segmentation. Some researchers [47] focus on the basic behaviors of vessels in scenarios where vessels come and go more frequently at sea, such as piers, anchorages, and waterways, and build a Semantic Model of vessel Behavior. The vessels are analyzed and classified to accurately obtain the basic and potential behaviors represented by sub-trajectory segments in all typical scenes. The advantage of the trajectory semantic-based method is that the sub-trajectory segments can be given complex semantic information based on manual tagging or domain expertise, allowing the segmentation results to provide the basis for subsequent semantic annotation or behavior analysis. However, the disadvantage is that it takes a lot of time to determine the appropriate thresholds or to label the trajectory dataset. Such methods rely heavily on the quality of the trajectory dataset and domain knowledge.

4 Unsupervised Vessel Trajectory Segmentation Method

The unsupervised trajectory segmentation can be mainly divided into two types according to the implementation methods, which are clustering-based segmentation and heuristic-based segmentation. The clustering-based segmentation method can identify consecutive trajectory points with similar characteristics, and the sub-trajectories formed by these consecutive trajectory points can be considered to have inherent characteristic consistency, thus realizing the segmentation of trajectories. Heuristic-based trajectory segmentation focuses on finding the feasible solution to the segmentation problem and then optimizing the loss function to search for the optimal segmentation method. Several effective methods are proposed in the field of time series segmentation that can well achieve segmentation of time series recorded by sensors and can also be extended to the task of trajectory segmentation.

4.1 Clustering-Based Segmentation

Sheikholeslami et al. [43] proposed WaveCluser, a grid-based clustering method. This multi-resolution clustering algorithm is capable of detecting arbitrarily shaped clusters at different scales and handling noise in the data. Furthermore, WaveCluster is noise-insensitive and suitable for segmenting vessel trajectory. Inspired by the DBSCAN clustering technique, Derya et al. proposed a density-based clustering method ST-DBSCAN [5]. This method, which adds distance parameters for spatial and non-spatial attributes, can cluster data according to their temporal and spatial attributes and assign different points to different clusters. ST-DBSCAN also solves the problem of noise points by assigning a density factor to each cluster. The segmentation produced by ST-DBSCAN is more fine-grained. In open areas, vessel trajectories mostly travel in a straight line, while ST-DBSCAN divides a straight-line trajectory into multiple sub-trajectory segments, which does not have better interpretability. Based on the Hausdorff distance, Tra-DBScan groups line segments based solely on the spatial characteristics of their trajectory segments [29]. Its limitation is that it ignores the temporal characteristics. Yan et al. [53] use a density-based method to analyze vessel trajectories and identity traffic patterns from historical AIS data, which also considers speed and direction. Some researchers [8,16] have also designed density clustering-based trajectory segmentation methods based on spatial density rules. Damiani et al. introduce a clustering-based segmentation method called SeqScan for dividing trajectories into successive subsequences based on spatial density criteria [8], similarly there is a spatial-temporal clustering-based trajectory segmentation method for detecting spatial-temporal clusters of any shape, all dividing the trajectories into internally similar segments [20]. Niu et al. [35] can identify important locations such as stopping points in trajectories by clustering algorithms.

There are also some segmentation methods, such as Tang et al. GMM [45] proposed a probabilistic generation model-based method EM-GMM. Applying EM-GMM to trajectory segmentation, it can obtain the probability that each trajectory point belongs to a certain segment, and it is applicable to data of various dimensions. However, the segmentation points obtained by EM-GMM cannot distinguish the complete turning process of vessel trajectory. Warped K-Means [26] is to segment and classify human activities. However, its main limitation is that it requires a given number of segments as input. In practice, the number of vessel trajectory segments is usually unknown and this type of method requires several iterations to determine the appropriate number of segments. Wu et al. [49] used the K-Means to design the trajectory segmentation criterion and adaptive distance measurement, but did not consider the clustering of temporal information. The method proposed by Izakian et al. [21] can take multiple motion parameters as input and detect trajectory subsequences with similar movement characteristics by clustering the extracted features such as position, direction and velocity. Varlamis et al. [46] identify the speed and heading patterns of specific types of vessels navigating on typical routes by clustering.

The advantage of the clustering-based method is that it can naturally discover trajectory points with similar behavioral and spatial-temporal characteristics in vessel trajectories and merge them into sub-trajectory segments. However, its disadvantage is that it is difficult to interpret, which may be solved if the sub-trajectory segments obtained by clustering are matched with predefined vessel behaviors. The clustering method can only focus on the similarity of trajectory point features, but cannot capture the semantics of the formed segments, and it needs more tuning parameters to determine the relatively suitable segments, otherwise it may produce too fragmented segmentation results.

4.2 Heuristic-Based Segmentation

TraClus is a well-known algorithm for trajectory segmentation [24]. According to the minimum description length (MDL) principle, TraClus segments trajectories into segments, and clusters those segments according to density, ensuring that as few sub-trajectory segments as possible are formed at the same time, so that the difference between the sub-trajectory segments and the trajectory itself is as small as possible. Based on the MDL principle, GRASP-UTS segments trajectories by optimizing a cost function that takes into account the homogeneity of the segments and the separation of their representative points. It is a metaheuristic algorithm that constructs segments by modifying the number and position of the representative points of the trajectory [44]. Semantic trajectory segmentation is focused by RGRASP-SemTS [22] discuss feature evaluation, non-monotonic criterion, semantic annotation, cost function and metaheuristic algorithms, which infer segments of trajectories in an unsupervised manner. RGRASP-SemTS also uses the MDL principle to measure uniformity within segments. TS-MF [52] is to measure the similarity based on the motion characteristics such as velocity, acceleration, differential position, and angle of two trajectory points by using Pearson coefficients, and then segmentation is performed. To solve the local over-segmentation of trajectories, the trajectories are merged by minimizing the cost function values.

With heuristic segmentation methods, detection of change points is also a concern. An error signal is first generated by using some interpolation to indicate the deviation of each point from its desired position. Then, the signal is used to segment the trajectory into segments. OWS [11] and SWS [12] belong to this type of unsupervised trajectory segmentation algorithms. The WS-II method, mentioned above, is an improvement of the SWS method. The framework of SEMANTIC-SEG [13] uses a movement pattern change detection algorithm to partition the original trajectory into uniform segments. Then a spatial-temporal probabilistic model is introduced for segmentation to achieve bottom-up trajectory segment inference. The Ramer-Douglas- Peucker [28] algorithm extracts the characteristic points of the trajectory to use them as change points. This method selects candidate points with spatial-temporal constraints and computes the optimal matching points from the candidate set by spatial-temporal distances. A limitation of this type of method is also the lack of good interpretation.

For time series, many unsupervised methods have been proposed for pattern discovery and sequence segmentation. Since vessel trajectory data contains rich temporal features, some of these methods can also be applied to trajectory segmentation. Adams et al. [2] proposed the Bayesian online change point detection method BOCD to find the change points of time series. Literatures [18,23,34,38,48] are based on a variant of the BOCD model. Matsubara et al. [31] proposed a multivariate time series segmentation method AutoPlait based on Hidden Markov Model in 2014. The method is parameter-free and well-scalable, and can automatically identify meaningful potential patterns in the series and segment the time series. Based on AutoPlait, Matsubara et al. also proposed ECOWEB [32] and REGIMECAST [30] to mine meaningful sequence patterns of large-scale co-evolutionary network data and predict long-term future events. In recent years Matsubara et al. proposed the CUBEMAKER [19] to segment large-scale multidimensional sensor data to find important time series patterns and change points, and also proposed a unified adaptive nonlinear method RFCast [9] to predict data flow change patterns and future evolutionary patterns in an online search. Gharghabi et al. [14] proposed a fast and general unsupervised semantic segmentation method FLOSS. FLOSS is a domain-agnostic method with only one easily determined parameter and can process data streams at high speed, but does not have the ability to handle multidimensional data. Recently Patrick et al. proposed a self-supervised time series segmentation method ClaSP [40] and its improved version [39], which can effectively discover the change points in time series. However, ClaSP can only achieve segmentation for one-dimensional time series and does not consider the co-evolutionary pattern of multidimensional features. These time series segmentation methods have good segmentation effect on the temporal characteristics of vessel trajectories, but they cannot pay attention to the spatial characteristics of trajectories. Moreover, BOCD, FLOSS and ClaSP can only handle one-dimensional feature sequences, which leads to the lack of segmentation capability for multidimensional data.

5 Applications

Through trajectory segmentation, we can discover different behaviors in the same vessel trajectory or the same behaviors of different vessels, and furthermore, we can mine the typical behavioral patterns or periodic movement patterns of vessels. Vessel trajectory segmentation provides the basis for vessel-related downstream tasks, and has many applications in vessel behavior analysis, water traffic management, and vessel route recommendation.

Shahir et al. [42] study vessel behaviors based on sub-trajectories for detecting and tracking illegal fishing activities. Chen et al. [7] analyze vessel collision avoidance behavior based on trajectory segmentation results. Li et al. [27] extract shape features from trajectory segments and then perform vessel pattern recognition. Abreu et al. [1] propose a visual analytics tool that uses spatial segmentation to divide trips into sub-trajectories and score them. The scores are displayed in a tabular visualization where users can rank trips by segment to identify local anomalies. Xiang et al. [51] use a grid-based method to segment trajectories to obtain the grid frequency diagram of trajectory distribution, and then use this to recommend safe routes. Yang et al. [55] segment the vessel sailing route map so as to achieve intelligent recommendations of vessel routes. Varlamis et al. [46] identify key areas in the monitored region for information extraction such as ports, platforms or areas where vessels change courses.

6 Challenges and Future Directions

Trajectory segmentation results have a significant impact on downstream tasks. Currently, vessel trajectory segmentation faces challenges in three aspects:

a) The uneven sampling frequency of on-board AIS equipment and the presence of errors in the AIS transmission and analysis process can result in a large number of error points in the data, which significantly affects the quality and refinement of the segmentation results. Therefore, trajectory segmentation methods require strong adaptability.
b) Unlike trajectory data from robots or road networks, vessel trajectories are a co-evolving multi-dimensional sequence with complex spatial topology relationships between sequence points. Vessel trajectories are influenced by a variety of factors in open water, complex marine waters, or encounter scenarios, which can result in different behavior patterns. Consequently, trajectory segmentation methods should be suitable for various situations.
c) Traditional methods rely on prior knowledge or specific rules to predefine basic vessel behavior. However, the core problem of vessel trajectory segmentation is designing an automatic method to discover meaningful trajectory segments for vessels based on the characteristics of the data itself. As trajectory points lack classification labels, identifying the start and end points of each subsequence from the spatio-temporal characteristics of the data itself is challenging. The difficulty lies in ensuring that the segments are meaningful and natural.

In light of the limitations of existing methods, several potential directions for future research are suggested.

Interpretability. Segmentation methods based on manual predefinition can explain the behavior of a vessel, but cannot exhaust all behaviors. Moreover, the domain-agnostic method lacks interpretability, leaving the rationale for segmentation unclear. Future research could aim to address these issues by developing methods that provide convincing explanations for the results of domain-agnostic segmentation.

Universality. Currently, most segmentation methods are designed for specific scenarios or a limited set of vessel behaviors, and therefore they are less adaptable. Therefore, future work could explore the development of universal trajectory segmentation methods that are adaptable to diverse navigation scenarios.

Multi-semanticity. For trajectory segmentation, hierarchical semantics could be combined with behavioral semantics for a single vessel, interaction semantics for multiple vessels, and environmental semantics. Using the multi-semantic segmentation results, downstream tasks can be studied more effectively.

7 Conclution

This survey presents an overview of AIS data and provides a description and definition of the vessel trajectory segmentation task. Furthermore, common spatio-temporal trajectory segmentation methods are examined and categorized into supervised and unsupervised methods. The supervised methods include threshold-based segmentation and trajectory semantic-based segmentation, while the unsupervised methods are clustering-based segmentation and heuristic-based segmentation, respectively.

Most supervised trajectory segmentation algorithms require labeled data or the priori information, including time thresholds, velocity thresholds, and the number of trajectory segments. In contrast, semi-supervised trajectory segmentation algorithms use a combination of labeled and unlabeled data for segmentation. Nevertheless, most vessel trajectory datasets do not contain labels and, thus, demand laborious manual labeling in advance. The unsupervised trajectory segmentation algorithm does not require labeled data. However, the existing unsupervised segmentation algorithms exhibit high time complexity and weak interpretability, making them unsuitable for analyzing large amounts of trajectory data. With clustering methods, it is difficult to obtain the semantics of the spatio-temporal trajectory segments. Although the existing time series segmentation methods applied to vessel trajectory segmentation can yield some obvious segmentation points, the segmentation results still lack satisfactory interpretation and fail to capture the spatial characteristics of the trajectory data.

This survey also introduces applications of vessel trajectory segmentation, highlights the challenges faced by existing methods, and suggests some future research directions.

Acknowledgements. This work is partially supported by the National Natural Science Foundation of China under Grants 52031009; by the Natural Science Foundation of Hubei Province, China, under Grant No. 2021CFA001.

References

1. Abreu, F.H., Soares, A., Paulovich, F.V., Matwin, S.: A trajectory scoring tool for local anomaly detection in maritime traffic using visual analytics. ISPRS Int. J. Geo Inf. **10**(6), 412 (2021)
2. Adams, R.P., MacKay, D.J.: Bayesian online changepoint detection. arXiv preprint arXiv:0710.3742 (2007)
3. Alvares, L.O., Bogorny, V., Kuijpers, B., de Macedo, J.A.F., Moelans, B., Vaisman, A.: A model for enriching trajectories with semantic geographical information. In: Proceedings of the 15th Annual ACM International Symposium on Advances in Geographic Information Systems, pp. 1–8 (2007)

4. Bermingham, L., Lee, I.: Mining place-matching patterns from spatio-temporal trajectories using complex real-world places. Expert Syst. Appl. **122**, 334–350 (2019)
5. Birant, D., Kut, A.: St-DBScan: an algorithm for clustering spatial-temporal data. Data Knowl. Eng. **60**(1), 208–221 (2007)
6. Buchin, M., Driemel, A., Kreveld, M.v., Sacristán Adinolfi, V.: Segmenting trajectories: a framework and algorithms using spatiotemporal criteria. J. Spatial Inf. Sci. **3**, 33–63 (2011)
7. Chen, P., Shi, G., Liu, S., Gao, M.: Collision avoidance situation matching with vessel maneuvering actions identification from vessel trajectories. Int. J. Perform. Eng. **15**(6), 1499 (2019)
8. Damiani, M.L., Hachem, F., Issa, H., Ranc, N., Moorcroft, P., Cagnacci, F.: Cluster-based trajectory segmentation with local noise. Data Min. Knowl. Disc. **32**, 1017–1055 (2018)
9. Do, T.M., Matsubara, Y., Sakurai, Y.: Real-time forecasting of non-linear competing online activities. J. Inf. Process. **28**, 333–342 (2020)
10. Etemad, M., Etemad, Z., Soares, A., Bogorny, V., Matwin, S., Torgo, L.: Wise sliding window segmentation: a classification-aided approach for trajectory segmentation. In: Goutte, C., Zhu, X. (eds.) Canadian AI 2020. LNCS (LNAI), vol. 12109, pp. 208–219. Springer, Cham (2020). https://doi.org/10.1007/978-3-030-47358-7_20
11. Etemad, M., Júnior, A.S., Hoseyni, A., Rose, J., Matwin, S.: A trajectory segmentation algorithm based on interpolation-based change detection strategies. In: EDBT/ICDT Workshops, p. 58 (2019)
12. Etemad, M., Soares, A., Etemad, E., Rose, J., Torgo, L., Matwin, S.: SWS: an unsupervised trajectory segmentation algorithm based on change detection with interpolation kernels. GeoInformatica **25**, 269–289 (2021)
13. Gao, Y., Huang, L., Feng, J., Wang, X.: Semantic trajectory segmentation based on change-point detection and ontology. Int. J. Geogr. Inf. Sci. **34**(12), 2361–2394 (2020)
14. Gharghabi, S., Ding, Y., Yeh, C.C.M., Kamgar, K., Ulanova, L., Keogh, E.: Matrix profile viii: domain agnostic online semantic segmentation at superhuman performance levels. In: 2017 IEEE International Conference on Data Mining (ICDM), pp. 117–126. IEEE (2017)
15. Guo, S., Li, X., Ching, W.K., Dan, R., Li, W.K., Zhang, Z.: GPS trajectory data segmentation based on probabilistic logic. Int. J. Approximate Reasoning **103**, 227–247 (2018)
16. Hachem, F., Damiani, M.L.: Periodic stops discovery through density-based trajectory segmentation. In: Proceedings of the 26th ACM SIGSPATIAL International Conference on Advances in Geographic Information Systems, pp. 584–587 (2018)
17. Han, J., Liu, M., Ji, G., Zhao, B., Liu, R., Li, Y.: Efficient semantic enrichment process for spatiotemporal trajectories in geospatial environment. In: Wang, X., Zhang, R., Lee, Y.-K., Sun, L., Moon, Y.-S. (eds.) APWeb-WAIM 2020. LNCS, vol. 12318, pp. 342–350. Springer, Cham (2020). https://doi.org/10.1007/978-3-030-60290-1_27
18. Hayashi, S., Kawahara, Y., Kashima, H.: Active change-point detection. In: Asian Conference on Machine Learning, pp. 1017–1032. PMLR (2019)
19. Honda, T., Matsubara, Y., Neyama, R., Abe, M., Sakurai, Y.: Multi-aspect mining of complex sensor sequences. In: 2019 IEEE International Conference on Data Mining (ICDM), pp. 299–308. IEEE (2019)

20. Hwang, S., VanDeMark, C., Dhatt, N., Yalla, S.V., Crews, R.T.: Segmenting human trajectory data by movement states while addressing signal loss and signal noise. Int. J. Geogr. Inf. Sci. **32**(7), 1391–1412 (2018)
21. Izakian, Z., Mesgari, M.S., Weibel, R.: A feature extraction based trajectory segmentation approach based on multiple movement parameters. Eng. Appl. Artif. Intell. **88**, 103394 (2020)
22. Junior, A.S., Times, V.C., Renso, C., Matwin, S., Cabral, L.A.: A semi-supervised approach for the semantic segmentation of trajectories. In: 2018 19th IEEE International Conference on Mobile Data Management (MDM), pp. 145–154. IEEE (2018)
23. Knoblauch, J., Jewson, J.E., Damoulas, T.: Doubly robust Bayesian inference for non-stationary streaming data with *beta*-divergences. In: Advances in Neural Information Processing Systems, vol. 31 (2018)
24. Lee, J.G., Han, J., Whang, K.Y.: Trajectory clustering: a partition-and-group framework. In: Proceedings of the 2007 ACM SIGMOD International Conference on Management of Data, pp. 593–604 (2007)
25. Lei, P.R.: A framework for anomaly detection in maritime trajectory behavior. Knowl. Inf. Syst. **47**(1), 189–214 (2016)
26. Leiva, L.A., Vidal, E.: Warped k-means: an algorithm to cluster sequentially-distributed data. Inf. Sci. **237**, 196–210 (2013)
27. Li, J., Liu, H., Chen, X., Li, J., Xiang, J.: Vessel pattern recognition using trajectory shape feature. In: 2021 5th International Conference on Computer Science and Artificial Intelligence, pp. 84–90 (2021)
28. Liu, C., Wang, J., Liu, A., Cai, Y., Ai, B.: An asynchronous trajectory matching method based on piecewise space-time constraints. IEEE Access **8**, 224712–224728 (2020)
29. Liu, L.X., Song, J.T., Guan, B., Wu, Z.X., He, K.J.: Tra-DBScan: a algorithm of clustering trajectories. In: Applied Mechanics and Materials, vol. 121, pp. 4875–4879. Trans Tech Publ (2012)
30. Matsubara, Y., Sakurai, Y.: Regime shifts in streams: real-time forecasting of co-evolving time sequences. In: Proceedings of the 22nd ACM SIGKDD International Conference on Knowledge Discovery and Data Mining, pp. 1045–1054 (2016)
31. Matsubara, Y., Sakurai, Y., Faloutsos, C.: Autoplait: automatic mining of co-evolving time sequences. In: Proceedings of the 2014 ACM SIGMOD International Conference on Management of Data, pp. 193–204 (2014)
32. Matsubara, Y., Sakurai, Y., Faloutsos, C.: The web as a jungle: non-linear dynamical systems for co-evolving online activities. In: Proceedings of the 24th International Conference on World Wide Web, pp. 721–731 (2015)
33. Mazzarella, F., Vespe, M., Damalas, D., Osio, G.: Discovering vessel activities at sea using AIS data: mapping of fishing footprints. In: 17th International Conference on Information Fusion (FUSION), pp. 1–7. IEEE (2014)
34. Mellor, J., Shapiro, J.: Thompson sampling in switching environments with Bayesian online change detection. In: Artificial Intelligence and Statistics, pp. 442–450. PMLR (2013)
35. Niu, X., Wang, S., Wu, C.Q., Li, Y., Wu, P., Zhu, J.: On a clustering-based mining approach with labeled semantics for significant place discovery. Inf. Sci. **578**, 37–63 (2021)
36. Palma, A.T., Bogorny, V., Kuijpers, B., Alvares, L.O.: A clustering-based approach for discovering interesting places in trajectories. In: Proceedings of the 2008 ACM symposium on Applied computing. pp. 863–868 (2008)

37. Rocha, J.A.M., Times, V.C., Oliveira, G., Alvares, L.O., Bogorny, V.: DB-SMOT: a direction-based spatio-temporal clustering method. In: 2010 5th IEEE International Conference Intelligent Systems, pp. 114–119. IEEE (2010)
38. Ruggieri, E., Antonellis, M.: An exact approach to Bayesian sequential change point detection. Comput. Stat. Data Anal. **97**, 71–86 (2016)
39. Schäfer, P., Ermshaus, A., Leser, U.: Clasp - time series segmentation. In: Proceedings of the 30th ACM International Conference on Information & Knowledge Management, pp. 1578–1587. CIKM 2021, New York, NY, USA. Association for Computing Machinery (2021). https://doi.org/10.1145/3459637.3482240
40. Schäfer, P., Ermshaus, A., Leser, U.: Clasp-time series segmentation. In: Proceedings of the 30th ACM International Conference on Information & Knowledge Management, pp. 1578–1587 (2021)
41. Shahir, A.Y., Charalampous, T., Tayebi, M.A., Glässer, U., Wehn, H.: Triptracker: unsupervised learning of fishing vessel routine activity patterns. In: 2021 IEEE International Conference on Big Data (Big Data), pp. 1928–1939. IEEE (2021)
42. Shahir, A.Y., Tayebi, M.A., Glässer, U., Charalampous, T., Zohrevand, Z., Wehn, H.: Mining vessel trajectories for illegal fishing detection. In: 2019 IEEE International Conference on Big Data (Big Data). pp. 1917–1927 (2019). https://doi.org/10.1109/BigData47090.2019.9006545
43. Sheikholeslami, G., Chatterjee, S., Zhang, A.: Wavecluster: a multi-resolution clustering approach for very large spatial databases. In: VLDB, vol. 98, pp. 428–439 (1998)
44. SoaresJúnior, A., Moreno, B.N., Times, V.C., Matwin, S., Cabral, L.D.A.F.: Grasp-UTS: an algorithm for unsupervised trajectory segmentation. Int. J. Geograph. Inf. Sci. **29**(1), 46–68 (2015)
45. Tang, H., Chu, S.M., Hasegawa-Johnson, M., Huang, T.S.: Emotion recognition from speech via boosted gaussian mixture models. In: 2009 IEEE International Conference on Multimedia and Expo, pp. 294–297. IEEE (2009)
46. Varlamis, I., Kontopoulos, I., Tserpes, K., Etemad, M., Soares, A., Matwin, S.: Building navigation networks from multi-vessel trajectory data. GeoInformatica **25**, 69–97 (2021)
47. Wen, Y., et al.: Semantic modelling of ship behavior in harbor based on ontology and dynamic Bayesian network. ISPRS Int. J. Geo Inf. **8**(3), 107 (2019)
48. Wilson, R.C., Nassar, M.R., Gold, J.I.: Bayesian online learning of the hazard rate in change-point problems. Neural Comput. **22**(9), 2452–2476 (2010)
49. Wu, H.R., Yeh, M.Y., Chen, M.S.: Profiling moving objects by dividing and clustering trajectories spatiotemporally. IEEE Trans. Knowl. Data Eng. **25**(11), 2615–2628 (2012)
50. Xiang, L., Gao, M., Wu, T.: Extracting stops from noisy trajectories: A sequence oriented clustering approach. ISPRS Int. J. Geo Inf. **5**(3), 29 (2016)
51. Xiang, Z., Chaojian, S., Hu, Q., Chun, Y.: Study on method of safe routes planning with massive AIS data. J. Safety Sci. Technol. **12**(10), 160–164 (2016)
52. Xu, W., Dong, S.: Application of artificial intelligence in an unsupervised algorithm for trajectory segmentation based on multiple motion features. Wirel. Commun. Mob. Comput. **2022**, 1–11 (2022)
53. Yan, W., Wen, R., Zhang, A.N., Yang, D.: Vessel movement analysis and pattern discovery using density-based clustering approach. In: 2016 IEEE International Conference on Big Data (Big Data), pp. 3798–3806. IEEE (2016)
54. Yan, Z., Chakraborty, D., Parent, C., Spaccapietra, S., Aberer, K.: Semantic trajectories: mobility data computation and annotation. ACM Trans. Intell. Syst. Technol. (TIST) **4**(3), 1–38 (2013)

55. Yang, L., Wang, Y., Ma, W.I.: Research on intelligent recommendation of ship sailing route by big data analysis. Ship Sci. Technol. **43**(14), 52–54 (2021)
56. Yoon, H., Shahabi, C.: Robust time-referenced segmentation of moving object trajectories. In: 2008 Eighth IEEE International Conference on Data Mining, pp. 1121–1126. IEEE (2008)
57. Zhao, B., Liu, M., Han, J., Ji, G., Liu, X.: Efficient semantic enrichment process for spatiotemporal trajectories. Wirel. Commun. Mob. Comput. **2021**, 1–13 (2021)

Deep Normalization Cross-Modal Retrieval for Trajectory and Image Matching

Xudong Zhang(✉) and Wenfeng Zhao

Wuhan University of Technology, No. 122 Luoshi Road, Wuhan 430070, Hubei, China
{248659,258960}@whut.edu.cn

Abstract. In response to the current state of urban traffic, this work introduces a brand-new challenge termed cross-modal retrieval of traffic trajectory and picture, which aims to achieve cross-modal retrieval between trajectory coordinates and traffic images. It is difficult to effectively capture the temporal-spatial properties of trajectory data, and there is still a challenge in narrowing the heterogeneity gap between the hash representations of these two modalities. Thus, we provide a brand-new end-to-end cross-modal retrieval technique for trajectory data that extracts features from both the picture and the trajectory and computes similarity. In order to confirm the efficacy of the suggested strategy, we also produced a traffic multimodal dataset with more than 20,000 traffic photos and trajectory data. The experimental findings on our own cross-modal dataset show that our model works better than other standard techniques.

Keywords: Trajectory coordinates · Cross-modal retrieval · Image

1 Introduction

Modern urban life is impossible without traffic, and as urbanisation has accelerated, issues like traffic congestion and frequent accidents have gotten worse. Among them, the cross-retrieval technology between traffic images and trajectory coordinates is one of the important means to solve traffic problems. Traffic images refer to images of traffic participants such as vehicles and pedestrians obtained through cameras or other sensors, while trajectory coordinates refer to trajectories formed by the position information of traffic participants such as vehicles and pedestrians. The cross-retrieval between traffic images and trajectory coordinates refers to the mutual matching and retrieval between these two types of data, so as to obtain more comprehensive and accurate traffic information. In traffic management, cross-retrieval between traffic images and trajectory coordinates can be used for traffic flow statistics, violation detection, traffic accident analysis, etc. Currently, the cross-retrieval technology between traffic images and trajectory coordinates has become an important research direction

A. El Abbadi et al. (Eds.): DASFAA 2023 Workshops, LNCS 13922, pp. 181–193, 2023.
https://doi.org/10.1007/978-3-031-35415-1_13

in traffic management, and relevant research has made some progress in domestic and international academic journals and conferences. However, due to problems such as large traffic data volume, complex data types, and serious noise interference, the cross-retrieval technology between traffic images and trajectory coordinates still faces many challenges and difficulties. Therefore, further research and exploration in this field, and improving the accuracy and efficiency of cross-retrieval between traffic images and trajectory coordinates, are of great significance for promoting traffic management and smart city construction.

As a result, a technique that can mutually retrieve trajectory coordinates and traffic photos is urgently needed. This technique is crucial for the intelligent processing and analysis of current traffic data. The heterogeneity between various traffic picture and trajectory coordinate modalities, as well as the semantic gap between low-level features and high-level semantics, make it difficult to design effective and efficient cross-modal retrieval algorithms. The majority of cross-modal retrieval techniques currently in use are deep learning-based, using deep neural networks to build two separate networks to learn the characteristics of each modality. In order to learn cross-modal correlations, the two models are then connected via a joint layer, as depicted in Fig. 1. 1 for the general multimodal retrieval architecture. The spatiotemporal properties of trajectory coordinate sequences prevent these approaches from effectively extracting features from trajectory data, despite the fact that they can model images. In addition to these cross-modal retrieval techniques, learning trajectory embeddings [1–6] are the subject of additional publications. These techniques, however, are limited to modelling coordinate trajectories-not pictures. In other words, there isn't a method available right now that can be employed for trajectory cross-modal retrieval.

In this study, we suggest a traffic trajectory multimodal retrieval model for multimodal data retrieval. The following are our key contributions:

- In this research, we define a novel multimodal retrieval task-traffic trajectory multimodal retrieval-and suggest a new multimodal retrieval technique for putting it into practise. This method will enable the retrieval of trajectory coordinates and traffic images. We believe that this is the first time that a cross-modal retrieval challenge for trajectory coordinates and traffic images has been proposed.
- We suggest a new model for carrying out trajectory cross-modal retrieval, attempt to employ a novel technique for learning the semantic information of images and trajectories, and carry out additional processing on the learned hash codes to bring the homogeneous modal data closer together in Hamming space.
- We manually created a dataset with more than 23,000 pieces of multimodal data to show the efficacy of the suggested approach. Thorough tests on the self-created dataset show that the suggested model is valid.

The remainder of this essay is structured as follows: The related study on quamodal retrieval is presented in Part II. Part III explains the formulation, loss

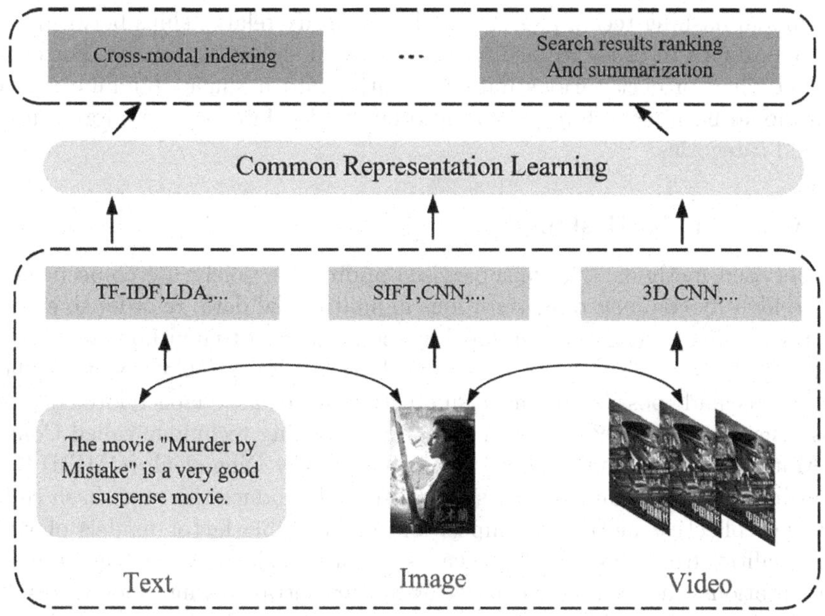

Fig. 1. A general multimodal retrieval framework.

function, and some implementation specifics for our suggested trajectory cross-modal retrieval model. The experimental analysis and results are presented in Sect. 5. The entire study is concluded in Sect. 6.

2 Related Work

We primarily introduce some traditional cross-modal hashing and trajectory feature extraction techniques in this section. Real-valued representation learning [7–9] and binary representation learning [10–12], commonly referred to as cross-modal hashing, are the two main categories under which cross-modal retrieval falls. Real-valued representation learning involves learning a common representation for diverse data modalities. Binary representation learning techniques try to convert data from many modalities into a single Hamming space in order to speed up cross-modal retrieval. However, because the representation is encoded as binary codes, information loss could result in a modest decline in retrieval precision.

Since the majority of real-valued cross-modal retrieval methods now in use are based on strong linear search, which takes a very long time for big amounts of data. Binary representation learning, commonly referred to as hashing, is a useful technique for expediting similarity searches. The approach suggested in this study, which is a cross-modal hashing approach, is one of the binary representation learning approaches suggested above. This group of techniques includes cross-modal hashing techniques that seek to identify relationships between various modal data in order to facilitate cross-modal simulation search. For quick retrieval, they project various data modalities into a single Hamming space. Cross-modal hashing techniques can also be divided into supervised and unsupervised categories.

2.1 Cross-Modal Hashing

Unsupervised methods: The unsupervised approaches solely use co-occurrence data to identify common representations in multimodal data. In order to extend spectral hashing from the conventional unimodal context to multimodal settings, Kumar et al. [13] devised a Cross-View Hashing (CVH) [14]. Cross-view similarity search is made possible by the hash function, which associates related objects across views with related codes. A brand-new hashing technique called Collective Matrix Factorization Hashing was introduced by Ding et al. (CMFH) [15]. According to CMFH, an instance's modalities all produce the same hash code. By using collective matrix decomposition and possible factor models of various modalities from an example, it learns uniform hash codes. To learn reliable representations, the technique can better capture intra- and intermodal correlations. In contrast to other approaches, Liu et al. [16] introduced fused similarity hashing, which involves creating undirected asymmetric graphs and directly embedding them in the shared Hamming space. To acquire accurate and compact multimodal representations that can learn intra- and inter-modal correlations as well as representations, Wang et al. [17] suggested a Deep Multimodal Hashing with Orthogonal Regularization (DMHOR). Moreover, an orthogonal regularizer is applied to the learned matrix to eliminate extraneous data in the algorithm. Nevertheless, none of the aforementioned techniques make use of labelling data, leading to subpar performance.

Supervised methods: To develop differentiated and compressed binary codes based on deep autoencoders, Cao et al. [18] introduced a brand-new supervised cross-modal hashing technique dubbed Correlated Autoencoder Hashing (CAH). By creating a framework for linear classification, xu et al. [19] created Discrete Cross-modal Hashing (DCH). Lin et al. [20] suggested SePH (Semantics Preserving hashing), a two-step supervised hashing technique for cross-view retrieval, to capture more complicated data structures. A semantic-preserving hashing paradigm was developed by Lin et al. [21] Before approximating semantic similarity to the hash code that will be learned by reducing the KL divergence, they turn semantic similarity into a probability distribution. The aforementioned method, however, makes heavy use of hand-crafted features and distinct algorithms for learning hash codes from features. Deep learning techniques for cross-modal hashing have also drawn attention because of how powerful deep neural networks are. Deep Cross-Modal Hashing (DCMH), a new cross-modal hashing technique proposed by Jiang et al. [10], is an end-to-end framework that combines feature learning and hash code learning into one framework. One deep neural network is used by DCMH, an end-to-end learning framework, to execute feature learning from scratch for each modality.

2.2 Trajectory Feature Learning

Feng et al. [5] proposed a method using Recurrent Neural Networks (RNN) and attention mechanism to predict human mobility behavior. In this method, trajectory data is represented as sequences, and modeled and predicted through attention mechanism and RNN. Li et al. [6] proposed a method using Convolutional Neural Networks (CNN) and sequence-to-sequence model to model and predict pedestrian trajectories. The method encodes trajectories using CNN and then predicts future trajectories using sequence-to-sequence model. Tong et al. [4] proposed a method based on Optimal Transport (OT) network for modeling cell motion and trajectories. The method models and analyzes trajectories by learning the flow distribution in the dynamic optimal transport network. However, although these methods can model and extract features of trajectories well, they do not involve cross-modal retrieval.

In order to build more accurate similarity matrices and boost retrieval accuracy, this study focuses on exploiting the classification information of multimodal trajectory data to direct the model to learn more discriminative feature representations of various classes of data.

3 Preliminary

We initially describe the terminology and formulas used in this work in this part before defining the problem in more depth.

3.1 Notation

Bold lowercase characters, such as a, denote vectors in this work while bold capital letters, such as A, denote matrices. In addition, X_{ij} signifies the element of the i-th row, j-th column of matrix X. The j-th column of the matrix X is denoted by $X_{*j}.O = \{o_i\}_{i=1}^{n}$denotes a set containing n elements.

3.2 Problem Definition

However, in this study, we only focus on the task of retrieval between two modalities. Our method can be extended to retrieve between several modalities. Consider that each of the n training samples has one image, coordinate and label, denoted as $O = \{o_i\}_{i=1}^{n}$. We use P and T to denote two modalities, image modality and coordinate modality. $P = \{p_i\}_{i=1}^{n}$ represents the modality of the image, where x_i can be the original pixels of the image or handcrafted features. $Y = \{y_i\}_{i=1}^{n}$ represents the modality of the coordinates, where y_i is the label information associated with image i, i.e., the coordinate information. $S_{N\times N}$ is the training similarity matrix of the data. If x_i and y_i are similar, $S_{ij} = 1$, otherwise $S_{ij} = 0$. For example, if image x_i and coordinate y_i belong to the same class, it means they are similar.

Only the inputs P, T, and the similarity matrix S of the training data are needed because the model is an end-to-end framework. The goal of the subsequent model training is to acquire knowledge of the $h_p(\cdot)$ and $h_t(\cdot)$ hash functions for the two modalities. For queries and database instances in both modalities, the learnt hash functions can be utilised to construct c-bit hash codes, with c designating the length of the final generated hash code. If $S_{ij} = 1$, then the distance between the binary hash codes of picture p_i and coordinate t_j in Hamming space should be minimal, and we use Hamming distance as the distance metric in this case to gauge how similar the modalities are. On the other hand, it should be far apart.

4 Proposed Model

This part focuses on the model's specifics, including the model and the formulation's introduction. Our suggested model is depicted in Fig. 2. It can be observed that the model is made up of two primary branching models that are trained end-to-end: the image feature extraction component and the coordinate feature extraction part.

4.1 Feature Extraction

Due to the enormous success of deep learning networks, we use them to extract reliable image and text hash representations for the image modality, continuously replacing the original image and text hash codes with the learnt representations. To extract picture features, we use traditional deep convolutional neural networks (CNNs), including ResNet [22] and VGG [23]. Because ResNet50 uses residual connections and converges more quickly during training than VGG, it enables training deeper networks without running into the gradient disappearance problem. This allows us to propagate gradients more effectively and prevents the gradient disappearance problem. With the help of Resnet50, which has been trained on ImageNet, we are able to classify images accurately. As a high-level semantic representation of the original image, we create a 2048-dimensional feature vector from the final Bottleneck block of Stage 4. We then change the number of hidden nodes in the final fully connected layer with c so that the length of its output vector equals the length of our final feature vector. We initialise only its parameters, leaving the parameters of the other layers unchanged, using a lower learning rate because the parameters of the other layers have already been trained to a higher level, and replacing the number of hidden nodes of the final fully connected layer with c so that the output vector length is the length of our final hash code. We specifically select the final layer's output to serve as the picture hash representation.

For the trajectory coordinate modalities, since the Word2Vec [24] model could not be used to obtain the feature vectors of the trajectory coordinates, we chose to use the k-dimensional binary code of the WGS84 coordinates directly for

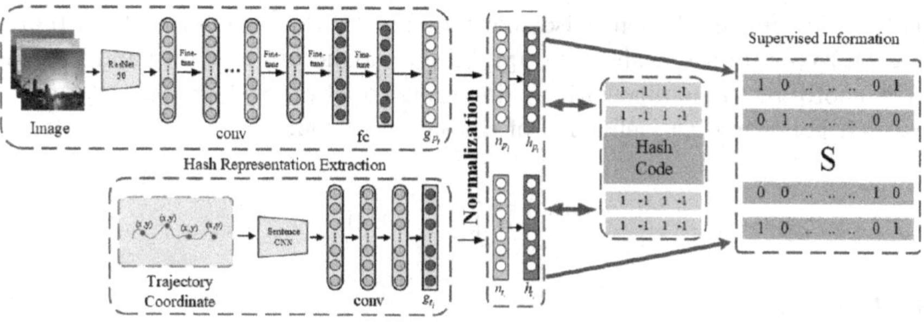

Fig. 2. According to the design of the model we've proposed, the hash representation extraction module's job is to collect a high-level semantic representation of coordinates and images before normalising them to produce the hash code.

the representation. Then, the trajectory coordinate matrix is fed to the convolutional layer, where a sentence CNN [25] is used to generate a high-level semantic representation of the original trajectory coordinates. The result is a hash code high-level semantic representation of the c bits of the trajectory coordinates. We do this by substituting c for the number of hidden nodes in the final fully connected layer. The process described above can be summed up as follows:

$$\begin{cases} g_{p_i} = f^p(p_i; \theta_p) \\ g_{t_i} = f^t(t_i; \theta_t) \end{cases}$$

where $f^p(\cdot)$, $f^t(\cdot)$ are the encoders of image and trajectory, respectively, and θt and θp are the parameters of image model and trajectory coordinate model learning, respectively. g_{p_i} and g_{t_i} are the c-bit hash representations of the i-th image and trajectory coordinates, respectively.

4.2 Hash Representation Normalization

Most studies [10,26] often employ inner product to gauge how comparable the hash representations of visual and textual modalities are to one another after collecting their c-bit hash representations. The hash representation with the greater value will be used to evaluate similarity if the size of these multimodal hash representations varies significantly. This could have a negative impact on how similar the target hash representations are computed, which would result in subpar retrieval performance. We normalise the learned hash representations by compressing them to a range of −1 to 1 in order to solve this issue. As a result, both modalities' hash representations serve the same purpose in successfully reducing the impact of modalities with bigger magnitudes. The following formulation represents the normalising operation:

$$\begin{cases} n_{p_i} = \dfrac{g_{p_i}}{\left\| g_{p_i} \right\|_2} \\ n_{t_i} = \dfrac{g_{t_i}}{\left\| g_{t_i} \right\|_2} \end{cases}$$

Finally, we adopt an element-wise sign function for the normalized hash representation to obtain the hash codes hpi, hti for the picture modality and the trajectory coordinate modality with the following equations: $h_{p_i} = sign(n_{p_i}), h_{t_i} = sign(n_{t_i})$. Following is a definition of the sign function:

$$\text{sign}(x) = \begin{cases} 1, & x \geq 0 \\ -1, & x < 0 \end{cases}$$

4.3 Loss Function

The objective of the model we propose is to learn high-level semantic representations of pictures and trajectories, i.e., to learn a common space in which modal samples from the same classes should be similar. We aim to minimise the discriminative loss in the common representation space in order to understand the characteristics of multimodal data. The objective function of the suggested model is then presented, and is defined as follows.

$$\begin{aligned} \mathfrak{J} = & \frac{1}{n^2} \sum_{i,j=1}^{n} (log(1 + e^{\Psi_{ij}}) - S_{ij}\Psi_{ij}) \\ & + \gamma \frac{1}{n^2} (\left\| B^{(p)} - P \right\|_F^2 + \left\| B^{(t)} - T \right\|_F^2) \\ & + \theta \frac{1}{n} \left\| P - T \right\|_F \\ s.t. \quad & B^{(p)} \in \{-1, 1\}^{c \times n} \\ & B^{(t)} \in \{-1, 1\}^{c \times n} \\ & B \in \{-1, 1\}^{c \times n} \\ & B = B^{(p)} = B^{(t)} \end{aligned} \tag{1}$$

where $\Psi_{ij} = \frac{1}{2} COS(P_{*i}, T_{*i})$, $P_{*i} = f^p(p_i; \theta_p)$, $T_{*j} = f^t(t_j; \theta_t)$, B^p is the binary hash code of the image, B^t is the binary hash code of the trajectory coordinates, and γ and θ are the hyperparameters. The first term $\frac{1}{n^2} \sum_{i,j=1}^{n} (log(1 + e^{\Psi_{ij}}) - S_{ij}\Psi_{ij})$ in (1) is the negative log likelihood of cross-modal similarity, and the likelihood function is defined as follows:

$$p(S_{ij} | p_i, t_j) = \begin{cases} \delta(\Psi_{ij}), & \text{if } S_{ij} = 1 \\ 1 - \delta(\Psi_{ij}), & \text{otherwise} \end{cases}$$

where $\delta(\Psi_{ij}) = \frac{1}{1+e^{-\Psi_{ij}}}$ is the sigmoid function, thus it is simple to discover that maximising the likelihood corresponds to minimising this negative log-likelihood function, which can make the similarity between p_i and t_j larger at $S_{ij} = 1$ and smaller at $S_{ij} = 0$. Optimizing the first term in (1) keeps the similarity between image features and trajectory features.

The second term $\left\| B^{(p)} - P \right\|_F^2 + \left\| B^{(t)} - T \right\|_F^2$ in (1), where $B^{(p)} = sign(n_p)$, $B^{(t)} = sign(n_t)$, n_p, n_t are the hash codes after feature normalization, and P and

T are the features after normalization process, i.e., n_p and n_t, respectively. Here, P and T are used as successive substitutions of $B^{(p)}$ and $B^{(t)}$. Thus, minimizing this term maximizes the modal similarity.

The third term $\|P - T\|_F$ in (1), We reduce the distance between the representations of all picture text pairs in order to eliminate cross-modal disparities.

4.4 Out-of-Sample Extension

During the training phase, text and images are each taught θ_p, θ_t in turn. After the model has been trained, the high-level semantic representation of the modalities is obtained directly using the $f^p(\cdot)$ and $f^t(\cdot)$ functions. Following normalisation, we then combine these two functions with the sign function to produce the corresponding modality's hash code and complete the retrieval task. As long as a point is observed, we may determine the hash code of one of its modalities (image or coordinate) even if it is not part of the training set. When an image modality p_i is provided, forward propagation can be used to determine its hash code b_{p_i}:

$$b_{p_i} = h_p(p_i) = sign(n_{p_i}) = sign(\frac{f^p(p_i;\theta_p)}{\|f^p(p_i;\theta_p)\|_2})$$

Similarly, when given a trajectory coordinate ti, we can obtain its binary hash code b_{t_i}:

$$b_{t_i} = h_t(t_i) = sign(n_{t_i}) = sign(\frac{f^t(t_i;\theta_t)}{\|f^t(t_i;\theta_t)\|_2})$$

As the query points have one modality and the points in the retrieved database have a different modality, our model may use it for cross-modal retrieval.

5 Experiment

Using a dataset of picture coordinates, we ran experiments to verify the efficacy of our suggested model. Using the open source deep learning framework Pytorch, the trials were carried out on four NVIDIA 1080 ti graphics cards.

5.1 Dataset

The experiments were performed on our self-made dataset, and a total of 23,000 pieces of data were obtained by calling the Baidu Maps api, of which 13,000 were from Beijing and 10,000 from Shanghai. Each data includes the traffic road condition picture and the corresponding coordinates. In order to make the images generalize better, we randomly color-lighten and darken the images in the range of 0.6 to 1.2 of the original image brightness, and normalize the images using the mean and variance learned from the ImageNet dataset. For the processing of the trajectory coordinates, we uniformly transform the WGS84 coordinates into 106-dimensional vectors and each point is labeled with a unique 24-dimensional vector label. This dataset is used as a criterion for performance evaluation of our proposed model.

5.2 Evaluation Metric

For our self-made dataset, we used 2500 of these points as the query dataset, 11000 points as the training dataset, and 20500 points as the retrieval dataset (the queried points).

Mean average precision (MAP) [27], a commonly used performance evaluation criterion in cross-modal retrieval studies, is a measure of the accuracy of the Hamming ranking technique for hash-based retrieval. The MAP metric jointly includes ranking information and precision. We assessed two distinct cross-modal retrieval tasks [7,8,28] in our experiments: 1) Using images to query trajectory coordinates (Image2Coord) and 2) Using trajectory coordinates to query images (Coord2Image).

5.3 Baselines

To evaluate the validity of our model, we compared our method with state-of-the-art methods that have been widely used as benchmarks in other literatures. The following four baselines are included: CCA [29], DCH [19], DCMH [10], SSAH [26].

- **CCA**: This is a crucial method for discovering a shared subspace for heterogeneous data. Through the use of a linear analysis technique, the correlation between two sets of data is established.
- **DCH**: By creating a linear classification framework, discrete cross-modal hashing is added in order to preserve discrete limitations and produce identifiable binary codes.
- **DCMH**: With the use of deep learning networks, a cross-modal retrieval framework from beginning to end has been developed.
- **SSAH**: To increase the semantic relevance and coherence across the representations of various modalities, the authors used two adversarial networks. Additionally, they used a self-supervised semantic network to find multi-label annotations that represent high-level semantic information.

Meanwhile, we applied pre-trained ResNet50 and VGG19 images for feature extraction, respectively. Notably, since all the proposed frameworks are not fully adaptable to our proposed task, we tried our best to adapt the models and report their best performance in the paper. Also, our model uses the ResNet50 model pre-trained on ImageNet [30].

5.4 Result Comparison

We examined the impact of various hash code lengths on the retrieval accuracy to illustrate the efficacy of our methodology. (i.e., 16-bit, 32-bit, 64-bit, and 128-bit) and two image feature extraction methods (i.e., ReNet50 and VGG19). Table 1 and Table 2 show the image feature extraction using ResNet50 and VGG19, respectively. We come to the following conclusions based on the comparison: 1) Deep learning models are more accurate than shallow learning models. The

Table 1. The MAP performance comparison on our dataset between our suggested model and the most recent baselines. The best outcomes of image feature extraction using ResNet50 are shown in bold.

Mothod	Manual Mult-modal DataSet							
	Image ⇒ Coordinate				Coordinate ⇒ Image			
	16b	32b	64b	128b	16b	32b	64b	128b
CAA	0.560	0.563	0.560	0.558	0.562	0.567	0.559	0.540
DCH	0.651	0.662	0.670	0.665	0.653	0.657	0.662	0.650
DCMH	0.741	0.745	0.755	0.749	0.748	0.758	0.752	0.743
SSAH	0.762	0.769	0.767	0.759	0.767	0.762	0.761	0.752
Ours	**0.772**	**0.774**	**0.771**	**0.779**	**0.775**	**0.775**	**0.779**	**0.766**

Table 2. The MAP performance comparison on our dataset between our suggested model and the most recent baselines. The best findings from image feature extraction using VGG19 are shown in bold.

Mothod	Manual Mult-modal DataSet							
	Image ⇒ Coordinate				Coordinate ⇒ Image			
	16b	32b	64b	128b	16b	32b	64b	128b
CAA	0.551	0.543	0.547	0.541	0.553	0.545	0.544	0.546
DCH	0.633	0.640	0.659	0.664	0.634	0.641	0.658	0.640
DCMH	0.728	0.731	0.735	0.726	0.726	0.734	0.740	0.737
SSAH	0.745	0.751	0.753	0.749	0.747	0.753	0.732	0.742
Ours	**0.750**	**0.757**	**0.767**	**0.760**	**0.751**	**0.755**	**0.765**	**0.756**

learned hash codes can be closely matched to the image features because the deep learning framework blends modal feature learning and hash code learning into an end-to-end framework, which lessens the accuracy loss caused by segmental learning. 2) In the deep learning framework, our model exhibits better accuracy compared to DCMH and SSAH because we build a model for trajectory coordinates a feature extraction model, and compared to DCMH and SSAH, we normalize the learned features in the process of transforming the modal high-level semantic representations learned by the model into hash codes, which attenuates the effect of larger numerical modalities on the computational similarity and further improves the performance.

6 Conclusion

In this article, we suggest a brand-new model and multimodal retrieval task for trajectory coordinates. This is the first paper that, to the best of our knowledge, suggests cross-modal retrieval of trajectory coordinates and traffic visuals.

Our suggested model is a complete framework that learns both a modality's feature representation and hash code. Research on our own traffic multimodal dataset demonstrates that other cutting-edge baselines provide the basis for our methodology.

References

1. Yao, D., Cong, G., Zhang, C., et al.: Computing trajectory similarity in linear time: a generic seed-guided neural metric learning approach. In: International Conference on Data Engineering (2019)
2. Yao, D., Zhang, C, Zhu, Z., et al.: Trajectory clustering via deep representation learning. In: International Joint Conference on Neural Network (2017)
3. Li, X., Zhao, K., Cong, G., et al.: Deep representation learning for trajectory similarity computation. In: International Conference on Data Engineering (2018)
4. Tong, A., Huang, J., Wolf, G., et al.: Trajectorynet: a dynamic optimal transport network for modeling cellular dynamics. In: Proceedings of the International Conference on Machine Learning. PMLR (2020)
5. Feng, J., Li, Y., Zhang, C., et al.: DeepMove: predicting human mobility with attentional recurrent networks. In: The Web Conference (2018)
6. Li, C., Zhang, Z., Lee, W.S., et al.: Convolutional sequence to sequence model for human dynamics. In: Proceedings of the Proceedings of the IEEE Conference on Computer Vision and Pattern Recognition (2018)
7. Wang, B., Yang, Y., Xu, X., et al.: Adversarial cross-modal retrieval. ACM Multimedia (2017)
8. Peng, Y., Huang, X., Qi, J.: Cross-media shared representation by hierarchical learning with multiple deep networks. In: Proceedings of the IJCAI (2016)
9. Peng, Y., Qi, J., Huang, X., et al.: CCL: cross-modal correlation learning with multigrained fusion by hierarchical network. IEEE Trans. Multimedia **20**(2), 405–20 (2017)
10. Jiang, Q.-Y., Li, W.-J.: Deep cross-modal hashing. In: Proceedings of the IEEE Conference on Computer Vision and Pattern Recognition (2017)
11. Cao, Y., Long, M., Wang, J., et al.: Deep visual-semantic hashing for cross-modal retrieval. In: Proceedings of the 22nd ACM SIGKDD International Conference on Knowledge Discovery and Data Mining (2016)
12. Zheng, F., Tang, Y., Shao, L.: Hetero-manifold regularisation for cross-modal hashing. IEEE Trans. Pattern Ana. Mach. Intell. **40**(5), 1059–71 (2016)
13. Sun, L., Ji, S., Ye, J.: A least squares formulation for canonical correlation analysis. In: Proceedings of the 25th International Conference on Machine Learning (2008)
14. Weiss, Y., Torralba, A., Fergus, R.: Spectral hashing. In: Advances in Neural Information Processing Systems, vol. 21 (2008)
15. Ding, G., Guo, Y., Zhou, J.: Collective matrix factorization hashing for multimodal data. In: Proceedings of the IEEE Conference on Computer Vision and Pattern Recognition (2014)
16. Liu, H., Ji, R., Wu, Y., et al.: Cross-modality binary code learning via fusion similarity hashing (2017)
17. Wang, D., Cui, P., Ou, M., et al.: Learning compact hash codes for multimodal representations using orthogonal deep structure. IEEE Trans. Multimedia **17**(9), 1404–16 (2015)

18. Cao, Y., Long, M., Wang, J., et al.: Correlation autoencoder hashing for supervised cross-modal search. In: Proceedings of the 2016 ACM on International Conference on Multimedia Retrieval (2016)
19. Xu, X., Shen, F., Yang, Y., et al.: Learning discriminative binary codes for large-scale cross-modal retrieval. IEEE Trans. Image Process. **26**, 2494–2507 (2017)
20. Lin, Z., Ding, G., Hu, M., et al.: Semantics-preserving hashing for cross-view retrieval. In: Proceedings of the IEEE Conference on Computer Vision and Pattern Recognition (2015)
21. Lin, Z., Ding, G., Han, J., et al.: Cross-view retrieval via probability-based semantics-preserving hashing. IEEE Trans. Cybernet. **47**, 4342–4355 (2017)
22. He, K., Zhang, X., Ren, S., et al.: Deep residual learning for image recognition. In: Proceedings of the IEEE Conference on Computer Vision and Pattern Recognition (2016)
23. Simonyan, K., Zisserman, A.: Very deep convolutional networks for large-scale image recognition. arXiv preprint arXiv:1409.1556 (2014)
24. Mikolov, T., Sutskever, I., Chen K., et al.: Distributed representations of words and phrases and their compositionality. arXiv: Computation and Language (2013)
25. Kim, Y.: Convolutional neural networks for sentence classification. Cornell University - arXiv, (2014)
26. Li, C., Deng, C., Li, N., et al.: Self-supervised adversarial hashing networks for cross-modal retrieval (2018)
27. Liu, W., Mu, C., Kumar, S., et al.: Discrete graph hashing. In: Neural Information Processing Systems (2014)
28. Wang, K., Yin, Q., Wang, W., et al.: A comprehensive survey on cross-modal retrieval. arXiv preprint arXiv:1607.06215 (2016)
29. Gong, Y., Lazebnik, S.: Iterative quantization: a procrustean approach to learning binary codes. In: Computer Vision and Pattern Recognition (2011)
30. Deng, J, Dong, W., Socher, R., et al.: ImageNet: a large-scale hierarchical image database. In: Computer Vision and Pattern Recognition (2009)
31. Castrejon, L., Aytar, Y., Vondrick, C., et al.: Learning aligned cross-modal representations from weakly aligned data. In: Proceedings of the IEEE Conference on Computer Vision and Pattern Recognition (2016)
32. Chen, Y., Wang, L., Wang, W., et al.: Continuum regression for cross-modal multimedia retrieval. In: Proceedings of the 2012 19th IEEE International Conference on Image Processing. IEEE (2012)
33. Feng, F., Wang, X., Li, R.: Cross-modal retrieval with correspondence autoencoder. In: Proceedings of the 22nd ACM International Conference on Multimedia (2014)
34. Hu, P., Peng, X., Zhu, H., et al.: Learning cross-modal retrieval with noisy labels. In: Proceedings of the IEEE/CVF Conference on Computer Vision and Pattern Recognition (2021)
35. Hu, Y., Jin, Z., Ren, H., et al.: Iterative multi-view hashing for cross media indexing. In: Proceedings of the 22nd ACM International Conference on Multimedia (2014)
36. Jia, Y., Salzmann, M., Darrell, T.: Learning cross-modality similarity for multinomial data. In: Proceedings of the 2011 International Conference on Computer Vision. IEEE (2011)
37. Li, D., Dimitrova, N., Li, M., et al.: Multimedia content processing through cross-modal association. In: Proceedings of the Eleventh ACM international Conference on Multimedia (2003)

C^3BR: Category-Aware Cross-View Contrastive Learning Framework for Bundle Recommendation

Fuyong Xu, Zhenfang Zhu, Peiyu Liu(✉), and Ru Wang(✉)

School of Information Science and Engineering,
Shandong Normal University, Jinan, China
liupy@sdnu.edu.cn, ruwang0929@gmail.com

Abstract. Bundle recommendation aims to suggest a series of items that users are interested in, and MealRec is a novel bundle recommendation dataset recently introduced. While current methods utilize user, bundle, and item interaction data to create user and bundle representations, the connection between item view and bundle view has not received sufficient attention. This study addresses this gap by examining the relationship between item view and bundle view. To capture interactive cooperative relations, we present the Category-aware Cross-view Contrastive learning framework (C^3BR). We participated in the workshop of BundleRS 2023 and the team name is Reset. Our experiments on MealRec dataset prove the superiority of C^3BR over baseline method.

Keywords: Bundle Recommendation · Contrastive Learning · Category · MealRec

1 Introduction

MealRec [17] is a bundle recommendation dataset that is based on meal data and user-meal interaction data. It collects recipes and related reviews from Allrecipes.com, which is one of the most popular food-sharing sites worldwide. A recipe is an essential representation of food, as it contains detailed descriptions such as food names, ingredients, instructions, pictures, and more. Bundle recommendation aims to suggest a series of items that users are interested in. Providing a series bundles in place of single products can vastly enhance the user's satisfaction in various online shopping websites. Additionally, platform that utilizes bundles as a marketing approach can elevate its sales revenue and entice more customers to the platform. Therefore, the development of effective bundle recommendation methods has garnered significant attention from both academic and industrial circles.

Existing studies that aim to model user preference can be classified into two main approaches: the bundle view and the item view. The user preferences in a bundle view are represented by the interactions between users and bundles, as shown in the user-bundle graph (U-B graph), while the item view describes the attribute of users and bundle from item level (U-I graph and B-I graph).

A. El Abbadi et al. (Eds.): DASFAA 2023 Workshops, LNCS 13922, pp. 194–203, 2023.
https://doi.org/10.1007/978-3-031-35415-1_14

By considering both the bundle view and the user-bundle graph, we can gain a comprehensive understanding of users' interests, which can be leveraged to build more effective recommendation systems from multiple perspectives. However, detailed differences exist between bundle and item views that have yet to be addressed in prior studies. The bundle view focuses on the behavior similarity of customers, while the item view concentrates on the relationship between the bundle and the user's preferences. Therefore, the collaboration of these two perspectives is essential to create effective bundle recommendations. BGCN [3] utilizes individual view representation and preference prediction, which are subsequently fused together to make predictions. While this approach gains better results than previous researches, it only used interactive cooperative relations to predict and does not integrate such relation signals directly for representation learning. CrossCBR [21] was proposed to capture the mutual enhancement between different views of meal items. However, it is important to note that the category feature of each meal is also crucial for bundle recommendation. To build an effective recommendation system, it is crucial to appropriately model the collaborative associations and category features across various item views and bundles.

Inspired by CrossCBR [21], Category-aware Cross-view Contrastive learning framework (C^3BR) is proposed to recommend varieties items. The cooperative associations between different item views can be effectively captured using C^3BR, which leverages cross-view contrastive learning and mutual fusion of category representations. We choose LightGCN [12] as the backbone for representation learning. Additionally, we incorporate the category features of meals in order to capture category-aware cross-view representations. In training process, Bayesian Personalized Ranking (BPR) [24] and contrastive loss [11] were used to optimize our method.

The main contributions of our work are:

- We propose the C^3BR, the Category-aware Cross-view Contrastive learning framework to learn the interactive cooperative relations from different views.
- The performance of our approach surpasses that of the baseline on MealRec [17], which is a newly introduced dataset for meal recommendations.

2 Methodology

This section provides an overview of the bundle recommendation task and presents our proposed model, C^3BR, as depicted in Fig. 1. Our model is designed to address the bundle recommendation task by taking into account both the user behaviors and the characteristics of the bundles and items. We will provide more details about our proposed model and its components in the following sections.

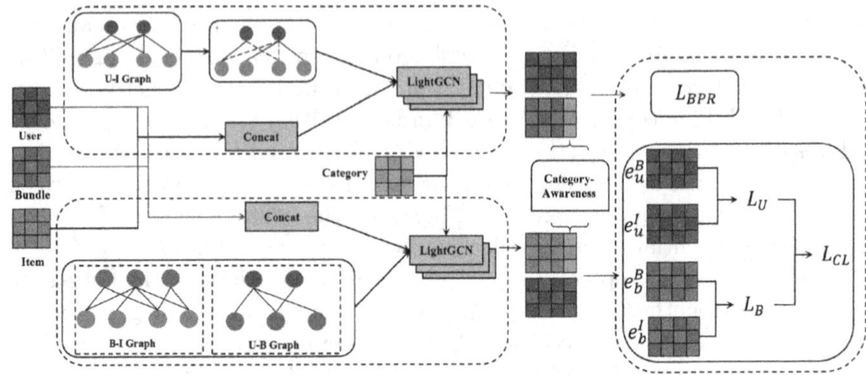

Fig. 1. The whole framework of C^{3}BR.

2.1 Problem Definition

Given a set of users $U = \{u_1, u_2, ..., u_J\}$, a set of bundles $B = \{b_1, b_2, ..., b_K\}$, a set of categories $C = \{c_1, c_2, ..., c_L\}$ and a set of items $I = \{i_1, i_2, ..., i_M\}$, where J, K, L, and M are the number of users, bundles, categories, and items, respectively. The user-bundle interactions, user-item interactions, and bundle-item interactions matrix are denotes as $\text{UB}_{J \times K}$, $\text{UI}_{J \times M}$, and $\text{BI}_{K \times M}$, respectively. The objective of the bundle recommendation task is to train a model using historical data $\{\text{UB}, \text{UI}, \text{BI}, C\}$ and make predictions on new user-bundle interactions that were not seen during training.

2.2 Representation Learning Module

This section discusses the representation learning module, which is designed to extract meaningful representations from the bundle and item views. The purpose of this module is to transform the raw data into a lower-dimensional representation that captures important features of the items and users in a recommendation system. By learning these representations, the recommendation system can better capture the preferences and behavior of users, leading to more accurate and effective recommendations.

To derive user and bundle representations features (bundle-views), the first step involves creating a U-B interaction graph, which is also referred to as the U-B. This graph is constructed using the user-bundle interaction matrix UB. Following that, LightGCN [12] was adopted to acquire the user and bundle representations. More precisely, we perform information propagation on the U-B interaction graph, with k-th layer's propagation represented as:

$$e_{u/b}^{B_k} = \sum_{b \in N_u} \frac{1}{\sqrt{|N_u|}\sqrt{|N_b|}} e_b^{B_{k-1}}. \tag{1}$$

Here, $e_u^{B_k}$ and $e_b^{B_k} \in \mathbb{R}^d$ represent the k-th layer's features that has been spread to u and b, where d denotes the hidden size and B_k refers to the bundle view features of k-th layer. N_u and N_b means the set of neighboring nodes connected to u and b in the user-bundle interaction graph.

In this paper, LightGCN [12] is employed to extract connection features from user-bundle interaction information. Then, we join all K-layers' embedding together the features obtained from the adjacent nodes with different depths. The final representations of bundle-view are denoted as:

$$e_{u/b}^{B} = \sum_{k=0}^{K} e_u^{B_k}. \tag{2}$$

We constructed the U-I and B-I interaction graphs from UI and BI to obtain the features of user and bundle. Furthermore, LightGCN [12] is applied to model the representation of user and item. Using the bundle-item interaction graph, we propagate information and refer to the information propagation at the k-th layer as follows:

$$e_{\{i/u\}}^{I_k} = \sum_{\{i,u\} \in N_{\{i,u\}}} \frac{1}{\sqrt{|N_u|}\sqrt{|N_i|}} e_b^{I_{k-1}}. \tag{3}$$

In this equation, $e_u^{I_k}$ and $e_i^{I_k} \in \mathbf{R}^d$ denote the k-th layer's features that has been spread to u and i, where d represents hidden size and I_k refers to the item view of the k-th layer. N_u and N_i are the neighbors of u and i in U-I interaction graph. We describe the process of information propagation using the following formulation:

$$\begin{aligned} e_u^I &= \sum_{k=0}^{K} e_u^{I_k} \\ e_i^I &= \{\sum_{k=0}^{K} e_i^{I_k}, e^c\}, \end{aligned} \tag{4}$$

where e_u^I and e_i^I are the user and item representations of item-view, e_c means category aware representations. Since each meal has its own category, the relationship between a meal's category and its item components can be leveraged to enhance the representation of the meal item. e_c can be formulated as:

$$e_c = \begin{bmatrix} e_{c_{00}} & \cdots & e_{c_{0M}} \\ \vdots & \ddots & \vdots \\ e_{c_{l0}} & \cdots & e_{c_{lM}} \end{bmatrix} \in \mathbb{R}^{l \times |C| \times d}, \tag{5}$$

In the formulation, l denotes the meal length, and d represents the size of the hidden category representation. Utilizing e_i^I and bundle-item interaction graph, we can obtain e_b^I by performing average pooling, which is represented as:

$$e_b^I = \frac{1}{|N_b|} \sum_{i \in N_b}^{K} e_i^I. \tag{6}$$

Items sets included in a particular bundle b is denoted by N_b.

2.3 Cross-View Contrastive Learning

The purpose of comparative learning is to narrow the distance between relevant features in different perspectives and to widen the distance between unrelated features. The contrast loss can be directly applied if multiple views are naturally present for each object. Proper data augmentation method can remove not only the limit of the comparison study but also improve the generalization ability of the model to latent random noise variable. Similar to CrossCBR [21], we also employ two simple ways to enhance the data: **graph-based and embedding-based** augmentations.

To obtain the extra graph data of graph structure, we use graph augmentation method to produce extra graph data by modifying the structure of raw graph data. This paper employ edge dropout to drop the edge of raw graph data randomly. Edge dropout operates on the principle of maintaining the fundamental local structure of a graph while removing certain edges. Thus, it is possible to increase the trained representations' robustness to cope with a certain amount of noise.

Unlike the above Augmentation method, which is only applicable to graphical data, embedding is more universal and applicable to all kinds of deep representation learning methods. The primary goal of our method is to modify the embedding, no matter how the embeddings are obtained. We make use of the message dropout technique, which randomly drops a large number of nodes in the network with a high dropout rate of ρ.

The contrast loss is employed to enhance the performance of a model by enabling comparison between various perspectives or viewpoints. Every viewpoint encapsulates a unique facet of the user's preferences, and effective coordination between these viewpoints is crucial to maximize the overall modeling capability. InfoNCE [11] is used to model the cross-view representation. To be more specific, the comparison loss facilitates the simultaneous alignment of the same user or bundle across multiple viewpoints, as well as the differentiation of individual users or bundles:

$$\begin{aligned} L_U &= -\frac{1}{|U|}\sum_{u\in U}\log\frac{e^{\text{sim}(e_u^B,e_u^I)/\tau}}{\sum_{v\in U}e^{\text{sim}(e_u^B,e_v^I)/\tau}} \\ L_B &= -\frac{1}{|B|}\sum_{b\in B}\log\frac{e^{\text{sim}(e_b^B,e_b^I)/\tau}}{\sum_{p\in B}e^{\text{sim}(e_b^B,e_p^I)/\tau}}, \end{aligned} \tag{7}$$

where L_U and L_B denote the contrastive loss of users and bundles respectively. Sim() means the similarity measure function, we employ cosine similarity function as shown in Eq. 8. τ is the temperature hyper-parameter. To construct negative pairs within a batch, we adopt the CrossCBR [21] approach. The final contrastive loss is $L_{\text{CL}} = L_U + L_B$.

$$Sim(a,b) = \frac{a \cdot b}{||a|| \times ||b||} \tag{8}$$

2.4 Predict&Train

We first calculate the item-view and bundle view predictions by taking the inner product, and then combine them by adding to produce the overall prediction.

$$\hat{y}_{u,b} = f(u,b) = {e_u^B}^\top e_b^B + {e_u^I}^\top e_b^I. \tag{9}$$

The primary loss employed is the standard Bayesian Personalized Ranking (BPR) [24] loss:

$$L_{\mathrm{BPR}} = - \sum_{(u,b,b') \in (U,B)} \ln \sigma(\hat{y}_{u,b} - \hat{y}_{u,b'}), \tag{10}$$

$\sigma()$ means sigmoid function. Final loss is formulated as:

$$L = \lambda_1 L_{\mathrm{BPR}} + \lambda_2 L_{\mathrm{CL}} + (1 - \lambda_1 - \lambda_2) \left|\left|\theta\right|\right|_2^2. \tag{11}$$

We use hyperparameters λ_1 and λ_2, which both belong to the interval $[0,1]$, to control the relative weights of three terms in our objective function.

3 Experiments

3.1 Experiment Setting

To demonstrate the effectiveness of C^{3}BR, we evaluate its performance on the MealRec dataset. The set of MealRec datasets that were released contains two datasets, MealRec-77 with a user food interaction density of 0.77%, and MealRec-17 with a user food interaction density of 0.17%. Recall@K and NDCG@K are used to evaluate our method, where $K \in \{20, 40, 80\}$. We can optimize our model using Xavier normal initialization [10] and the Adam optimizer [15]. All models are trained with Pytorch 1.7.1 or NVIDIA RTX3090. The hyperparameters of our experiments show in Table 1.

Table 1. Hyperparameters of our methods.

Parameter	MealRec77	MealRec17
batch_size	1024	512
embedding_size	128	64
num_layers	2	4
learning_rate	1e−4	1e−4
item_ratios	0.1	0.1
bundle_ratios	0.3	0.3
bundle_agg_ratios	0.1	0.1
l2_regs	1e−4	1

Table 2. Experimental results of our method.

	MealRec-17					
	Recall@20	NDCG@20	Recall@40	NDCG@40	Recall@80	NDCG@80
BGCN	0.0841	0.0402	0.12	0.0483	0.181	0.0599
Our	**0.1474**	**0.0868**	**0.1881**	**0.0962**	**0.2465**	**0.1073**
Improve	0.0630	0.047	0.068	0.048	0.066	0.047
	MealRec-77					
	Recall@20	NDCG@20	Recall@40	NDCG@40	Recall@80	NDCG@80
BGCN	0.2267	0.1415	0.3131	0.1629	0.4095	0.1854
Our	**0.3654**	**0.2342**	**0.4601**	**0.2604**	**0.5812**	**0.2886**
Improve	0.1387	0.0927	0.1470	0.0975	0.1717	0.1032

3.2 Results

We compare C^{3}BR's overall recommendation performance on MealRec as shown in Table 2:

The best performing results are bold; Improve measures the relative improvements of C^{3}BR over BGCN. From the above Table 2, our method has achieved good results, and has improved objectively compared with the baseline method, especially on the MealRec-77 dataset. In the process of our experiment, we found that our method is not sensitive to the MealRec-17 dataset. When the density of interactive data decreases, our method improvement is limited. Graph-based methods have shown promising results in bundle recommendation. This is because they are able to effectively capture the features of graph structure and multi-hop cooperative information. Our method is also based on the graph, and it leverages graph characteristics and multi-hop collaborative information dissemination to achieve favorable performance.

4 Related Work

In recent years, graph-based approaches have gained significant popularity owing to their excellent performance in modeling complex interactions among users and items, particularly with the use of graph neural networks. Wang et al. proposed NGCF [26], which is based on the user-item interaction matrix, and uses graph convolutional networks as graph learning modules. Following NGCF, He proposed to eliminate some of the redundant modules (such as the non-linear characteristic transform and the activation level) from NGCF, leading to LightGCN [12]. LightGCN has been successful in a variety of recommendation tasks [8].

Bundle recommendation is specifically tailored to address a specific type of recommendation scenario in which the recommended object consists of a collection of items that are linked by a common topic. The original researches neglect the associated entries in bundle, only represent the bundle with an id [25]. Several studies have recognized the significance of related items, and as a result, various techniques have been proposed by researchers to capture additional user-item interactions. These methods include EFM [2] and DAM [5].With the growing popularity of GNN-style recommendation approaches, BundleNet [7] was proposed by Deng et al., while Chang et al. introduced a new method called BGCN [3,4]. In BundleNet [7], several types of relationships among users, bundles, and items are integrated, whereas The user's preferences are separated into two distinct types in BGCN [3,4]: item view and bundle view. Both kinds of preferences are captured efficiently by the two-view representation, which leads to better performance. Our work is built on the basis of these two perspectives, and we also stress the importance of collaborative association modeling between them. Certain related issues, for example, set, basket, or bundle recommendation [14,16,23], and bundle generation [1,4], differ depending on whether it is a suggested subject (loose/random cooccurrence/basket/pack versus a predefined set of topic related items) or a task (creating a bundle of entries).

In recent years, contrastive learning has found applications in various domains such as computer vision [6,13,22], natural language processing [9,20], and graph learning [18,29]. This tendency has been taken advantage of by the community of recommender systems, which include generic CF [28,34,36], sequence and session recommendation [19,30–32,35], multimedia and social recommendation [27,33]. The effective construction of comparative pairs is crucial to the implementation of contrastive learning. Several current methods utilize diverse data augmentations to produce multiple views from the original data. For instance, edge drop is one of the various graph augmentation methods employed in SGL [3], while CL4SRec [32] and CoSeRec [19] adopt diverse sequence augmentation techniques, such as inserting, deleting, and re-ordering. An alternative group of methods focuses on extracting multiple views present in the data. For example, CLCRec [27] constructs contrastive pairs by treating different modes and users/items as distinct views. COTREC [30] utilizes two views, namely the Item View and Session View, to learn session representations from two distinct data sources, namely the session transition graph and session similarity graph. The method leverages contrastive learning to perform the learning process. CrossCBR, described in [21], employs cross-view interactive contrastive learning to capture the collaborative relationship between bundles and achieve improved performance in bundle recommendation.

5 Conclusion

We introduced the Category-aware Cross-view Contrastive learning framework named C^{3}BR to regularize the cross-view representations. Although C^{3}BR has achieved good results, comparative learning-based research for bundle or universal recommendation is still in its early stages. Several directions are promising in the future. For example, the representation learning of user and bundle, the data augmentation methods of contrastive learning. Our code will be released in https://github.com/fuyongxu0908/C3BR

Acknowledgement. This research received support from the Key R&D Project of Shandong Province under Grant 2019JZZY010129, and the Shandong Provincial Social Science Planning Project through Awards 19BJCJ51, 18CXWJ01, and 18BJYJ04.

References

1. Bai, J., et al.: Personalized bundle list recommendation. In: WWW, pp. 60–71. ACM (2019)
2. Cao, D., Nie, L., He, X., Wei, X., Zhu, S., Chua, T.: Embedding factorization models for jointly recommending items and user generated lists. In: SIGIR, pp. 585–594. ACM (2017)
3. Chang, J., Gao, C., He, X., Jin, D., Li, Y.: Bundle recommendation with graph convolutional networks. In: SIGIR, pp. 1673–1676. ACM (2020)
4. Chang, J., Gao, C., He, X., Jin, D., Li, Y.: Bundle recommendation and generation with graph neural networks. IEEE Trans. Knowl. Data Eng. **35**, 2326–2340 (2021)
5. Chen, L., Liu, Y., He, X., Gao, L., Zheng, Z.: Matching user with item set: collaborative bundle recommendation with deep attention network. In: IJCAI, pp. 2095–2101. ijcai.org (2019)
6. Chen, T., Kornblith, S., Norouzi, M., Hinton, G.E.: A simple framework for contrastive learning of visual representations. In: ICML. Proceedings of Machine Learning Research, vol. 119, pp. 1597–1607. PMLR (2020)
7. Deng, Q., et al.: Personalized bundle recommendation in online games. In: CIKM, pp. 2381–2388. ACM (2020)
8. Ding, Y., Ma, Y., Wong, W.K., Chua, T.: Leveraging two types of global graph for sequential fashion recommendation. In: ICMR, pp. 73–81. ACM (2021)
9. Gao, T., Yao, X., Chen, D.: SimCSE: simple contrastive learning of sentence embeddings. In: EMNLP, pp. 6894–6910. Association for Computational Linguistics (2021)
10. Glorot, X., Bengio, Y.: Understanding the difficulty of training deep feedforward neural networks. In: Teh, Y.W., Titterington, D.M. (eds.) AISTATS. JMLR Proceedings, vol. 9, pp. 249–256. JMLR.org (2010)
11. Gutmann, M., Hyvärinen, A.: Noise-contrastive estimation: a new estimation principle for unnormalized statistical models. In: AISTATS. JMLR Proceedings, vol. 9, pp. 297–304. JMLR.org (2010)
12. He, X., Deng, K., Wang, X., Li, Y., Zhang, Y., Wang, M.: LightGCN: simplifying and powering graph convolution network for recommendation. In: SIGIR, pp. 639–648. ACM (2020)
13. Hjelm, R.D., et al.: Learning deep representations by mutual information estimation and maximization. In: ICLR. OpenReview.net (2019)
14. Hu, H., He, X., Gao, J., Zhang, Z.: Modeling personalized item frequency information for next-basket recommendation. In: SIGIR, pp. 1071–1080. ACM (2020)
15. Kingma, D.P., Ba, J.: Adam: a method for stochastic optimization. In: Bengio, Y., LeCun, Y. (eds.) ICLR (2015)
16. Li, C., et al.: Package recommendation with intra- and inter-package attention networks. In: SIGIR, pp. 595–604. ACM (2021)
17. Li, M., Li, L., Xie, Q., Yuan, J., Tao, X.: MealRec: a meal recommendation dataset. CoRR abs/2205.12133 (2022)
18. Li, S., Wang, X., Zhang, A., Wu, Y., He, X., Chua, T.: Let invariant rationale discovery inspire graph contrastive learning. In: ICML. Proceedings of Machine Learning Research, vol. 162, pp. 13052–13065. PMLR (2022)

19. Liu, Z., Chen, Y., Li, J., Yu, P.S., McAuley, J.J., Xiong, C.: Contrastive self-supervised sequential recommendation with robust augmentation. CoRR abs/2108.06479 (2021). arxiv.org/abs/2108.06479
20. Logeswaran, L., Lee, H.: An efficient framework for learning sentence representations. In: ICLR. OpenReview.net (2018)
21. Ma, Y., He, Y., Zhang, A., Wang, X., Chua, T.: CrossCBR: cross-view contrastive learning for bundle recommendation. In: KDD, pp. 1233–1241 (2022)
22. van den Oord, A., Li, Y., Vinyals, O.: Representation learning with contrastive predictive coding. CoRR abs/1807.03748 (2018). arxiv.org/abs/1807.03748
23. Qin, Y., Wang, P., Li, C.: The world is binary: contrastive learning for denoising next basket recommendation. In: SIGIR, pp. 859–868. ACM (2021)
24. Rendle, S., Freudenthaler, C., Gantner, Z., Schmidt-Thieme, L.: BPR: bayesian personalized ranking from implicit feedback. In: Bilmes, J.A., Ng, A.Y. (eds.) UAI, pp. 452–461. AUAI Press (2009)
25. Rendle, S., Freudenthaler, C., Schmidt-Thieme, L.: Factorizing personalized Markov chains for next-basket recommendation. In: WWW, pp. 811–820. ACM (2010)
26. Wang, X., He, X., Wang, M., Feng, F., Chua, T.: Neural graph collaborative filtering. In: SIGIR, pp. 165–174. ACM (2019)
27. Wei, Y., et al.: Contrastive learning for cold-start recommendation. In: MM, pp. 5382–5390. ACM (2021)
28. Wu, J., et al.: Self-supervised graph learning for recommendation. In: SIGIR, pp. 726–735. ACM (2021)
29. Wu, Y., Wang, X., Zhang, A., He, X., Chua, T.: Discovering invariant rationales for graph neural networks. In: ICLR. OpenReview.net (2022)
30. Xia, X., Yin, H., Yu, J., Shao, Y., Cui, L.: Self-supervised graph co-training for session-based recommendation. In: CIKM, pp. 2180–2190. ACM (2021)
31. Xia, X., Yin, H., Yu, J., Wang, Q., Cui, L., Zhang, X.: Self-supervised hypergraph convolutional networks for session-based recommendation. In: AAAI, pp. 4503–4511. AAAI Press (2021)
32. Xie, X., et al.: Contrastive learning for sequential recommendation. In: ICDE, pp. 1259–1273. IEEE (2022)
33. Yu, J., Yin, H., Gao, M., Xia, X., Zhang, X., Hung, N.Q.V.: Socially-aware self-supervised tri-training for recommendation. In: KDD, pp. 2084–2092. ACM (2021)
34. Zhou, C., Ma, J., Zhang, J., Zhou, J., Yang, H.: Contrastive learning for debiased candidate generation in large-scale recommender systems. In: KDD, pp. 3985–3995. ACM (2021)
35. Zhou, K., et al.: S3-Rec: self-supervised learning for sequential recommendation with mutual information maximization. In: CIKM, pp. 1893–1902. ACM (2020)
36. Zhou, X., Sun, A., Liu, Y., Zhang, J., Miao, C.: SelfCF: a simple framework for self-supervised collaborative filtering. CoRR abs/2107.03019 (2021). arxiv.org/abs/2107.03019

GDMA

Threat Action Extraction Based on Coreference Resolution

Dengtian Mao(✉), Ruike Zhao, Rui He, Pengyi He, Fanghui Ning, and Lingqi Zeng

CNOOC EnerTech-Safety and Environmental Protection Co., Tianjin 300452, China
maodengtian@163.com, {ex_zhaork,herui,hepy3,ex_ningfh,zenglq}@cnooc.com.cn

Abstract. The knowledge of attacks contained in Cyber Threat Intelligence (CTI) reports is very important to effectively identify and quickly respond to cyber threats. However, CTI reports are usually described in natural language, and the existence of the phenomenon of coreference affects threat action extraction, which leads to the absence of some threat actions. In order to further investigate this phenomenon, in this paper the original CTI text is fed into the threat action extraction model after coreference resolution. The experimental results show that reference resolution can effectively improve the performance of models that use NLP technology, especially those that rely on POS tagging technology.

Keywords: CTI · Threat action extraction · Coreference resolution

1 Introduction

With the continuous development of computer science and technology, the number and complexity of cyber attacks have increased [1], and their stealthiness and other characteristics. Therefore, the sharing of cyber threat intelligence (CTI) becomes critical to identify and respond to cyber-attacks in a timely, cost-effective manner.

CTI reports contain a wealth of knowledge about cyber-attacks, such as information about the sequence of actions, contextual mechanisms, impact on the attacked system, mitigation measures and countermeasures, and compromise indicators [2], and a simple example of a CTI report is shown in Fig. 1. This knowledge is critical for network operations and response personnel, system administrators, and intrusion detection and prevention product vendors. However, most of this knowledge is described using natural language, and it is very time-consuming for security practitioners to analyze and utilize multi-source and unstructured CTIs. Therefore it is important to extract information from unstructured CTI automatically and efficiently.

Information extraction refers to the extraction of valuable information from unstructured CTI reports. The extracted information mainly includes entities related to cybersecurity and the relationships between entities. Cybersecurity entity identification is used to identify named entities in the cybersecurity

A. El Abbadi et al. (Eds.): DASFAA 2023 Workshops, LNCS 13922, pp. 207–221, 2023.
https://doi.org/10.1007/978-3-031-35415-1_15

domain, mainly including names of people, names of organizations, names of locations and some security terms. Entity relationship extraction is to extract the relationships between security entities in unstructured CTI reports. Besides, it is also important to identify threat actions. Threat action consists of a subject, verb and object. Threat actions not only describe the attack behavior during the attack, but also include non-predefined entities and their contextual semantic relationships. The subject and object in the action correspond to a pair of security entities, and the verb describes the semantic relationship between the entity pairs.

The malware connects to the Command & Control (CnC) server.

The "Authorization.exe" malware has keylogger functionality.

It stores the logged keystrokes in the following file: [CWD].tmp

When the "Authorization.exe" malware is executed it :

Creates a copy of itself in the following locations: %APPDATA% %USERNAME%

Tries to open the following file: [CWD]\Authorization.exe.config

Entrenches in the system for persistence in the following registry locations:

HKCU...\bf7a7ffda58092e10 HKLM...\bfda58092e10

Beacons to the following C2 node IP:.* over TCP port 1177:"217.66.231.245"

Makes the following modification to the registry to bypass the Windows Firewall:

HKLM...\msnco.exe

The downloaded file is decoded, written to disk as %APPDATA%...\ccSvcHst

The following files created when the Authorization.exe malware executed:
msnco.exe authorization.EXE-0AD199D6.pf

Msnco.exe and Authorization.EXE-0AD199D6.pf are created by Authorization.exe.

Fig. 1. A simple example of a CTI report.

Most of the current work uses natural language processing techniques for threat action extraction, and unstructured CTI reports are processed by syntactic analysis, semantic dependency [3], and relationship extraction. Since CTI reports are usually described using natural language, in which pronouns inevitably appear. Only the subject, predicate and object are generally considered when performing threat action extraction. In this case, there will be a situation where the pronoun is extracted as the subject, and if only the pronoun is considered as the subject and the specific content to which the pronoun refers is ignored, it may lead to the loss of some action information. For example, in the third sentence of the example in Fig. 1, without coreference resolution, the extracted subject is 'it', while in fact 'it' refers to the 'Authorization.exe' above. Therefore, in order to extract the relevant information from the text more accurately and without omission, it is necessary to resolve the referential phenomenon in the text.

Referentiality is a common linguistic phenomenon widely found in natural language. Generally speaking, there are 2 types of referents: anaphora (also known as indicative referents): the current referent is closely semantically related to the word, phrase or sentence above, and the referent is dependent on the contextual semantics and may refer to different entities in different linguistic contexts, which is asymmetric and non-transmissive; coreference (also known as homonymous): coreference mainly refers to 2 nouns (including pronouns and noun-phrases) pointing to the same referent in the real world, and this kind of reference is still valid out of context. Coreference resolution is the process of dividing different denotations representing the same entity into a set of equivalents (coreference chain). There are two main steps in coreference resolution: the first step is mention identification, finding all the mentions in a sentence, and the second is performing coreference resolution.

This paper investigates the role of coreference resolution in threat action extraction. The original CTI reports and the reports processed by coreference resolution were input into the threat action extraction model respectively, and the extraction results were compared. The experimental results show that in the case that there are more pronouns in the article, the coreference resolution process of CTI reports can effectively improve the accuracy of threat action extraction.

The remainder of the paper is organized as follows: in Sect. 2, we list the work related to the extraction of key information from CTI reports and coreference resolution; in Sect. 3, we present the baselines we use for coreference resolution and threat action extraction; in Sect. 4, we present the relevant experimental results, and in Sect. 5, we summarise our work.

2 Related Work

2.1 Threat Information Extraction

The fragmentation of information in the era of big data gives unstructured CTI reports the characteristics of diversification, fragmentation, and heterogeneity. For these characteristics of unstructured CTI reports, Liao et al. [4] proposed a method that extracts indicators of compromise(IOC)(e.g. malware signatures, botnet IPs) from public resources(e.g. blogs, forums, tweets, etc.)and converts them into a machine-readable OpenIOC format.SaraQamar et al. [5] proposed a threat analytics framework based on Web Ontology Language(OWL)for formal specification, semantic reasoning, and contextual analysis. By constructing the Structured Threat Information eXpression(STIX), the framework provides an automated mechanism to investigate cyber threats targeting the network under question by classifying the threat relevance, determining threat likelihood, and identifying the affected and exposed assets through formulated rules and inferences. Xun et al. [6] proposed an automatic identification model of threat intelligence (TI) based on convolutional neural network (CNN) called AITI. This model can automatically identify TI from unstructured semantic text such as security articles. These studies reduce the noise data in the CTI report by reorganizing unstructured threat report knowledge to identify cyber threat information in an effective manner.

In addition to this, many studies have used graphical modes to reconstruct CTI knowledge. Shu et al. [7] introduced threat intelligence computing as a new methodology that models threat discovery as a graph computation problem, which formalizes cyber threat intelligence computing into a new security paradigm. Ya et al. [8] proposed an attack entities recognition method based on a neural network CNN-BiLSTM-CRF model combined with a feature template (FT) to construct a CTI knowledge graph. Jia et al. [9] used existing machine learning technology to extract entities and build ontology to obtain a cybersecurity knowledge base, then new rules are deduced by calculating formulas and using the path-ranking algorithm. Du et al. [10] proposed a knowledge graph for People-Readable Threat Intelligence recommendation (PRTIRG) and incorporates knowledge graph representation into the PRTI recommender system for click-through prediction. The threat intelligence knowledge graph helps security practitioners understand cyber threats in a timely and rapid manner.

For threat action extraction from unstructured CTI reports, Husari et al. [11] proposed a method named TTPDrill to extract actions based on semantic dependence from the unstructured text of threat reports and an ontology database, which is used to map actions to different attack patterns. However, TTPDrill will neglect part of threat actions in clause structure and parallel sentences. And it used ontology structure to identify threat actions, which will lose some undefined threat actions in the ontology structure. Husari et al. [12] developed an approach named ActionMiner, which used NLP technology based on the metrics of entropy and mutual information from Information Theory, to extract low-level cyber threat actions from publicly available CTI sources in an automated manner to enable timely defense decision making. However, ActionMiner has relied on the syntactic analysis to extract low-level threat actions. It lacks a behavioral subject, and the information content is difficult to guarantee. Zhang et al. [2] proposed a framework called EX-Action. It extracts actions based on the syntactic structure and rules mapping and identifies them by a multimodal learning algorithm. However, EX-Action has relied on part-of-speech and semantic analysis, which may lose part of threat actions and fail to recognize the pronoun referent.

2.2 Coreference Resolution

There are three main approaches to the development of coreference resolution, which are the rule-based approach, themention-pair-based approach and themention-ranking-based approach.

In 1976, Hobbs proposed the rule-based plain algorithm, which is known as Hobbs' algorithm. Themention-pair-based approach transforms the coreference resolution problem into a binary classification problem. The sentence is traversed from left to right, and each mention is found as anmention pair with each of the previously found mentions. A classifier is used to determine whether the two mentions point to the same entity. If so, the two mentions are connected, using mainly features such as vocabulary, distance, consistency, syntax, and semantics. In the mention ranking method, each mention is scored with all previous mentions at the same time, and the prior word with the highest probability is found after normalization with softmax, and a concatenated edge is added between the prior word and the mention. The probability calculation is usually performed using deep learning methods.

For using deep learning to solve the coreference resolution problem, Clark et al. [13] presented a coreference system that captures entity-level information with distributed representations of coreference cluster pairs. These learned, dense, high-dimensional feature vectors provide a cluster-ranking coreference model with a strong ability to distinguish beneficial cluster merges from harmful ones. Lee et al. [14] introduced the first end-to-end coreference resolution model and showed that it significantly outperforms all previous work without using a syntactic parser or hand-engineered mention detector. The key idea of the model is to directly consider all spans in a document as potential mentions and learn distributions over possible antecedents for each. Then using a span-ranking model to decide which of the previous spans (if any) is a good antecedent for each span. The core of this model is vector embeddings representing spans of text in the document, which combine context-dependent boundary representations with a head-finding attention mechanism over the span. Joshi et al. [15] apply BERT to coreference resolution, achieving strong improvements on the OntoNotes. Kirstain et al. [16] proposed a modification of the mention ranking model, which does not use span representations. This model is a lightweight end-to-end coreference model that removes the dependency on span representations, handcrafted features, and heuristics.

3 Approach

In this paper, two models, coreference resolution and threat action extraction are used. The coreference resolution model uses the word-level coreference resolution model proposed by Vladimir Dobrovolskii [17]. By considering coreference links between individual words rather than word spans and then reconstruct the word spans, this work reduces the complexity of the coreference model to O(n2) and allows it to consider all potential mentions without pruning any of them out. Threat action extraction uses EXTRACTOR proposed by Satvat et al. [18], which is able to extract attack behaviors from unstructured CTI reports and map the sources by performing a series of processing on the CTI reports.

3.1 Coreference Resolution

Token Representation. After obtaining contextual embeddings of all the subtokens of a text, the model computes the weighted sums of their respective subtokens as token representations T. The weights are obtained by applying the softmax function to the raw scores of subtokens of each token. The scores are calculated by multiplying the matrix of subtoken embeddings X by a matrix of learnable weights W_a:

$$A = W_a * X \tag{1}$$

Coarse-to-Fine Antecedent Pruning. This model uses a bilinear scoring function to compute k most likely antecedents for each token, which helps to further reduce the computational complexity of the model.

Then, it constructs a matrix of n × k pairs, where each pair is represented as a concatenation of two token embeddings, the product of their elements and the feature embeddings (distance, genre and same/different speaker embeddings). The fine antecedent scores are obtained using feedforward neural networks:

$$s_a(i,j) = FFNN_a\left([\mathbf{T}_i, \mathbf{T}_j, \mathbf{T}_i \odot \mathbf{T}_j, \phi]\right) \tag{2}$$

The resulting coreference score is defined as the sum of the two scores:

$$s(i,j) = s_c(i,j) + s_a(i,j) \tag{3}$$

The candidate antecedent with the highest positive score is assumed to be the predicted antecedent of each token. If there are no candidates with positive coreference scores, the token is concluded to have no antecedents.

Span Extraction. The tokens that are found to be coreferent to some other tokens are further passed to the span extraction module. For each token, the module reconstructs the span by predicting the most probable start and end tokens in the same sentence. To reconstruct a span headed by a token, all tokens in the same sentence are connected to that head token and then passed through a feed-forward neural network followed by a convolution block with two output channels(for start and end scores). The kernel size is three. The intuition captured by this approach is that the best span boundary is located between a token that is likely to be in the span and a token that is unlikely to be in the span. During inference, tokens to the right of the head token are not considered as potential start tokens, while tokens to the left are not considered as potential end tokens.

3.2 Threat Action Extraction

EXTRACTOR simplifies the text by performing a series of processes on the original unstructured CTI report to extract the threat actions and transform them into the form of a source map.EXTRACTOR consists of four main components:

normalization, resolution, summarization and graph generation. Normalization is used to simplify the original text and process each sentence into a normalized form. Resolution is used to deal with the ambiguities in the sentences and further simplify the text. Summarization is used to remove parts of the text that are not strictly related to the attack behavior. Finally, the graph generation part presents the extracted actions in a graph form. Figure 2 is the framework of EXTRACTOR.

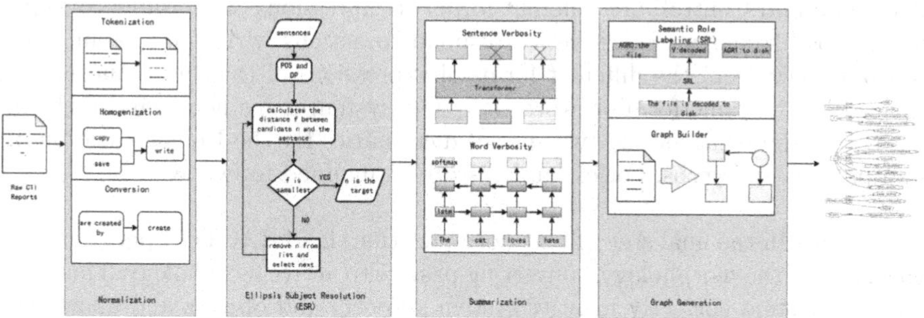

Fig. 2. The framework of EXTRACTOR

Normalization. Due to the complex structure of CTI text sentences, it is necessary to convert the complex sentences into sentences with a canonical form. Through normalization, long and complex sentences are decomposed into shorter sentences with standardized forms to ensure that each sentence expresses only a single action as much as possible, thus facilitating the presentation of the action and the subject and object of the action. Tokenization, Homogenization, and Conversion are the three main parts of normalization, which are the division of sentences, the homogenization of words, and the conversion of passive voice into active voice.

Tokenization. Since EXTRACTOR extracts actions at the sentence level, it is important to identify the correct sentence boundaries. Currently available clause splitting tools mainly use ('. ,!, ?') to perform clause splitting, but in CTI reports, clause splitting with classical punctuation only performs poorly, as shown in the example in Fig. 1, where there are multiple '.exe' words, and the clause splitting tool can easily slice sentences with such phenomena, resulting in the identification of inaccurate sentences. To solve such problems, in addition to the classic sentence separator, line breaks, enumeration numbers and headings are used as sentence separators to divide longer sentences into shorter sequences. A short sequence is considered to be a complete sentence if it satisfies one of the following conditions: 1). The sequence begins with an initial capitalized subject, the sentence contains all the necessary parts of a complete sentence (subject, predicate, and object), and the preceding and following sequences of the sequence are also complete sentences; 2). The sentence begins with a dictionary verb, contains all the necessary parts except the subject, and the preceding and following sequences of the sequence are also complete sentences. The second case usually occurs when the subject is omitted. The result of the tagger is a set of shorter sentences that are more likely to describe a single action.

Homogenization. In CTI reports, it is common to include synonyms that can cause ambiguity and affect the final result. For example, 'C2', 'C & C' and 'Command and Control' are different forms of expressions of the same concept, while verbs like 'copy' and 'save' correspond to actions 'write' in system calls. By Homogenization, the different forms of expressions of the same concept are replaced by the same form of expression. The system maps nouns and verbs in CTI reports by constructing two special dictionaries for the Homogenization of noun phrases and verbs. For example, 'store' is mapped to 'write.'

Conversion. In the final step of text normalization, EXTRACTOR performs the conversion of the morphology, converting passive to active morphology. This step helps the system correctly identify system subjects and objects and allows for a more accurate inference of causality. The passive clause is first detected using lexical annotation and dependent syntactic analysis to perform the morphological transformation. For example, Fig. 3 shows the syntactic dependency tree of the sentence 'Msnco.exe and Authorization.exe-0AD199D6.pf are created by Authorization.exe.'. In this sentence, 'are' is marked as an auxiliary verb, 'created' is marked as a verb, and is the ROOT node of the dependency syntax tree. 'Msnco.exe' is marked as a noun phrase, which is the subject of the passive sentence (nsubjpass), and indicates parallelism, 'Authorization.exe-0AD199D6.pf' is marked as a parallel (conj), and 'Authorization.exe' is marked as a prepositional object (pobj). After identifying the passive clause, the system switches the subject and object and converts the passive verb to an active verb, thus converting the passive clause into an active one.

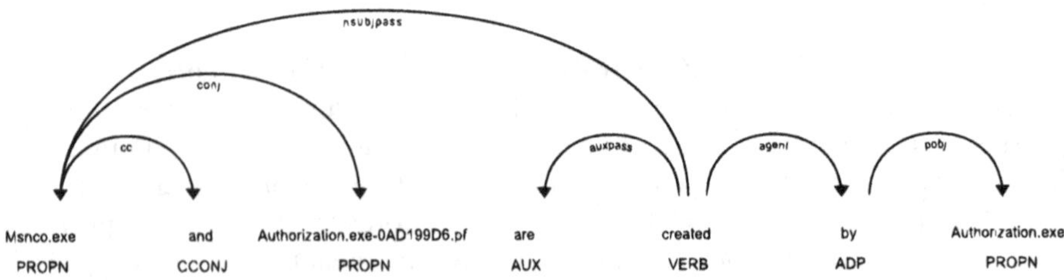

Fig. 3. The syntactic dependency tree of the sentence 'Msnco.exe and Authorization.exe-0AD199D6.pf are created by Authorization.exe.'

The end result of the normalization is that long sentences are converted into short sentences in the active voice, each containing only one action if possible.

Resolution. After normalizing the original text, Resolution reconciles implicit references that refer to the same entity into the actual referent.

Ellipsis Subject Resolution (ESR). The omission of the subject is a common linguistic construction, such that the subject is not present in the sentence. This phenomenon is present in a large number of CTI reports and is used to describe a series of actions performed by the same actor (process or attacker). The presence of this phenomenon leads to the loss of narrative order and story relationships (subject and object of the action). As in the example in Fig. 1, the subject is omitted from lines 5-10. This system solves the problem of omitted subjects through ESR. The first step is to detect sentences with omitted subjects by POS and DP together with the system calls dictionary, as was described in the discussion about the Tokenizer, on the sentences. In the second step, after detecting the sentences with omitted subjects, ESR takes the entities appearing in the sentences before the current sentence as a list of candidates, calculates the distance between each candidate and the distance of the sentence with omitted subjects separately, and selects the candidate with the closest distance as the subject of the current sentence. In particular, the closer candidate is more likely to be selected. As in the example in Fig. 1, for lines 5-10, where there is an omitted subject, the ESR module detects the subject and other objects in the preceding sentences and selects the pronoun 'it,' which appears before the colon, as the subject.

After resolution, the sentence in the text consists of a clear subject, object and verb, which facilitates the extraction of the action.

Summarization. In order to reduce Verbosity and obtain concise descriptions that can be used to directly detect attack behavior, the redundant text needs to be removed. To do this, EXTRACTOR needs to know which sentences describe the attack behavior and which ones do not. Previous related work [4, 10] uses topic clustering to identify topic-related text from context (e.g., advertising text and technical text). This approach is able to distinguish advertising from technical texts but cannot further distinguish texts with attack behaviors from other technical texts (e.g., introductions or contextual descriptions, etc.) EXTRACTOR wants to remove redundant sentences and keep only texts about attack behaviors. Another issue is word Verbosity. Some word structures, such as gerunds and adjective phrases, are not helpful for describing actions, and these words can be safely removed (e.g., but, also, able).

To address these issues, EXTRACTOR takes a two-step approach, using a BERT classifier to deal with sentence Verbosity and then a BiLSTM to deal with word Verbosity. To deal with sentence Verbosity requires a deeper understanding of the text; intuitively, sentences about attack behavior express a more "direct" connection between subject and object than other sentences. Therefore, to classify these connections, the language model of the text must construct a fine-grained representation of the word context, and BERT is currently better able to do this fine-grained representation. The second part of performing text summarization deals with word Verbosity, and for the output of the first part, the semantic components of the words in the sentence are first tagged using BiLSTM, and then the unnecessary sentence components are removed.

The text summary is one of the main components of the extractor. It is responsible for significantly reducing the complexity and volume of the text while retaining the most important sentences describing observable behavior.

Graph Extraction. After the above steps, most of the text exists in a form with a clear subject, object, and verb, and most of the redundant text is eliminated, the final step of EXTRACTOR is to extract the actions in the text and convert them into graph form. Since several roles and relations between subject and object may be expressed in the same sentence, EXTRACTOR uses Semantic Role Labeling (SRL) and a set of rules to extract the causality and direction of the information flow.

Semantic Role Labeling (SRL). SRL is a technique for labeling semantic roles in sentences. To provide a visual overview of SRL's capabilities, SRL is able to extract two roles from each sentence (represented by Raw SRL) and understand which noun is the object (i.e., the one on which the action falls, represented by ARG1) and which is the subject (the noun with the action, represented by ARG0.) SRL is able to correctly associate each component of the sentence with a semantic tag.

Graph Builder (GB). The last step of our method constructs a graph from the output of SRL. GB operates in two steps. First, it merges SRL parameters with the same text into the same node. Next, GB constructs the graph using the following method: 1) Node-Edge-Node triples. For each sentence, GB generates edge and node pairs if it has at least three roles, including a verb role (as a system call representation of a connector) and two nodes. 2) Edge orientation. EXTRACTOR determines the orientation of edges by using a small mapping of edge orientations associated with the system called dictionary.

4 Experiment

4.1 Experiment Setup

We selected 50 CTI reports from different sources for the experiment, and there were roughly 637 sentences in these reports. We used our pool of CTI reports scraped from various sources. We used different sources of namely APT report repository, Microsoft Threat Center, Symantec Security Center, Threat Encyclopedia, and Virus Radar to ensure diversity and coverage.

4.2 Evaluation Metrics

In this study, accuracy, recall, precision and F1-score are used as performance metrics. The recall, precision, and F1-score reflect the quantitative difference of threat actions between machine identification and the ground truth. They can be calculated by Eqs. 4–6.

$$Recall = \frac{TP}{TP + FN} \tag{4}$$

$$Precision = \frac{TP}{TP + FP} \tag{5}$$

$$F1 - score = \frac{2 * Precision * Recall}{Precision + Recall} \tag{6}$$

4.3 Results and Analysis

For our collection of CTI reports, we input the original text and the text that has undergone coreference resolution into EXTRACTOR for extraction separately and manually judge the extraction results. Table 1 shows the recall, precision, and F1-score of the number of actions extracted by EXTRACTOR when different texts were input.

According to the data in the table, the recall, precision, and F1-score of threat actions extracted can be improved to a certain extent after referring to the original text with 2.65% increase in recall, 4.27% increase in precision, and 3.45% increase in F1-score. Not all of the original CTI reports have as many pronouns as the example in Fig. 1. When there are fewer pronouns, the performance gain from coreference resolution is smaller. Still, in general, coreference resolution of the original CTI reports before threat action extraction helps to further extract more accurate threat actions.

Table 1. Experimental results of different methods.

Method	Recall (%)	Precision (%)	F1-score (%)
EXTRACTOR	82.56	83.23	82.89
coreference resolution + EXTRACTOR	85.21	87.50	86.34

For the CTI report example in Figs. 1, 4 and 5 show the threat action graphs extracted by EXTRACTOR before and after the coreference resolution was performed. By comparing the images, it is obvious to observe that before the coreference resolution was performed, the 'it' node was present in the threat action graph extracted by EXTRACTOR, except for the 'it' present in lines 3-4 of that report itself, and lines 5-10 10 lines, ESR chose the nearest pronoun 'it' as a supplementary subject due to the lack of a subject, which resulted in this node being the subject of most of the actions. In the absence of context, we cannot

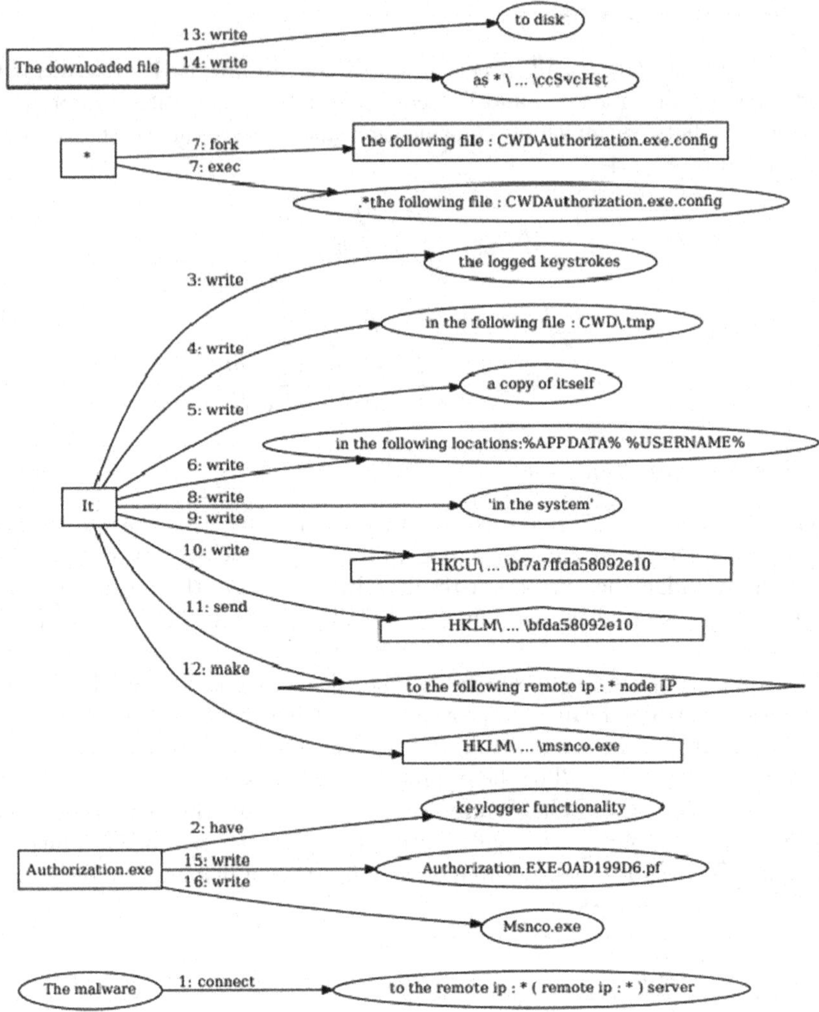

Fig. 4. Threat action graph extracted by EXTRACTOR without coreference resolution.

know exactly which entity in the graph the 'it' node refers to, thus reducing the performance of EXTRACTOR in extracting threatening actions. After the coreference resolution, the model replaces 'it' and 'itself' with the entity it really refers to, 'Authorization.exe'. As shown in Fig. 5, most of the actions have a real subject, and EXTRACTOR extracts more accurate threat actions.

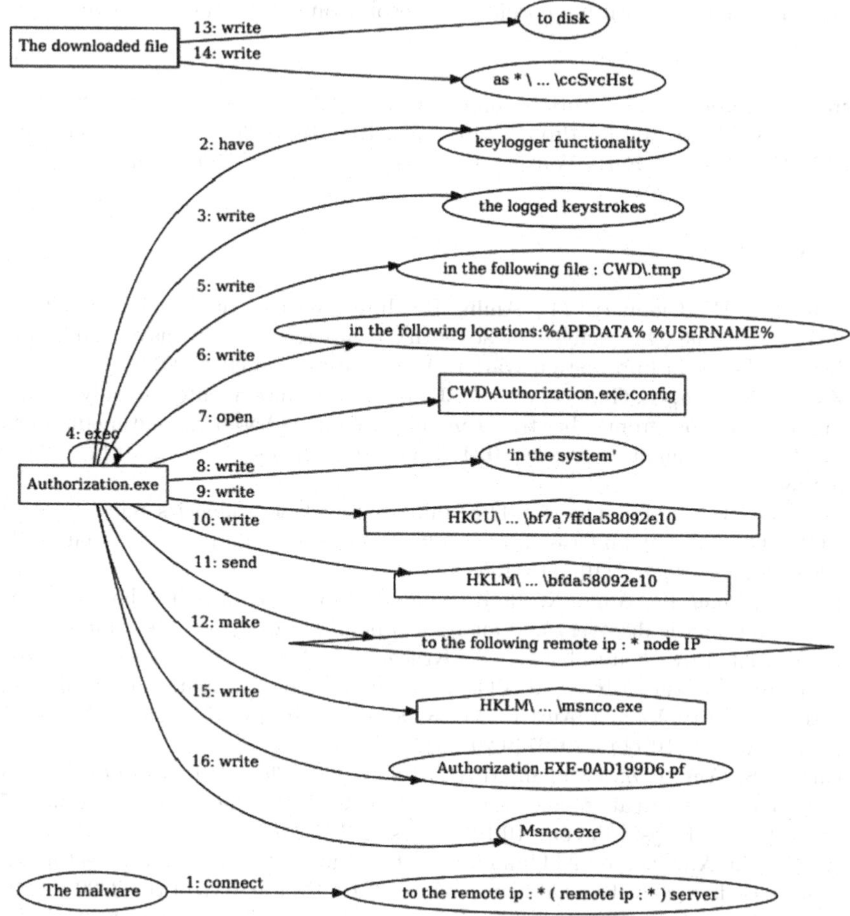

Fig. 5. Threat action graph extracted by EXTRACTOR with coreference resolution.

5 Conclusion

Pronouns are inevitably present in the original CTI reports, and ignoring pronouns when performing threat action extraction can result in some actions being missing. As the results are shown in Fig. 4, the subjects of most actions are pronouns, and the entities to which the pronouns refer cannot be clearly identified. Therefore, it is necessary to perform coreference resolution before performing threat action extraction. The performance of threat action extraction can be improved by effectively reducing the cases of pronouns as subjects through coreference resolution. In this paper, the original text and the text after coreference resolution are input into EXTRACTOR for threat action extraction, and the extraction results are evaluated. The experimental results show that for EXTRACTOR, which uses NLP techniques, especially POS tagging for threat action extraction, performing coreference resolution can further improve its performance.

Acknowledgement. This work is supported by CNOOC Energy Technology & Services Limited Major Special Project Sub-topic: Key Technology Research on Safety Risk Identification and Early Warning for On-site Production Operations.

References

1. Hutchins, E., Cloppert, M., Amin, R.: Intelligence-driven computer network defense informed by analysis of adversary campaigns and intrusion kill chains. Leading Issues in Information Warfare & Security Research 1 (2011)
2. Zhang, H., Shen, G., Guo, C., Cui, Y., Jiang, C.: EX-Action: automatically extracting threat actions from cyber threat intelligence report based on multimodal learning. Secur. Commun. Networks **2021**, 1–12 (2021). https://doi.org/10.1155/2021/5586335
3. Marneffe, M.C., Manning, C.: The stanford typed dependencies representation. COLING Workshop on Cross-framework and Cross-domain Parser Evaluation (01 2008). https://doi.org/10.3115/1608858.1608859
4. Liao, X., Yuan, K., Wang, X., Li, Z., Xing, L., Beyah, R.A.: Acing the IOC game: toward automatic discovery and analysis of open-source cyber threat intelligence. In: Weippl, E.R., Katzenbeisser, S., Kruegel, C., Myers, A.C., Halevi, S. (eds.) Proceedings of the 2016 ACM SIGSAC Conference on Computer and Communications Security, Vienna, Austria, 24–28 October 2016, pp. 755–766. ACM (2016). https://doi.org/10.1145/2976749.2978315
5. Qamar, S., Anwar, Z., Rahman, M.A., Al-Shaer, E., Chu, B.-T.: Data-driven analytics for cyber-threat intelligence and information sharing. Comput. Secur. **67**, 35–58 (2017). https://doi.org/10.1016/j.cose.2017.02.005
6. Xun, S., Li, X., Gao, Y.: AITI: an automatic identification model of threat intelligence based on convolutional neural network. In: Proceedings of the 2020 the 4th International Conference on Innovation in Artificial Intelligence, pp. 20–24. ICIAI 2020, Association for Computing Machinery, New York, NY, USA (2020). https://doi.org/10.1145/3390557.3394305
7. Shu, X., et al.: Threat intelligence computing. In: Proceedings of the 2018 ACM SIGSAC Conference on Computer and Communications Security, pp. 1883–1898. CCS 2018, Association for Computing Machinery, New York, NY, USA (2018). https://doi.org/10.1145/3243734.3243829
8. Qin, Y., Shen, G.W., Zhao, W.B., Chen, Y.P., Yu, M., Jin, X.: A network security entity recognition method based on feature template and CNN-BiLSTM-CRF. Front. Inf. Technol. Electr. Eng. **20**(6), 872–884 (2019). https://doi.org/10.1631/FITEE.1800520
9. Jia, Y., Qi, Y., Shang, H., Jiang, R., Li, A.: A practical approach to constructing a knowledge graph for cybersecurity. Engineering **4**(1), 53–60 (2018). https://doi.org/10.1016/j.eng.2018.01.004
10. Du, M., Jiang, J., Jiang, Z., Lu, Z., Du, X.: PRTIRG: a knowledge graph for people-readable threat intelligence recommendation. In: Douligeris, C., Karagiannis, D., Apostolou, D. (eds.) KSEM 2019. LNCS (LNAI), vol. 11775, pp. 47–59. Springer, Cham (2019). https://doi.org/10.1007/978-3-030-29551-6_5

11. Husari, G., Al-Shaer, E., Ahmed, M., Chu, B., Niu, X.: TTPDrill: automatic and accurate extraction of threat actions from unstructured text of CTI sources. In: Proceedings of the 33rd Annual Computer Security Applications Conference, pp. 103–115. ACSAC 2017, Association for Computing Machinery, New York, NY, USA (2017). https://doi.org/10.1145/3134600.3134646
12. Husari, G., Niu, X., Chu, B., Al-Shaer, E.: Using entropy and mutual information to extract threat actions from cyber threat intelligence. In: 2018 IEEE International Conference on Intelligence and Security Informatics, ISI 2018, Miami, FL, USA, 9–11 November 2018, pp. 1–6. IEEE (2018). https://doi.org/10.1109/ISI.2018.8587343
13. Clark, K., Manning, C.D.: Improving coreference resolution by learning entity-level distributed representations. In: Proceedings of the 54th Annual Meeting of the Association for Computational Linguistics, ACL 2016, 7–12 August 2016, Berlin, Germany, Volume 1: Long Papers. The Association for Computer Linguistics (2016). https://doi.org/10.18653/v1/p16-1061
14. Lee, K., He, L., Lewis, M., Zettlemoyer, L.: End-to-end neural coreference resolution. In: Palmer, M., Hwa, R., Riedel, S. (eds.) Proceedings of the 2017 Conference on Empirical Methods in Natural Language Processing, EMNLP 2017, Copenhagen, Denmark, 9–11 September 2017, pp. 188–197. Association for Computational Linguistics (2017). https://doi.org/10.18653/v1/d17-1018
15. Joshi, M., Levy, O., Zettlemoyer, L., Weld, D.: BERT for coreference resolution: baselines and analysis. In: Proceedings of the 2019 Conference on Empirical Methods in Natural Language Processing and the 9th International Joint Conference on Natural Language Processing (EMNLP-IJCNLP), pp. 5803–5808. Association for Computational Linguistics, Hong Kong, China (2019). https://doi.org/10.18653/v1/D19-1588
16. Kirstain, Y., Ram, O., Levy, O.: Coreference resolution without span representations. CoRR abs/2101.00434 (2021). arXiv: 2101.00434
17. Dobrovolskii, V.: Word-level coreference resolution. In: Moens, M., Huang, X., Specia, L., Yih, S.W. (eds.) Proceedings of the 2021 Conference on Empirical Methods in Natural Language Processing, EMNLP 2021, Virtual Event / Punta Cana, Dominican Republic, 7–11 November 2021, pp. 7670–7675. Association for Computational Linguistics (2021). https://doi.org/10.18653/v1/2021.emnlp-main.605
18. Satvat, K., Gjomemo, R., Venkatakrishnan, V.N.: EXTRACTOR: extracting attack behavior from threat reports. CoRR abs/2104.08618 (2021). arxiv.org/abs/2104.08618

An Improved Method for Constructing Domain-Agnostic Knowledge Graphs

Yuzhou Han[1(✉)], Wenqing Deng[1], Zhe Wang[2], and Kewen Wang[2]

[1] College of Intelligence and Computing, Tianjin University, Tianjin, China
{yuzhou_han,2019218039}@tju.edu.cn

[2] School of Information and Communication Technology, Griffith University, Brisbane, Australia
{zhe.wang,k.wang}@griffith.edu.au

Abstract. Texts contain a vast amount of useful information. Converting unstructured text into structured data enables machines to better understand and mine useful information within texts. Knowledge graphs provide a natural method for data representation and can be applied in downstream tasks. However, existing knowledge graph construction (KGC) methods are mostly domain-specific, which heavily rely on specific external knowledge bases. In this paper, we focus on the domain-agnostic knowledge graph construction task to decouple the reliance on external knowledge. For this purpose, we propose a new method for constructing knowledge graphs from texts, which improves the accuracy of the baseline method Seq2KG. Specifically, we employ a state-of-the-art coreference resolution module in the pre-pocessing stage to provide more priori information for subsequent network learning. Moreover, we introduce a self-attention layer to dynamically model the weights of tokens, thereby enhancing the token representation in the sequence. Our experiments on three public datasets demonstrate that our new model outperforms Seq2KG in both triple extraction and entity typing accuracy.

Keywords: Knowledge Graph · Self-Attention · Coreference Resolution

1 Introduction

With the advent of the digital age, numerous texts are generated in the form of news, social media, and reports. Knowledge graphs provide an intuitive way to represent such unstructured information as RDF triples and can provide support for downstream tasks, such as image caption [46,49], recommender systems [13], and question answering [2,47]. A text contains large amount of information. Converting an unstructured text into a structured format allows machines to better understand the text and mine useful information, which has attracted extensive attention in both academic and industrial community.

Early methods proposed for knolwedge graph construction from texts (KGC) are mainly pipeline-based, consisting of two stages, i.e., information extraction

A. El Abbadi et al. (Eds.): DASFAA 2023 Workshops, LNCS 13922, pp. 222–237, 2023.
https://doi.org/10.1007/978-3-031-35415-1_16

and entity linking. In the information extraction stage, most of existing methods [24,29] first perform named entity recognition (NER) to detect entities and then relation classification to extract relations between entities. In the entity linking stage, KGC methods link entities and relationships to the corresponding concepts in existing knowledge bases like DBpedia [1] and Freebase [4]. Such separated framework makes the KGC task easy to deal with, and each component to be more flexible. However, pipeline-based methods suffer from severe error propagation [42] since they omit the evident correlation between each steps.

To avoid unintuitive error propagation, a number of joint learning-based methods have recently been proposed to jointly extract entities and relations from text using a single model. Different from pipeline-based methods, joint learning can implicitly simulate the correlation between tasks and alleviate the error propagation to a certain extent. Generally, joint learning-based methods can be divided into three categories [32]: sequence labeling method [27,34,35, 40,48], table filling method [12,39] and generative method [25,42–44].

Although significant progress has been made in joint learning-based KGC, existing approaches still suffer from some shortcomings, including: 1) Some domain feature entities cannot be properly mapped to existing knowledge bases. 2) Domain-agnostic KGC are less studied. Existing KGC systems are mainly built for specific domain such as healthcare [30], education [8] and finance [11]. The domain-specific KGC systems need complex pipelines purposely built for target domain, which is not generalisable. 3) Based on named entity recognition, only named entities will be regarded as nodes, but some crucial concepts that are not represented as named entities will be missed out. Therefore, it is beneficial to construct domain-agnostic knowledge graphs without existing knowledge bases. The head and tail of each triple should be any terms carrying important information, whether they are named entities or not. The predicates can be determined through the text without mapping to existing knowledge bases. In this sense, a KGC system can still excel when data is stored within "non-entities", and is generalisable to any domain even in the absence of the corresponding domain knowledge base.

Among efforts for addessing the above issues, a KGC model named Seq2KG is proposed [34] in order to construct knowledge graphs from scratches. However, the accuracy of Seq2KG for triple extraction and entity typing is still limited. For instance, in the benchmark dataset BBN, if the input text is the sentence: *'It's just like weather forecasting,' says Energy Transportation trial attorney Harry Reasoner of Vinson & Elkins.* The ground truth expects the following two triples:

(*Harry Reasoner*, *of*, *Vinson & Elkins*) and

(*Harry Reasoner*, *of*, *Energy Transportation trial attorney*).

But Seq2KG can only extract the triple (*Harry Reasoner*, *trial*, *attorney*).

In this paper, we propose a self-attention-based end-to-end network to achieve domain-agnostic knowledge graph construction from texts. Specifically, as a preprocessing step, we introduce a advanced coreference resolution module to improve the accuracy of knowledge graph construction. In addition, we introduce a self-attention mechanism to enable the tokens in the sequence to have better semantic interactions with the sequence. The better representation for each token will ultimately lead to improved performance in KGC task. In the above example, our method extracts exactly the two ground truth triples. We used three public datasets and compared our proposed model with Seq2KG. Our experiments show that the proposed model of KGC is more effective than Seq2KG.

2 Related Work

Knowledge graph construction (KGC) is the process of automatically extracting structured information from unstructured data [23,31]. KGC can provide services of reasoning, analysis and decision making for users in various fields, such as social media [6] and scientific literature [36]. Typically, current KGC methods fall into two categories, i. e., pipeline-based and joint learning-based.

2.1 Pipeline-Based Approaches for KGC

Most traditional KGC methods are pipeline-based. The pipeline-based methods mainly consist of two stages, *i.e.*, information extraction and entity linking. Triples are extracted from the text in the information extraction stage. Most models first perform named entity recognition to detect entities [24] and after that, perform relation classification on pairs of obtained entities [29] to extract relations. The entity linking stage aims to map entities and relationships to existing knowledge bases, such as DBpedia [1] and Wikidata [38].

There are many off-the-shelf toolkits available for pipeline method, including:(1) applications with GUI support, such as GATE [5], (2) software packages that require some programming to customize, such as DeepDive [26], (3) software packages that require proficiency in NLP and programming to build software libraries suitable for purposeful text processing pipelines, such as NLTK [3], Stanford CoreNLP [21], and Spacy [37].

Different systems [6,7,16,22] may include multiple tasks such as named entity recognition, relation extraction, coreference resolution, and entity linking [33]. Diverse pipeline-based systems have radically different component organizations. T2KG [16] consists of five modules: entity mapping, coreference resolution, triple extraction, triple integration and predicate mapping. Martinez *et al.* [22] complete the task with following modules: sentence segmentation, POS tagging, syntax tree parsing, entity recognition and linking, relation extraction and triple filtering. Their system uses OpenIE to extract relationships. Many pipeline-based methods [9,20,45] have achieved great progress.

The pipeline-based approaches make the task easy to handle, and each component to be flexible. While these approaches have achieved good results, but they ignore the correlation among different stages. In these approaches, each stage is an independent model which can lead to error propagation [18].

2.2 Joint Learning-Based Approaches for KGC

Joint learning-based methods can implicitly simulate the correlation between tasks, effectively integrate the information of entities and relationships, and alleviate the shortcomings of pipeline-based methods to a certain extent. The methods based on neural networks can be divided into three types [32]: Sequence labeling method [27,34,35,40,48], table filling method [12,39] and generative method [25,42–44].

The Sequence labeling method formulates the triple extraction task as a sequence labeling problem. Zheng *et al.* [48] propose a strong neural end-to-end joint model of entities and relations based on an LSTM sequence tagger. Takanobu et al. [35] develope a hierarchical reinforcement learning (HRL) framework for KGC, in which the related entities are regarded as the arguments of a relation. CasRel models relations as functions that map subjects to objects in a sentence [40]. An end-to-end sequence to KG neural model (Seq2KG) jointly learns to generate triples and resolves entity types as a multi-label classification task [34]. Ren *et al.* [27] develops a KGC bidirectional extraction framework that extracts triples based on the entity pairs extracted from two complementary directions.

Table filling methods regard the triple extraction task as a table made up of the Cartesian product of the input sentence and itself. Fu *et al.* [12] used a graph convolution network (GCN) where they treated each token in a sentence as a node in a graph and edges were considered as relations. TPLinker [39] formulates joint extraction as a token pair linking problem and introduces a novel handshaking tagging scheme that aligns the boundary tokens of entity pairs under each relation type.

Generative methods usually employ an encoder-decoder framework to generate triples. CopyRE is an end-to-end model based on sequence-to-sequence learning with copy mechanism [44]. It is able to extract triples with overlapping entities but it cannot predict multi-token entities (e.g. Steve Jobs). Then the authors propose CopyMTL, a multi-task learning framework equipped with copy mechanism [43]. Two methods proposed by Nayak *et al.* [25] employ encoder-decoder architecture for jointly extracting entities and relations having multiple tokens. Ye *et al.* [42]proposes a contrastive triple extraction method with a generative transformer to address the long-term dependence and unfaithful issues.

However, existing KGC systems rely on complex, domain-specific pipeline-based approaches that typically include a range of natural language processing tasks. A KGC system designed this way is usually applicable to only a specific domain of expertise and not well suited for direct application to other domains. Meanwhile, joint learning-based approach naturally alleviate the shortcomings of pipeline-based method. Thus, it would be useful to develop techniques for constructing domain-agnostic knowledge graphs without using existing knowledge bases. Seq2KG is developed to tackle this challenge [34] but their model is still limited in several aspects. In particular, its accuracy for both triple extraction and entity typing is still limited. Our work aims to improve Seq2KG on the accuracy and develop a new method for KGC.

3 Proposed Method

3.1 Overview

In this section, we propose a self-attention-based end-to-end network. The overall architecture of the new model is dipicted in Fig. 1.

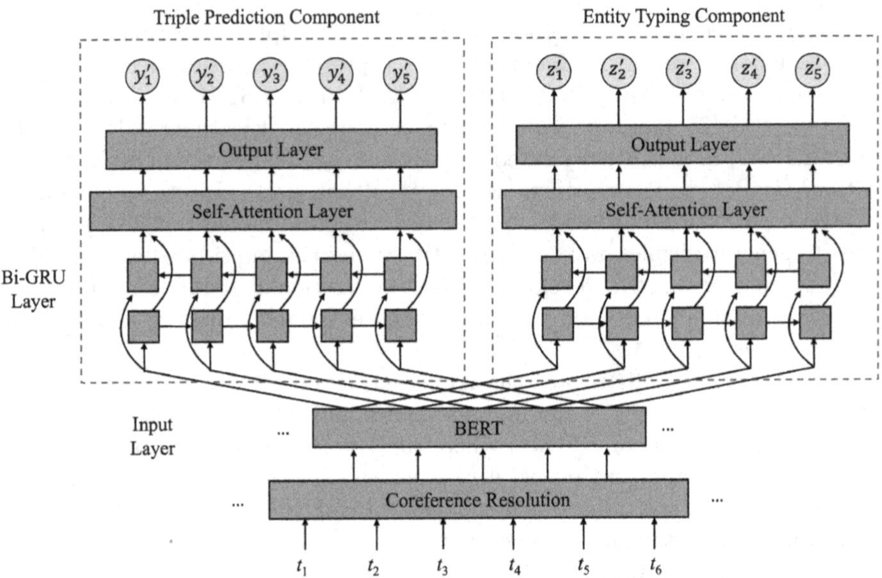

Fig. 1. Our method contains two main parts, *i.e.*, the coreference resolution which performed as preprocessing and an end-to-end model which performs triple prediction. Specifically, the end-to-end model consists of an input layer, a triple prediction component, and an entity typing component, where the two components share the same architecture, *i.e.*, a Bi-GRU layer, a self-attention layer, and an output layer.

Given an input raw text, we first introduce an improved coreference resolution as a step of preprocessing, which is explained in Sect. 3.2. This step aims to detect the mentions in the text that refer to the same entity, which can provide more priori information for subsequent network learning. Then we design an end-to-end model in Sect. 3.3 to generate triples from the input text. We enhance the network architecture for Seq2KG [34] and introduce a self-attention mechanism to improve the learning capability of the model by focusing on the relationship between tokens in the text.

3.2 Coreference Resolution

Coreference resolution identifies the mentions in the text that refer to the same entity. People often have many different ways of referring to the same entity, such as pronouns, abbreviations, and aliases. Usually, coreference resolution group the same entity to the same group and replace the pronouns, abbreviations and aliases with the main mention of the entity, which is called the antecedent. For instance, "Li Hua is a student and he likes reading books", the coreference resolution will find "Li Hua" and "he" referring to the same entity, and then group these two mentions into a cluster. After that, "he" will be replaced with "Li Hua". And finally we get "Li Hua is a student and Li Hua likes reading books.".

Coreference resolution plays an important role in KGC tasks [10,34,41]. SpanBERT [15] is a state-of-the-art coreference resolution model [14], which consists of a mention extraction module and a mention link module. The former extracts all possible mentions from the original text, while the latter clusters mentions into different groups.

In this paper, we use pre-trained SpanBERT [15] as the coreference resolution module to improve the performance of our system, instead of NeuralCoref[1] used in Seq2KG [34]. SpanBERT excels in span-related tasks, such as question answering and coreference resolution. Therefore, we use SpanBERT as a pre-processing step when extracting triples from documents comprised of multiple sentences.

3.3 End-to-End Model

The end-to-end model extracts triples directly from the input text by treating the task as multi-label sequence labelling problem. The model labels each token of the text with its membership and the role in zero or more triples, as well as its entity type(s) through neural networks. It is domain-independent since it does not need to be customized for a specific domain. In fact, when generating triples, the model Seq2KG [34] only uses Bi-GRU to handle the relationship within an input text. This is far from sufficient because tokens in the text can only interact with their neighbors. Therefore, in order to enable each token in the text to interact with other tokens, our work introduce a self-attention mechanism.

In Fig. 1, the input layer uses BERT to encode the sequence from coreference resolution. A Bi-GRU layer is used to further represent the embedding of each token in the sequence. Then these embeddings are fed into a self-attention layer. Finally, we use a linear layer as the output layer to get the probability of each label.

Input Layer. Given a text T, we first utilize coreference resolution as a pre-processing step. The coreference resolution is performed for the input text containing more than one sentence. The sequence $X = \{x_1, x_2, ..., x_m\}$ obtained from coreference resolution is then encoded by BERT and the embeddings $E = \{e_1, e_2, ..., e_m\}$ are fed to the triple prediction component and the entity typing component, respectively.

[1] https://github.com/huggingface/neuralcoref.

Bi-GRU Layer. The embeddings E obtained from BERT are used as the input of Bi-GRU. The multiple layers Bi-GRU extract the sequence features from both forward and backwards directions, providing a more comprehensive understanding of the sequence. The update gate z and the reset gate r are calculated as follows:

$$\begin{aligned} z_t &= \sigma\left(W_z \cdot \left[h_{t-1}, e_t\right] + b_z\right), \\ r_t &= \sigma\left(W_r \cdot \left[h_{t-1}, e_t\right] + b_r\right). \end{aligned} \tag{1}$$

The hidden layer h_t in the GRU is calculated as:

$$\begin{aligned} \tilde{h}_t &= \tanh\left(W_h \cdot \left[r_t * h_{t-1}, e_t\right] + b_h\right) \\ h_t &= (1 - z_t) * h_{t-1} + z_t * \tilde{h}_t \end{aligned} \tag{2}$$

where σ represents Sigmoid activation function, W_z, W_r and W_h represent the weight matrices that can be learned, []means that the two vectors are connected to each other, e_t is the current input at time t, h_t is the hidden state at time t, b_z,b_r and b_h are the bias vectors, and $*$represents the element-wise multiplication.

Self-Attention Layer. In order to enhance the learning capability of the model, we introduce self-attention mechanism into the Knowledge Graph Construction task. Specifically, by considering the internal correlation between tokens, we aim to assign weights to each token through the self-attention mechanism, allowing for the selective enhancement of useful embeddings and suppression of secondary ones.

The input of this layer is the output of the Bi-GRU layer. Let H be the matrix consisting of the output vectors $[h_1, h_2, ...h_m]$ by Bi-GRU layer, where m is the text length.

The training process consists of three main steps. First, a query matrix Q, a key matrix K and a value matrix V are obtained through linear transformations:

$$\begin{aligned} Q &= HW_q, \\ K &= HW_k, \\ V &= HW_v, \end{aligned} \tag{3}$$

where dim is the representation vector dimension of each token in the output of H, and $W_q \in \mathbb{R}^{dim*dim}$, $W_k \in \mathbb{R}^{dim*dim}$ and $W_v \in \mathbb{R}^{dim*dim}$ represent three learned matrices. Then, the matrix multiplication of Q and K^T is calculated to obtain the correlation between them, and then obtains attention weight matrix α through the $softmax$ function:

$$\alpha = \text{softmax}(\frac{QK^T}{\sqrt{d_k}}), \tag{4}$$

where d_k represents the dimension of embedding for each token.

Finally, the matrix multiplication of α and V is used to characterise the interaction between each token and the input sentence. The resulting sentence representation r is obtained as follow:

$$r = \alpha V. \tag{5}$$

Output Layer. The embeddings r transformed by the self-attention are then fed into the output layer to calculate the probability of each label. The output layer linearizes the embedding r followed by the sigmoid activation function:

$$y' = \frac{1}{1 + e^{-(rB^{\top}+b)}}. \tag{6}$$

The output vector y' has a weight for each label in the set N, where N represents the set of all labels. B is a matrix that can be learned.

Loss Function. Each component's loss is calculated using categorical cross entropy:

$$\mathcal{L} = \frac{1}{|N|} \sum_{n \in N} - \left(y_n \log y'_n + (1 - y_n) \log (1 - y'_n) \right), \tag{7}$$

where N is the set of labels, $y_n \in \{0, 1\}$ is the ground truth label of the class of index n, and $\{y'_n \in \mathbb{R} \mid 0 \leq y'_n \leq 1\}$ is the prediction probability associated with the class of index n. The joint loss is calculated as the average of the two component loss values. We denote that $\mathcal{L}_t$ and $\mathcal{L}_e$ are the loss of the triple prediction component and the entity typing component, respectively. In order to co-train the triple prediction and entity typing components, the joint loss is employed to back-propagate the weights throughout the entire model:

$$\mathcal{L}_{joint} = \frac{1}{2}(\mathcal{L}_t + \mathcal{L}_e). \tag{8}$$

4 Experiments

Our experiments are intended to evaluate the performance of the proposed model. Our experimental results show that the new model improves the accuracy of the baseline system Seq2KG [34] for both triple extraction and entity typing on three benchmark datasets.

4.1 Experimental Settings

Datasets. We conducted experiments to evaluate the proposed models on three practical datasets: Catering Service (CS), Automotive Engineering (AE), and BBN. CS and AE contain news articles related to the catering and automotive industries from 2012 to 2014, and the data are captured from catering and automotive news websites respectively. Seq2KG [34] took the first n paragraphs of each article until the total number of characters of the document exceeded 800. It will not be included in the dataset if the article content is fewer than 800 characters. NLTK is used to tokenize the text in the dataset.

The BBN dataset contains single sentences taken from the Wall Street Journal articles. BBN dataset is the entity typing benchmark dataset. We followed the settings in [34] for BBN, which is a subset of the BBN dataset, which was already tokenized and annotated for entity typing [28].

We used 80% of each dataset as the training set, 10% as the validation set, and the remaining 10% as the test set. Table 1 displays the statistics of the three datasets.

Table 1. Statistics of datasets CS, AE and BBN.

Name	#Train	#Dev	#Test
CS	162	20	20
AE	646	81	81
BBN	1639	183	166

The BBN dataset is the only dataset that includes both entity typing and triple membership annotations. In contrast, the CS and AE datasets only annotate triple membership.

Implementation Details. The experiments are designed to evaluate the performance of the proposed method.

We utilized Pytorch to implement the model. The experiments are carried out on a server equipped with a single NVIDIA Tesla V100 GPU. The hyperparameters are set as follows: the batch size is 10, the learning rate is 0.001, the dropout rate is 0.5, the number of max epochs is 1000, and the maximum length of the input sequence is 100. The dimension of the hidden state is 768. And the model is optimized by Adam [17].

Comparison Method. To achieve the comprehensive and comparative analysis of our model, we compared it with Seq2KG [34]. Seq2KG jointly learns to generate triples and resolves entity types as a multi-label classification task through deep neural networks. Seq2KG (no ET) learns to predict triples from text, but not the entity types of each head and tail, Seq2KG (ET) jointly learns to perform both triple extraction and entity typing. We follow the setting and have "Ours (no ET)" and "Ours (ET)".

Besides, we also conducted ablation experiments, including 1) "Ours (no ET) w/o SpanBERT", which does not utilize SpanBERT and only performs triple extraction, 2) "Ours (no ET) w/o Self-Attention Mechanism" which does not utilize self-attention layer and only performs triple extraction.

Our work is based on Seq2KG, and our model accepts the same data format as Seq2KG but is different from other joint learning-based models. Therefore, we only compare ours with Seq2KG instead of other models.

4.2 Evaluation Metrics

Following previous work [19,34], we adopted Embedding Similarity (ESim) to evaluate the triple extraction performance, and Strict Accuracy, Loose Macro, and Loose Micro scores to evaluate the entity typing performance.

Embedding Similarity. In order to combine semantic information with structural information of the knowledge graph, Embedding Similarity (ESim) [34] was used to evaluate the extracted triples.

For each document, two directed acyclic graphs P and G are created from the predicted triples and the ground truth triples, respectively. Then k random walks on two graphs are performed in order to obtain two sets PR and GR of token sequences. For example, if PR_1 is a sub-graph containing nodes $\{N_1,N_2,N_3\}$ and edges $\{E_1,E_2\}$ connecting N_1 to N_2 and N_2 to N_3, we can obtain a sequence N_1, E_1, N_2, E_2, N_3. We then embed the sequence to get pr_1. And perform the same procedure to GR_1 to get gr_1. Each pair of pr_i and gr_i is then compared using cosine similarity to obtain a score s_i. The final score s is the average of s_i:

$$\cos(\mathbf{a}, \mathbf{b}) = \frac{\mathbf{ab}}{\|\mathbf{a}\|\|\mathbf{b}\|} = \frac{\sum_{i=1}^{n} \mathbf{a}_i \mathbf{b}_i}{\sqrt{\sum_{i=1}^{n} (\mathbf{a}_i)^2} \sqrt{\sum_{i=1}^{n} (\mathbf{b}_i)^2}}, \tag{9}$$

$$s = \frac{\sum_{i=1}^{k} \cos\left(f\left(PR_i\right), f\left(GR_i\right)\right)}{k}, \tag{10}$$

where f is a function that computes the average embedding of a sequence. The score of the corpus is the average of each document score s.

Evaluating Entity Typing Performance. We denote $\hat{t_e}$ and t_e are predicted label and ground truth label, respectively. Denote the set of predicted entity labels P and the set of ground truth entity labels T. F1 scores of each metric are then calculated.

Strict Accuracy:

$$\begin{aligned} precision &= \frac{1}{|P|} \sum_{e \in P \cap T} \delta\left(\hat{t_e} = t_e\right), \\ recall &= \frac{1}{|T|} \sum_{e \in P \cap T} \delta\left(\hat{t_e} = t_e\right). \end{aligned} \tag{11}$$

Loose Macro:

$$\begin{aligned} precision &= \frac{1}{|P|} \sum_{e \in P} \frac{\left|\hat{t}_e \cap t_e\right|}{\left|\hat{t_e}\right|}, \\ recall &= \frac{1}{|T|} \sum_{e \in T} \frac{\left|\hat{t}_e \cap t_e\right|}{\left|t_e\right|}. \end{aligned} \tag{12}$$

Loose Micro:

$$precision = \frac{\sum_{e \in P} \left|t_e \cap \hat{t_e}\right|}{\sum_{e \in P} \left|\hat{t_e}\right|},$$
$$recall = \frac{\sum_{e \in T} \left|t_e \cap \hat{t_e}\right|}{\sum_{e \in T} \left|t_e\right|}. \tag{13}$$

F1 score:

$$F1 = \frac{2 \times precision \times recall}{precision + recall}. \tag{14}$$

4.3 Results and Analysis

Results. Table 2 display the comparison results of our model with Seq2KG [34] on the CS, AE and BBN datasets according to the ESim metric. Compared with the baseline model, "Ours (no ET)" outperforms "Seq2KG (no ET)" by 0.065, 0.004 and 0.004 on CS, AE and BBN, respectively. And "Ours (ET)" outperforms "Seq2KG (ET)" by 0.10 on BBN. Our method achieves higher ESim score. Such results demonstrate the effectiveness of our proposed method.

Table 2. The results of our model's triple extraction performance on the three datasets according to the ESim metric. Each score represents the average score obtained from evaluating the same model 36 times, with six different seeds used during the training period and six different seeds used during the evaluating period.

Methods	CS	AE	BBN
Seq2KG (no ET)	0.612	0.895	0.767
Ours (no ET)	**0.677**	**0.899**	0.771
Seq2KG (ET)	–	–	0.779
Ours (ET)	–	–	**0.789**

As mentioned in Sect. 4.1, the BBN dataset is labeled with triple extraction information and entity typing at the same time. The following experiments are designed to verify whether the joint training of entity typing tasks and triple extraction can improve system performance with the inclusion of self-attention mechanism. The results are shown in Table 3.

The result of the Seq2KG is 0.779, and ours can reach 0.789 on ESim. For the Strict Accuracy, Loose Macro and Loose Micro, our method also achieves better performance. The experimental results convey that the self-attention mechanism improves the performance for both the entity typing tasks and triple extraction tasks.

Table 3. The results of our method jointly performing triple extraction and entity typing on BBN according to the ESim, Micro, Macro and Strict metrics. Each score represents the average score obtained from evaluating the same model 36 times, with six different seeds used during the training period and six different seeds used during the evaluating period.

	ESim	Micro	Macro	Strict
Seq2KG (ET)	0.779	0.585	0.655	0.655
Ours (ET)	**0.789**	**0.592**	**0.657**	**0.657**

Table 4. The ablation study results for our model's triple extraction performance on the three datasets are presented based on the ESim metric. Each score represents the average score obtained from evaluating the same model 36 times, with six different seeds used during the training period and six different seeds used during the evaluating period.

Variants	CS	AE	BBN
Ours (no ET)	**0.677**	**0.899**	**0.771**
Ours (no ET) w/o SpanBERT	0.627	0.893	0.771
Ours (no ET) w/o Self-Attention Mechanism	0.631	0.898	–

Ablation Study. To evaluate the effectiveness of advanced coreference resolution and self-attention mechanism, we also conducted the ablation study.

Each item of the BBN dataset contains only one sentence, which will not be pre-processed by the coreference resolution, according to the model settings. Therefore, "Ours (no ET)" and "Ours (no ET) w/o SpanBERT" have the same result on the BBN dataset. Without using SpanBERT as co-reference resolution and without adding the self-attentive mechanism, the results on the BBN dataset will be the same as in "Seq2KG (no ET)" and will not be presented as a variant of our method.

Table 4 displays the results of ablation study. Compared with "Ours (no ET)", "Ours (no ET) w/o SpanBERT" decreases by 0.05 and 0.006 on CS and AE datasets, respectively, which shows the necessity of using SpanBERT to perform coreference as pre-processing. And compared with "Ours (no ET)","Ours (no ET) w/o Self-Attention Mechanism" decreases by 0.046 and 0.001 on CS and AE datasets, respectively. This result shows that the self-attention mechanism plays an important role in our model.

5 Conclusion

In this paper, we propose a self-attention-based end-to-end neural model for achieving domain-agnostic knowledge graph construction through texts. Specifically, our proposed model employs a coreference resolution strategy in the preprocessing stage and a self-attention mechanism in the network learning stage, both of which effectively enhance the token representation in the sequence. Our experiments on three public datasets demonstrate that our model outperforms the baseline Seq2KG in terms of accuracy.

It would be useful to develop more efficient KGC models that is scalable to larger KGs. However, the current training process for our proposed model and Seq2KG relies on manually annotated data, which may not be convenient for some practical applications. Therefore, it is of great interest to explore KGC models that require less annotated data, while maintaining high performance and scalability.

Acknowledgements. This work was partially supported by the National Natural Science Foundation of China under grant 61976153.

References

1. Auer, S., Bizer, C., Kobilarov, G., Lehmann, J., Cyganiak, R., Ives, Z.: DBpedia: a nucleus for a web of open data. In: Aberer, K., et al. (eds.) ASWC/ISWC -2007. LNCS, vol. 4825, pp. 722–735. Springer, Heidelberg (2007). https://doi.org/10.1007/978-3-540-76298-0_52
2. Bao, J., Duan, N., Yan, Z., Zhou, M., Zhao, T.: Constraint-based question answering with knowledge graph. In: Proceedings of COLING 2016, the 26th International Conference on Computational Linguistics: Technical Papers, pp. 2503–2514 (2016)
3. Bird, S., Loper, E.: NLTK: the natural language toolkit. In: Proceedings of the 42nd Annual Meeting of the Association for Computational Linguistics, Barcelona, Spain, 21–26 July 2004 - Poster and Demonstration. ACL (2004). https://aclanthology.org/P04-3031/
4. Bollacker, K., Evans, C., Paritosh, P., Sturge, T., Taylor, J.: Freebase: a collaboratively created graph database for structuring human knowledge. In: Proceedings of the 2008 ACM SIGMOD International Conference on Management of Data, pp. 1247–1250 (2008)
5. Bontcheva, K., et al.: Gate teamware: a web-based, collaborative text annotation framework. Lang. Resour. Eval. **47**(4), 1007–1029 (2013)
6. Bontcheva, K., Derczynski, L., Funk, A., Greenwood, M.A., Maynard, D., Aswani, N.: TwitIE: an open-source information extraction pipeline for microblog text. In: Proceedings of the International Conference Recent Advances in Natural Language Processing RANLP 2013, pp. 83–90 (2013)
7. Chan, Y.S., Roth, D.: Exploiting syntactico-semantic structures for relation extraction. In: Proceedings of the 49th Annual Meeting of the Association for Computational Linguistics: Human Language Technologies, pp. 551–560 (2011)
8. Chen, P., Lu, Y., Zheng, V.W., Chen, X., Yang, B.: KnowEdu: a system to construct knowledge graph for education. IEEE Access **6**, 31553–31563 (2018)
9. Chen, X., et al.: KnowPrompt: knowledge-aware prompt-tuning with synergistic optimization for relation extraction. In: Proceedings of the ACM Web Conference 2022, pp. 2778–2788 (2022)
10. Eberts, M., Ulges, A.: An end-to-end model for entity-level relation extraction using multi-instance learning. In: Proceedings of the 16th Conference of the European Chapter of the Association for Computational Linguistics: Main Volume, pp. 3650–3660. Association for Computational Linguistics (2021). https://doi.org/10.18653/v1/2021.eacl-main.319. www.aclanthology.org/2021.eacl-main.319

11. Elhammadi, S., et al.: A high precision pipeline for financial knowledge graph construction. In: Proceedings of the 28th International Conference on Computational Linguistics, pp. 967–977 (2020)
12. Fu, T.J., Li, P.H., Ma, W.Y.: GraphRel: modeling text as relational graphs for joint entity and relation extraction. In: Proceedings of the 57th Annual Meeting of the Association for Computational Linguistics, pp. 1409–1418 (2019)
13. Guo, Q., et al.: A survey on knowledge graph-based recommender systems. IEEE Trans. Knowl. Data Eng. **PP**, 1 (2020)
14. Hovy, E., Marcus, M., Palmer, M., Ramshaw, L., Weischedel, R.: OntoNotes: the 90% solution. In: Proceedings of the Human Language Technology Conference of the NAACL, Companion Volume: Short Papers, pp. 57–60 (2006)
15. Joshi, M., Chen, D., Liu, Y., Weld, D.S., Zettlemoyer, L., Levy, O.: SpanBERT: improving pre-training by representing and predicting spans. Trans. Assoc. Computat. Linguist. **8**, 64–77 (2020)
16. Kertkeidkachorn, N., Ichise, R.: T2KG: an end-to-end system for creating knowledge graph from unstructured text. In: Workshops at the Thirty-First AAAI Conference on Artificial Intelligence (2017)
17. Kingma, D.P., Ba, J.: Adam: a method for stochastic optimization. arXiv preprint arXiv:1412.6980 (2014)
18. Li, Q., Ji, H.: Incremental joint extraction of entity mentions and relations. In: ACL (1), pp. 402–412 (2014)
19. Ling, X., Weld, D.S.: Fine-grained entity recognition. In: Twenty-Sixth AAAI Conference on Artificial Intelligence (2012)
20. Liu, K., et al.: Noisy-labeled NER with confidence estimation. arXiv preprint arXiv:2104.04318 (2021)
21. Manning, C.D., Surdeanu, M., Bauer, J., Finkel, J.R., Bethard, S., McClosky, D.: The stanford coreNLP natural language processing toolkit. In: Proceedings of 52nd Annual Meeting of the Association for Computational Linguistics: System Demonstrations, pp. 55–60 (2014)
22. Martinez-Rodriguez, J.L., López-Arévalo, I., Rios-Alvarado, A.B.: OpenIE-based approach for knowledge graph construction from text. Expert Syst. Appl. **113**, 339–355 (2018)
23. Mooney, R.J., Bunescu, R.: Mining knowledge from text using information extraction. ACM SIGKDD Explor. Newsl. **7**(1), 3–10 (2005)
24. Nadeau, D., Sekine, S.: A survey of named entity recognition and classification. Lingvisticae Investigationes **30**(1), 3–26 (2007)
25. Nayak, T., Ng, H.T.: Effective modeling of encoder-decoder architecture for joint entity and relation extraction. In: Proceedings of the AAAI Conference on Artificial Intelligence, vol. 34, no. 05, pp. 8528–8535 (2020)
26. Niu, F., Zhang, C., Ré, C., Shavlik, J.W.: DeepDive: web-scale knowledge-base construction using statistical learning and inference. VLDS **12**, 25–28 (2012)
27. Ren, F., Zhang, L., Zhao, X., Yin, S., Liu, S., Li, B.: A simple but effective bidirectional framework for relational triple extraction. In: Proceedings of the Fifteenth ACM International Conference on Web Search and Data Mining, pp. 824–832 (2022)
28. Ren, X., He, W., Qu, M., Huang, L., Ji, H., Han, J.: AFET: automatic fine-grained entity typing by hierarchical partial-label embedding. In: Proceedings of the 2016 Conference on Empirical Methods in Natural Language Processing, pp. 1369–1378 (2016)

29. Rink, B., Harabagiu, S.: UTD: classifying semantic relations by combining lexical and semantic resources. In: Proceedings of the 5th International Workshop on Semantic Evaluation, pp. 256–259 (2010)
30. Rotmensch, M., Halpern, Y., Tlimat, A., Horng, S., Sontag, D.: Learning a health knowledge graph from electronic medical records. Sci. Rep. **7**(1), 1–11 (2017)
31. Sarawagi, S.: Information extraction. Now Publishers Inc (2008)
32. Shang, Y.M., Huang, H., Sun, X., Wei, W., Mao, X.L.: Relational triple extraction: one step is enough. In: Raedt, L.D. (ed.) Proceedings of the Thirty-First International Joint Conference on Artificial Intelligence, IJCAI-22, pp. 4360–4366 (July 2022). https://doi.org/10.24963/ijcai.2022/605. Main Track
33. Stewart, M., Enkhsaikhan, M., Liu, W.: ICDM 2019 knowledge graph contest: team UWA. In: 2019 IEEE International Conference On Data Mining (ICDM), pp. 1546–1551. IEEE (2019)
34. Stewart, M., Liu, W.: Seq2KG: an end-to-end neural model for domain agnostic knowledge graph (not text graph) construction from text. In: Calvanese, D., Erdem, E., Thielscher, M. (eds.) Proceedings of the 17th International Conference on Principles of Knowledge Representation and Reasoning, KR 2020, Rhodes, Greece, 12–18 September 2020, pp. 748–757 (2020). https://doi.org/10.24963/kr.2020/77
35. Takanobu, R., Zhang, T., Liu, J., Huang, M.: A hierarchical framework for relation extraction with reinforcement learning. In: Proceedings of the AAAI Conference on Artificial Intelligence, vol. 33, no. 01, pp. 7072–7079 (2019)
36. Torii, M., Arighi, C.N., Li, G., Wang, Q., Wu, C.H., Vijay-Shanker, K.: RLIMPS-P 2.0: a generalizable rule-based information extraction system for literature mining of protein phosphorylation information. IEEE/ACM Trans. Comput. Biol. Bioinform. **12**(1), 17–29 (2014)
37. Vasiliev, Y.: Natural language processing with python and SpaCy: a practical introduction. No Starch Press (2020)
38. Vrandečić, D.: Wikidata: a new platform for collaborative data collection. In: Proceedings of the 21st International Conference on World Wide Web, pp. 1063–1064 (2012)
39. Wang, Y., Yu, B., Zhang, Y., Liu, T., Zhu, H., Sun, L.: TPLinker: single-stage joint extraction of entities and relations through token pair linking. In: Proceedings of the 28th International Conference on Computational Linguistics, pp. 1572–1582. International Committee on Computational Linguistics (2020)
40. Wei, Z., Su, J., Wang, Y., Tian, Y., Chang, Y.: A novel cascade binary tagging framework for relational triple extraction. arXiv preprint arXiv:1909.03227 (2019)
41. Xue, Z., Li, R., Dai, Q., Jiang, Z.: CorefDRE: document-level relation extraction with coreference resolution (2022)
42. Ye, H., et al.: Contrastive triple extraction with generative transformer. In: Thirty-Fifth AAAI Conference on Artificial Intelligence, AAAI 2021, Thirty-Third Conference on Innovative Applications of Artificial Intelligence, IAAI 2021, The Eleventh Symposium on Educational Advances in Artificial Intelligence, EAAI 2021, Virtual Event, 2–9 February 2021, pp. 14257–14265. AAAI Press (2021). https://www.ojs.aaai.org/index.php/AAAI/article/view/17677
43. Zeng, D., Zhang, H., Liu, Q.: CopymMTL: copy mechanism for joint extraction of entities and relations with multi-task learning. In: The Thirty-Fourth AAAI Conference on Artificial Intelligence, AAAI 2020, The Thirty-Second Innovative Applications of Artificial Intelligence Conference, IAAI 2020, The Tenth AAAI Symposium on Educational Advances in Artificial Intelligence, EAAI 2020, New

York, NY, USA, 7–12 February 2020, pp. 9507–9514. AAAI Press (2020). https://www.ojs.aaai.org/index.php/AAAI/article/view/6495
44. Zeng, X., Zeng, D., He, S., Liu, K., Zhao, J.: Extracting relational facts by an end-to-end neural model with copy mechanism. In: Proceedings of the 56th Annual Meeting of the Association for Computational Linguistics (Volume 1: Long Papers), pp. 506–514 (2018)
45. Zhang, N., Deng, S., Sun, Z., Chen, J., Zhang, W., Chen, H.: Relation adversarial network for low resource knowledge graph completion. In: Proceedings of The Web Conference 2020, pp. 1–12 (2020)
46. Zhang, Y., Shi, X., Mi, S., Yang, X.: Image captioning with transformer and knowledge graph. Pattern Recogn. Lett. **143**, 43–49 (2021)
47. Zhang, Y., Dai, H., Kozareva, Z., Smola, A.J., Song, L.: Variational reasoning for question answering with knowledge graph. In: Thirty-second AAAI Conference on Artificial Intelligence (2018)
48. Zheng, S., Wang, F., Bao, H., Hao, Y., Zhou, P., Xu, B.: Joint extraction of entities and relations based on a novel tagging scheme. arXiv preprint arXiv:1706.05075 (2017)
49. Zhou, Y., Sun, Y., Honavar, V.: Improving image captioning by leveraging knowledge graphs. In: 2019 IEEE Winter Conference on Applications of Computer Vision (WACV), pp. 283–293 (2019). https://doi.org/10.1109/WACV.2019.00036

Zero-Shot Entity Typing in Knowledge Graphs

Shengye Zhou[1(✉)], Zhe Wang[2], Kewen Wang[2], and Zhiqiang Zhuang[1]

[1] College of Intelligence and Computing, Tianjin University, Tianjin, China
{zsyc,zhuang}@tju.edu.cn

[2] School of Information and Communication Technology, Griffith University, Nathan, Australia
{zhe.wang,k.wang}@griffith.edu.au

Abstract. Knowledge graphs are often highly incomplete due to their large sizes and one major task for knowledge graph completion is entity typing, that is to predict missing types of entities or vice versa. It is especially challenging to perform entity typing when the type is new, i.e., unseen during training, which is known as the zero-shot entity typing problem. Existing entity typing models cannot handle the zero-shot case as it requires the models to be retrained to embed the unseen types, and other zero-shot knowledge graph completion approaches cannot be applied to the entity typing task either. In this paper, we propose a novel zero-shot entity typing approach based on a generation architecture, and introduce a novel feature distribution and semantic encoding method that combines both ontological and textual knowledge. We also construct the first zero-shot entity typing datasets based on commonly used benchmarks. Our experiment evaluation shows the effectiveness of our approach.

Keywords: Knowledge graph · Entity typing · Zero-shot learning

1 Introduction

Knowledge graphs (KGs) are an effective way to represent and store knowledge, which are often expressed as a collection of triples of the form (*head*, *relation*, *tail*) with *head* and *tail* are entities corresponding to vertices and *relation* corresponding to edges in the KG. The triple says the *head* and *tail* entities are associated with the *relation*. Many KGs also contains information about *types* of entities, for example in a triple (*Jay Chou*, *type*, *person*), which says the entity *Jay Chou* has a type of *person*. KGs have been widely used in a range of applications, such as visual question answering [3], relation classification [17] and question answering [4]. Many large-scale KGs have been constructed in the past few decades, such as Freebase [1], YAGO [15] and DBpedia [10]. However, most existing KGs are highly incomplete, which limits their benefits to the downstream applications due to missing important facts. Hence, KG completion is an

A. El Abbadi et al. (Eds.): DASFAA 2023 Workshops, LNCS 13922, pp. 238–250, 2023.
https://doi.org/10.1007/978-3-031-35415-1_17

important research topic, which concerns the auto-inference of missing facts in KGs. And one major KG completion task is *entity typing*, which aims to predict missing types of entities or vice versa.

Knowledge graphs (KGs) are an effective way to represent and store knowledge, which are often expressed as a collection of triples of the form (*head*, *relation*, *tail*) with *head* and *tail* are entities corresponding to vertices and *relation* corresponding to edges in the KG. The triple says the *head* and *tail* entities are associated with the *relation*. Many KGs also contains information about *types* of entities, for example in a triple (*Jay Chou*, *type*, *person*), which says the entity *Jay Chou* has a type of *person*. KGs have been widely used in a range of applications, such as visual question answering [3], relation classification [17] and question answering [4]. Many large-scale KGs have been constructed in the past few decades, such as Freebase [1], YAGO [15] and DBpedia [10]. However, most existing KGs are highly incomplete, which limits their benefits to the downstream applications due to missing important facts. Hence, KG completion is an important research topic, which concerns the auto-inference of missing facts in KGs. And one major KG completion task is *entity typing*, which aims to predict missing types of entities or vice versa.

For practical KGs, new entity types are often added to the KGs, and downstream applications, such as question answering, often requires to refer to types that may not exist in KGs. This is known as the challenge of *zero-shot learning*, which aims to classify a object to new classes, called unseen classes, without examples available during the training. The current entity-typing models, such as ConnectE [18] and CET [12], performs well on classic entity-typing scenarios but cannot handle the zero-shot scenario without re-training the models. Recently, research has been done for entity typing in few-shot settings [20], which cannot be applied to the zero-shot scenario either where no example (i.e., entities as member) of the unseen types is available. While zero-shot KG completion has been studied [6], they focus on tasks like link prediction that involve unseen relations and cannot be directly applied to the entity-typing task where unseen types are involved instead of relations.

In this paper, we propose a zero-shot KG entity-typing approach for knowledge graph. Inspired by how human learns new concepts through building connections with prior concepts to identify familiar features of the new concepts, we use generative adversarial learning to generate feature encoding for unseen entity types based on their semantic connections with seen types. This allows us to transfer the zero-shot learning task into traditional supervised learning. In particular, we use conditional GAN to generate feature encodings of unseen types from their semantics encodings. We design a feature encoder that combines information of entities and their neighbours so that the generated encoding captures the cluster and structure feature distributions of the entities. For semantic embeddings of entity types, we combine ontological and textual knowledge about the types, which allows us to address the issue with long tail distributions of types. We also construct the first zero-shot entity typing datasets based on commonly used benchmarks and evaluate the effectiveness of our approach.

Our main contributions are summarized as below:

- We introduce the first zero-shot learning approach for KG entity-typing, through a generative adversarial framework.
- We propose a novel encoding methods for entities that preserve their feature distributions related to their types and for types that capture their semantic connections by combining both ontological and textual information of the types.
- We construct three new datasets for zero-shot entity-linking and show that our method achieves superior performance than both existing and adapted based methods.

2 Related Work

We are unaware of any existing work on KG entity-typing in zero-shot settings. In this section, we discuss research work that are most closely related to our work.

2.1 Entity Typing

ConnectE [18] is a classic embedding-based method for KG entity-typing, and it constructs embeddings with two sub-models, E2T and TRT. E2T captures the connection between entities and their types, and TRT captures the relation between types. They cannot handle the zero-shot scenario as the type embeddings rely on the embeddings of the entities belong to the types. ConnectE-MRGAT [19] improves ConncetE by addressing the issue that ConnectE consider each entity, relation and entity type independently, by making better use of structure information provided by KG with a self-attention graph neural network. AttET [21] divides neighbours of an entity into two groups, the outgoing neighbourhoods and incoming neighbourhoods, and use an attention-based neural network to calculate the representation of the entity. In this way, each neighbour has a different contribution to the entity embedding based on the given entity type. None of these approaches can be used in our zero-shot setting.

JOIE [9] sees a KG as a two-layered graph, an instance graph about entities and their inter-relations and an ontology graph about entity types and their relations, and the two layers are connected by entity-type associations. Two layers are separately embedded and the entity and type embeddings are aligned through a neural network mapping entity in instance graph to corresponding entity types in ontology graph. CORE [5] uses complex space to represent instance and ontology graph, the method is similar to JOIE. MET [20] also adopts the two-layered graph idea but focuses on entity-typing under a few-shot setting. They first use GCN to encode instance graph and ontology graph and use MAML from meta-learning to train a good initial weight to make the model able to quick adapt to other entity types with few instances. With a separate embedding obtained from the ontology graph, these embedding approaches can be applied in a zero-shot setting, but they are not designed for the zero-shot learning task.

2.2 Zero-Shot KG Completion

Zero-shot KG completion works mostly focus on link prediction [6,7,14], which aims to predict tail entity given head entity and relation. The zero-shot setting here refers to the scenario where relations are new. In [14], the authors design a feature encoder to use entity pairs to represent features of relations. The semantic embeddings of a relation are generated based on their text descriptions. It also uses a conditional GAN to generate feature encodings for unseen relations. OntoZSL [6] uses an ontology to capture semantic connections between seen and unseen relations and demonstrates the usefulness of ontological knowledge in semantic representation. K-ZSL [7] proposed a general method for construct ontology for zero-shot KG completion. None of these approaches can be directly used or easily adapted to the zero-shot entity-typing task.

3 Background

In this section, we define the problem and introduce some useful notations.

A *knowledge graph* (KG) consists of triples of the form (e_1, r, e_2), where e_1, e_2 are entities and r is a relation, and entity-type instances of the form (e, t), where e is an entity and t is an entity type. Each triple (e_1, r, e_2) represents that the head entity e_1 and tail entity e_2 are connected with the relation r, and each entity type instance (e, t) represents the entity e has the type t. $\mathcal{E}$, $\mathcal{R}$, and $\mathcal{T}$ denote the sets of all the entities, relations, and entity types, respectively. For a KG $\mathcal{K}$, let $ins(t)$ denote the set of entities belonging to a type t, i.e., $ins(t) = \{e \in \mathcal{E} \mid (e, t) \in \mathcal{K}\}$.

Each entity in a KG may have multiple types, and often such type information is (highly) incomplete in KGs. The task of *entity typing* (ET) in KGs is to infer the types t of a given entity e such that (e, t) do not occur in the KG, and *zero-shot entity typing* (ZSET) aims to infer types t for the entity that is new, i.e., not exist in the training data. In this setting, the set of entity types is split into two parts $\mathcal{T} = \mathcal{T}_s \cup \mathcal{T}_u$, which are the *seen* types $\mathcal{T}_s$, i.e., at least an instance (e, t) occur in the training data for each $t \in \mathcal{T}_s$, and the set of *unseen* types $\mathcal{T}_u$, i.e., no instance (e, t) occur in the training data for any $t \in \mathcal{T}_u$. The task of ZSET is to estimate for a given KG $\mathcal{K}$ the matching probability between an entity e and an unseen type $t \in \mathcal{T}_u$, i.e., $p(e, t|\mathcal{K})$.

Both ontological and textual information has been used to establish semantic connections between seen and unseen types for ZSET. A common form of ontological knowledge is the connection of types t with relations r, i.e., whether the entities of type t have relation r. An *ontological triple* is of the form (t_1, r, t_2) with $t_1, t_2 \in \mathcal{T}$ and $r \in \mathcal{R}$, which says entities of type t_1 have relation r with entities of type t_2. Such ontological triples can be automatically extracted from many KGs. $\mathcal{O}$ denote the set of all ontological triples in the KG. Moreover, types in KGs also have meaningful names or textual labels, which we also use as prior semantic knowledge.

4 Our Method

Figure 1 gives the overview of our approach, which is a common architecture for generation-based zero-short learning.

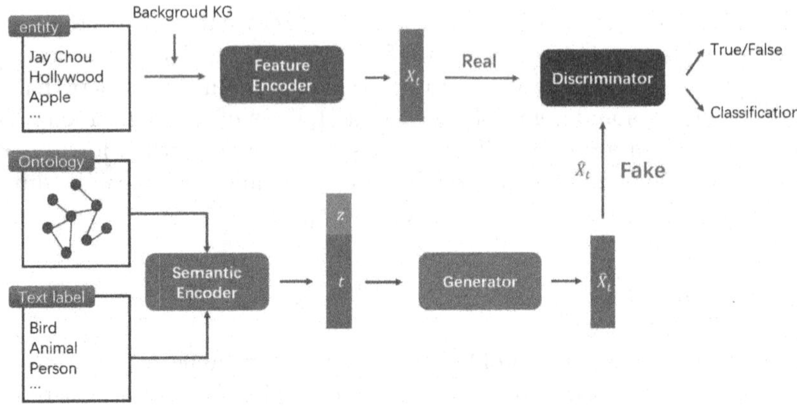

Fig. 1. An overview of our model.

During the training, for each seen type t, our model extracts its *feature encoding* $\mathcal{X}_t$ that represents all the entities belonging to t. The model also feeds the *semantic embedding* $\mathbf{t}$ of t to a GAN generator to generate a fake feature encoding $\hat{\mathcal{X}}_t$ while a GAN discriminator tries to distinguish $\hat{\mathcal{X}}_t$ from $\mathcal{X}_t$. After we train the generator, it can generate feature embeddings $\hat{\mathcal{X}}_t'$ for unseen types t' to predict the entities of t'. Such an architecture utilizes the generalizability of (conditional) GAN to predict the entities of unseen types based on their semantic connections with seen types.

The prediction performance largely depends on the feature encodings and semantic embeddings of types. In particular, the feature encodings $\mathcal{X}_t$ (or $\hat{\mathcal{X}}_t$ after training) need to accurately represent the cluster and structure features of all the entities of a seen (resp., unseen) type t, and the semantic embeddings $\mathbf{t}$ should capture the semantic connections between the seen and unseen types. In what follows, we introduce our feature and semantic embeddings.

4.1 Feature Encoder

The feature encodings need to capture the cluster and structure feature distribution of the entities of a type. We start with entity embeddings and adopt the TransE embedding $\mathbf{e}$ for each entity e due to its simplicity and effectiveness in capturing KG structure features. In particular, the TransE embedding enforces that $\mathbf{e}_1 + \mathbf{r} \approx \mathbf{e}_2$ for each triple (e_1, r, e_2) in the KG. Yet the collection $\{\mathbf{e} \mid e \in ins(t)\}$ is not a good feature encoding of type t, as TransE embeddings do not reflect the cluster features; that is, in the embedding space, those embeddings of entities belonging to the same type should be close to each other. Moreover, it is desirable that the distribution is not significantly dependent on the number of entities in each type. Hence, we use a feed-forward layer to encode each entity as in [14].

$$f_1(e) = \sigma(\mathbf{W}_1\mathbf{e} + \mathbf{b}_1), \tag{1}$$

where σ is the active function tanh, and $\mathbf{W}_1$ and $\mathbf{b}_1$ are learnable parameters.

To enhance the above encoding with structure features, we further encode the features of (direct) neighbours of entities. For each entity e, let $in(e) = \{(r, e') \in \mathcal{R} \times \mathcal{E} \mid (e', r, e) \in \mathcal{K}\}$ and $out(e) = \{(r, e') \in \mathcal{R} \times \mathcal{E} \mid (e, r, e') \in \mathcal{K}\}$. Then, for each entity e,

$$f_2(e) = \sigma(\frac{1}{|in(e)|} \sum_{(r,e') \in in(e)} \mathbf{W}_2(\mathbf{e}' \oplus \mathbf{r}) + \mathbf{b}_2), \tag{2}$$

$$f_3(e) = \sigma(\frac{1}{|out(e)|} \sum_{(r,e') \in out(e)} \mathbf{W}_2(\mathbf{e}' \oplus -\mathbf{r}) + \mathbf{b}_2), \tag{3}$$

where $\mathbf{W}_2$ and $\mathbf{b}_2$ are learnable parameters, and $\oplus$ is the concatenation operation.

The encoding of an entity e is the concatenation of $f_1(e)$, $f_2(e)$, and $f_3(e)$ as follows,

$$\mathbf{x}_e = \mathbf{W}_3(f_1(e) \oplus f_2(e) \oplus f_3(e)) + \mathbf{b}_3, \tag{4}$$

where $\mathbf{W}_3$ and $\mathbf{b}_3$ are learnable parameters. For scalability, we can set an upper limit for the number of neighbours.

The feature encoding of a type t is the collection of the entity encodings of all the entities belonging to t, i.e.,

$$\mathcal{X}_t = \{\mathbf{x}_e \mid e \in ins(t)\}. \tag{5}$$

To train the parameters $\omega = \{\mathbf{W}_1, \mathbf{W}_2, \mathbf{W}_3, \mathbf{b}_1, \mathbf{b}_2, \mathbf{b}_3\}$, for each entity type t, we follow the approach in [14] to randomly take out k reference entity type instances (e^*, t) from the training set, and the remaining are positive instances (e^+, t). Also, we generate a set of negative instances (e^-, t) by corrupting the entities. Based on the assumption that the distribution is invariant of the entity numbers, we calculate the cluster center using the reference instances, i.e., $\mathbf{x}^c = \frac{1}{k}\sum_{e^*} \mathbf{x}_{e^*}$. Then we calculate the dot similarity respectively with the feature encodings of positive instances $\mathbf{x}_{e^+}$ and negative instances $\mathbf{x}_{e^-}$, denoted as score_ω^+ and score_ω^-. We use margin-based ranking loss and it can be described as below,

$$L_\omega = \max(0, \gamma + \text{score}_\omega^+ - \text{score}_\omega^-), \tag{6}$$

where γ denotes the margin.

4.2 Semantic Encoder

To encode type semantics and establish semantic connections between the seen and unseen types, we use two sources of knowledge. One is the ontological knowledge expressed as ontological triples (t_1, r, t_2) and the other is type names or textual labels.

The encoding of ontological triples is based on TransE, by enforcing $\mathbf{t}_1^o + \mathbf{r}^o \approx \mathbf{t}_2^o$ for each ontological triple (t_1, r, t_2). Such an embedding captures ontological features, as semantically alike types are likely to be associated with similar relations and other types. It is also consistent with how we embed KG triples, but it is worth noting that the embedding of a relation $\mathbf{r}^o$ is different from its KG embedding $\mathbf{r}$, as the embedding space of the entities and that of the types are different. Unlike [9,18], we do not map the entity embeddings into the embedding space of their types, because the unseen types do not have entities and considering entities does not help to capture semantic similarities between seen and unseen types.

Hence, the score function based on ontological knowledge is as follows:

$$L_o = \max(0, \gamma + \sum_{(t_1,r,t_2)\in\mathcal{O}} -\|\mathbf{t}_1^o + \mathbf{r}^o - \mathbf{t}_2^o\|_2 - \sum_{(t_3,r,t_4)\notin\mathcal{O}} -\|\mathbf{t}_3^o + \mathbf{r}^o - \mathbf{t}_4^o\|_2), \quad (7)$$

where $\|\cdot\|_2$ denotes L2-norm.

A limitation of semantic encoding purely based on ontological knowledge is that such knowledge is often very incomplete. This is well recognized for KG completion and especially true for entity typing, as a majority of types have limited information about their entities and also lack ontological knowledge, which is known as the *long-tail* problem. To address such an issue, we use textual knowledge to complement the ontological knowledge, and the semantic encoding of a type combines its ontological and textual semantics. Most of the types come with meaningful names or textual labels. Hence, we use the language encoding $\mathbf{t}^l$ of a type t to represent its textual semantics. Unlike many existing approaches that enforce the language encoding to satisfy ontological constraints [8], e.g., $\mathbf{t}_1^l + \mathbf{r}^l \approx \mathbf{t}_2^l$, which could potentially change the initial textual semantics, we instead map the ontological embeddings $\mathbf{t}^o$ into the textual embedding space as follows.

$$L_l = \sum_{t\in\mathcal{T}} -\|\sigma(\mathbf{W}_l \cdot \mathbf{t}^o + \mathbf{b}_l) - \mathbf{t}^l\|, \quad (8)$$

where $\mathbf{W}_l$ and $\mathbf{b}_l$ are learnable parameters and $\sigma(\cdot)$ is a non-linear activation function such as `tanh`.

Then, we combine the semantic loss functions as follows:

$$L_s = L_o + L_l, \quad (9)$$

Finally, the semantic embedding of type t is $\mathbf{t} = \mathbf{t}^o$, which incorporates both the ontological and textual connections with other types.

4.3 Generative Adversarial Model

Generator. The GAN Generator takes the semantic embeddings $\mathbf{t}$ for types with a random vector $\mathbf{z}$ sampled from Gaussian distribution, and generates fake feature encodings $\hat{\mathcal{X}}_t$, i.e., $\hat{\mathcal{X}}_t = G_\theta(\mathbf{t}, \mathbf{z})$ with parameters θ, which is implemented by two fully-connected (FC) layers and a layer of normalization as in [14]. To avoid mode collapse and improve diversity, we adopt Wasserstein loss $\mathbb{E}[\cdot]$ and an additional classification loss $L_{cls}(\cdot)$, which is formulated as the margin ranking loss as Eq. (6). In addition, visual pivot regularization L_p is also applied to provide enough inter-class discrimination.

$$L_{G_\theta} = -\mathbb{E}[D_\phi(G_\theta(\mathbf{t}, \mathbf{z}))] + \lambda_1 L_{cls}(G_\theta(\mathbf{t}, \mathbf{z})) + \lambda_2 L_P, \tag{10}$$

where $D_\phi(\cdot)$ is the Discriminator which we will introduce in detail below, and ϕ, λ_1, and λ_2 are hyperparameters.

Discriminator. The first component is an FC layer that acts as a binary classifier to separate real data from fake data, and we utilize the Wasserstein loss again. The second part is the classification loss. In order to stabilize training behaviour and eliminate mode collapse, we also adopt the gradient penalty L_{GP} to enforce the Lipschitz constraint. It penalizes the model if the gradient norm moves away from its target norm value 1. In summary, the loss function of the Discriminator is formulated as:

$$L_{D_\phi} = \mathbb{E}[D_\phi(\hat{\mathcal{X}}_t)] - \mathbb{E}[D_\phi(\mathcal{X}_t)] + \frac{1}{2}L_{cls}(\hat{\mathcal{X}}_t) + \frac{1}{2}L_{cls}(\mathcal{X}_t) + L_{GP}. \tag{11}$$

Unseen Types Prediction. After the GAN is trained, the Generator can generate a feature encoding $\hat{\mathcal{X}}'_t$ for a given unseen type t'. For an entity e, its similarity score $score(e, t')$ can be calculated by cosine similarity between $\hat{\mathcal{X}}'_t$ and $\mathbf{x}_e$. Since $\hat{\mathcal{X}}'_t$ preserves the feature distribution in a way that is insensitive to the number of samples, we can sample a relatively small number of n samples from $\hat{\mathcal{X}}'_t$, i.e., $\{\hat{\mathbf{x}}_1, \ldots, \hat{\mathbf{x}}_m\}$.

We utilize the average cosine similarity value as the ultimate ranking score,

$$score(e, t') = \frac{1}{m}\sum_{i=1}^{m} \frac{\hat{\mathbf{x}}_i \cdot \mathbf{x}_e}{\|\hat{\mathbf{x}}_i\|_2 \cdot \|\mathbf{x}_e\|_2}. \tag{12}$$

5 Experiments

We have conducted two sets of experiments to evaluate the performance of our model.

5.1 Datasets

Since there is no available zero-shot entity typing dataset, we modified three commonly used entity typing datasets FB15KET [18], YAGO43KET [18] and DBpedia111K-174 [9] to zero-shot scenarios. In particular, we randomly selected k unseen types in each dataset and removed their entity-type instances from the training data. The number k of unseen types however are carefully selected, because if k is too small, the selected unseen types may not be representative; yet if k is too large, there will be a significant number of long-tail types. Also, we only removed entity-type instances but not triples as otherwise the resulting KG may not be a connected graph.

Besides separating seen and unseen types, we have also refined the names and text labels of the types for language models to encode their semantics. In FB15KET, the entity and type names are often composed of 3 phrases, such as "/base/tmv112233/topic", and some of the phrases are abbreviations or in Latin, which cannot be utilized by a language model like Glove [13] or does not provide much textual semantics. Hence, we filtered the entities based on their Glove encodings, and only kept those entities and types whose names only contain meaningful phrases (i.e., are contained in the dictionary). And we divided the remaining entity-type instances by 80:20 for train:test.

For YAGO43KET, many entity types are very specific, such as "*wikicat Footballers at the 2000 Summer Olympics*", which includes a specific year, making it difficult to obtain semantic representation which can distinguish different years. Thus, we substitute these types with their parent entity types with the help of relation *rdfs:subClassOf* which describes the hierarchical structure in ontology schema. The way we divide them into seen and unseen types in the same way as FB15KET.

As we discussed before, the *long-tail* issue is rather common in KGs, where a relatively small number of types have large numbers of entities belonging to them (i.e., having many entity-type instances), whereas the majority of types only have a few or none entities. To see how our approach address the long-tail issue, we also evaluated on long-tail datasets, by removing the unseen types with the largest numbers of entities, such as *person*, till the entity-type instances of the remaining unseen types are around 20% of the initial datasets.

The statistics of the modified datasets are presented in Table 1, including the number of unique entities (#Entity), triples (#Triples), seen types (#Seen) for training, unseen types (#Unseen) and long-tail unseen types (#Unseen') for testing, as well as entity-type instances for training (#Train), testing (#Test), and long-tail testing (#Test').

5.2 Ontological Knowledge

We obtain ontological knowledge by extracting ontological triples from Freebase and YAGO3 for FB15KET-ZS and YAGO43KET-ZS, respectively, by identifying triples whose subjects and objects correspond to the entity types in FB15KET-ZS and YAGO43KET-ZS. The ontological knowledge for DBpedia-ZS is from [9].

Table 1. Statistics of the constructed datasets.

Dataset	#Entity	#Triple	#Seen	#Unseen	#Unseen'	#Train	#Test	#Test'
FB15KET-ZS	14951	483142	1450	236	196	78236	20594	987
YAGO43KET-ZS	41723	331687	97	30	27	23493	5797	824
DBpedia-ZS	98336	658505	143	30	24	68272	17487	447

To control the numbers of ontological triples and to reduce the noise introduced, we selected certain relations, such as *equivalentType* (FB15K), *subClassOf* (YAGO), and *similarTo* (DBpedia), and sampled those types associated the selected relations.

The statistics of the ontologies are presented in Table 2, where we record the numbers of triples (#Triple), types (#Type), and relations (#Relation).

Table 2. Statistics of the ontologies.

Dataset	#Triple	#Type	#Relation
FB15KET-ZS	56852	27789	8
YAGO43KET-ZS	35021	14387	3
DBpedia-ZS	763	174	20

5.3 Baselines

We compare our model with the classic KG entity typing model ConnectE [18] and JOIE [9] that is specially designed to capture ontological knowledge. ConnectE cannot handle zero-shot learning and for the comparison, we adapted it for the zero-shot scenario inspired by [14], which we call ZS-ConnectE. In particular, for each type t, we use our Generator without random noise to initialize the type embedding $G_\theta(\mathbf{t})$ where $\mathbf{t}$ is the semantic embedding. This allows us to generate type embeddings for unseen types purely based on ontological and textual knowledge. JOIE can handle zero-shot entity typing tasks, as it generates type embeddings from ontological knowledge. We evaluate JOIE with various base KG embeddings TransE (TE), DisMULT (DM) [16] and HOLE (HL) [11]. Other entity typing models like CET [12] cannot handle nor be easily adapted for zero-shot learning, and thus are not included in the evaluation.

5.4 Main Results

We compare with the baselines and evaluate our model under various configurations, including using only ontological knowledge (O), using only textual knowledge (T), and both (OT). We use the standard metrics of the mean reciprocal

Table 3. Zero-shot entity typing on FB15KET-ZS, YAGO43KET-ZS, and DBpedia-ZS. The best results are in bold and the second best are underlined.

Dataset	FB15KET-ZS				YAGO43KET-ZS				DBpedia-ZS			
	MRR	H@1	H@3	H@10	MRR	H@1	H@3	H@10	MRR	H@1	H@3	H@10
JOIE(TE)	0.20	10.82	21.46	37.93	0.33	14.04	35.35	87.94	0.34	22.94	33.76	61.39
JOIE(DT)	0.02	0.05	0.23	0.67	0.28	17.18	25.60	48.44	0.33	21.97	36.72	45.14
JOIE(HE)	0.14	6.10	14.76	30.55	0.22	7.00	20.49	65.36	0.20	9.04	19.84	41.67
ZS-ConnectE	0.10	5.77	8.74	16.12	0.41	32.07	41.54	52.46	0.17	5.51	14.21	44.78
Ours (O)	0.13	9.16	13.08	19.46	0.58	43.54	61.89	94.24	0.35	19.43	39.11	73.65
Ours (T)	0.26	13.30	31.34	50.50	0.46	28.03	50.42	94.89	0.65	52.08	75.00	91.51
Ours (OT)	**0.36**	**27.51**	**38.70**	**51.59**	**0.68**	**50.04**	**83.32**	**98.15**	**0.75**	**68.63**	**75.26**	**92.61**

rank (MRR) and the percentage of correct entities ranked in the top 1, 3, or 10 (H@1, H@3, H@10). And we use the filtered setting as in [2].

From the results in Table 3, our model significantly outperforms both JOIE and ZS-ConnectE. Note that our model with only ontological knowledge already outperforms ZS-ConnectE and in many cases, JOIE as well, which shows the effectiveness of our generative model for zero-shot entity typing. The performance of our model is significantly improved with textual knowledge. The ontological and textual knowledge are complementary, as the benefit of ontological knowledge is more obvious where textual labels are rather simple, like YAGO43KET-ZS, whereas the benefit of textual knowledge is demonstrated when ontological triples are relatively limited, like DBpedia-ZS. Overall, our model has the best performance when combining ontological and textual knowledge.

5.5 Long-Tail Results

We evaluate our model on long-tail versions of the datasets, obtained by removing "popular" unseen types whose entity-type instances constitute of 80% of the total entity-type instances.

Table 4. Zero-shot entity typing on long-tail datasets. The best results are in bold and the second best are underlined.

Dataset	FB15KET-ZS*				YAGO43KET-ZS*				DBpedia-ZS*			
	MRR	H@1	H@3	H@10	MRR	H@1	H@3	H@10	MRR	H@1	H@3	H@10
JOIE(TE)	0.23	13.06	25.00	47.17	0.33	15.39	37.82	74.91	0.25	12.72	24.33	56.25
JOIE(DM)	0.03	0.30	1.01	4.05	0.40	28.00	39.64	76.36	0.22	6.47	22.77	54.46
JOIE(HL)	0.08	2.53	7.59	17.11	0.24	8.36	25.58	68.24	0.23	8.26	23.21	57.37
ZS-ConnectE	0.07	2.13	6.58	13.06	0.16	3.27	10.18	64.85	0.17	3.35	14.06	51.34
Ours (O)	0.06	2.23	4.45	10.43	0.33	20.61	31.27	62.91	0.21	12.28	18.53	27.90
Ours (T)	**0.28**	**16.50**	**31.28**	**47.67**	**0.52**	**41.33**	**52.48**	76.00	**0.44**	**31.03**	**50.00**	66.07
Ours (OT)	0.23	13.46	23.38	43.62	0.46	31.52	47.15	**86.91**	0.42	**31.03**	41.29	**68.53**

From the results in Table 4, we can see that focusing on the long-tail portion of the datasets highlights the usefulness of textual knowledge, and our models using only textual knowledge shows the best performance in most of the cases. Our model using ontological knowledge, including the combined ontological and textual knowledge, is impacted by the reduced ontological knowledge.

6 Conclusion

In this paper, we proposed as far as we know the first zero-shot entity-typing approach based on generative adversarial learning. To this end, we have proposed novel feature and semantic encoding methods to generate feature encodings for unseen types based on their semantic connections with seen types, and the semantic connections are captured by combining ontological and textual knowledge of types. Experiments on three zero-shot entity-typing benchmarks adapted from existing datasets demonstrate the superior performance of our model.

Acknowledgements. This work was partially supported by the National Natural Science Foundation of China under grant 61976153.

References

1. Bollacker, K., Evans, C., Paritosh, P., Sturge, T., Taylor, J.: Freebase: a collaboratively created graph database for structuring human knowledge. In: Proceedings of the 2008 ACM SIGMOD International Conference on Management of Data, pp. 1247–1250 (2008)
2. Bordes, A., Usunier, N., Garcia-Duran, A., Weston, J., Yakhnenko, O.: Translating embeddings for modeling multi-relational data. In: Advances in Neural Information Processing Systems 26 (2013)
3. Chen, Z., Chen, J., Geng, Y., Pan, J.Z., Yuan, Z., Chen, H.: Zero-shot visual question answering using knowledge graph. In: Hotho, A., et al. (eds.) ISWC 2021. LNCS, vol. 12922, pp. 146–162. Springer, Cham (2021). https://doi.org/10.1007/978-3-030-88361-4_9
4. Elsahar, H., Gravier, C., Laforest, F.: Zero-shot question generation from knowledge graphs for unseen predicates and entity types. arXiv preprint arXiv:1802.06842 (2018)
5. Ge, X., Wang, Y.C., Wang, B., Kuo, C.J.: Core: a knowledge graph entity type prediction method via complex space regression and embedding. Pattern Recogn. Lett. **157**, 97–103 (2022)
6. Geng, Y., et al.: OntoZSL: ontology-enhanced zero-shot learning. In: Proceedings of the Web Conference 2021, pp. 3325–3336 (2021)
7. Geng, Y., Chen, J., Chen, Z., Pan, J.Z., Yuan, Z., Chen, H.: K-ZSL: resources for knowledge-driven zero-shot learning. arXiv preprint arXiv:2106.15047 (2021)
8. Gesese, G.A., Biswas, R., Alam, M., Sack, H.: A survey on knowledge graph embeddings with literals: Which model links better literal-ly? Semantic Web **12**(4), 617–647 (2021)
9. Hao, J., Chen, M., Yu, W., Sun, Y., Wang, W.: Universal representation learning of knowledge bases by jointly embedding instances and ontological concepts. In: Proceedings of the 25th ACM SIGKDD International Conference on Knowledge Discovery & Data Mining, pp. 1709–1719 (2019)

10. Lehmann, J., et al.: DBpedia-a large-scale, multilingual knowledge base extracted from wikipedia. Semantic web **6**(2), 167–195 (2015)
11. Nickel, M., Rosasco, L., Poggio, T.: Holographic embeddings of knowledge graphs. In: Proceedings of the AAAI Conference on Artificial Intelligence, vol. 30 (2016)
12. Pan, W., Wei, W., Mao, X.L.: Context-aware entity typing in knowledge graphs. arXiv preprint arXiv:2109.07990 (2021)
13. Pennington, J., Socher, R., Manning, C.D.: Glove: Global vectors for word representation. In: Proceedings of the 2014 Conference on Empirical Methods in Natural Language Processing (EMNLP), pp. 1532–1543 (2014)
14. Qin, P., Wang, X., Chen, W., Zhang, C., Xu, W., Wang, W.Y.: Generative adversarial zero-shot relational learning for knowledge graphs. In: Proceedings of the AAAI Conference on Artificial Intelligence, vol. 34, pp. 8673–8680 (2020)
15. Suchanek, F.M., Kasneci, G., Weikum, G.: YAGO: a core of semantic knowledge. In: Proceedings of the 16th International Conference on World Wide Web, pp. 697–706 (2007)
16. Yang, B., Yih, W., He, X., Gao, J., Deng, L.: Embedding entities and relations for learning and inference in knowledge bases. arXiv preprint arXiv:1412.6575 (2014)
17. Zeng, D., Liu, K., Lai, S., Zhou, G., Zhao, J.: Relation classification via convolutional deep neural network. In: Proceedings of COLING 2014, the 25th International Conference on Computational Linguistics: Technical Papers, pp. 2335–2344 (2014)
18. Zhao, Y., Zhang, A., Xie, R., Liu, K., Wang, X.: Connecting embeddings for knowledge graph entity typing. arXiv preprint arXiv:2007.10873 (2020)
19. Zhao, Y., Zhou, H., Zhang, A., Xie, R., Li, Q., Zhuang, F.: Connecting embeddings based on multiplex relational graph attention networks for knowledge graph entity typing. IEEE Trans. Knowl. Data Eng. **35**, 4608–4620 (2022)
20. Zhu, G., Zhang, Z., Su, S.: Few-shot knowledge graph entity typing. In: Gama, J., Li, T., Yu, Y., Chen, E., Zheng, Y., Teng, F. (eds.) Advances in Knowledge Discovery and Data Mining. PAKDD 2022. LNCS, vol. 13280. Springer, Cham (2022). https://doi.org/10.1007/978-3-031-05933-9_26
21. Zhuo, J., Zhu, Q., Yue, Y., Zhao, Y., Han, W.: A neighborhood-attention fine-grained entity typing for knowledge graph completion. In: Proceedings of the Fifteenth ACM International Conference on Web Search and Data Mining, pp. 1525–1533 (2022)

Causal MRC: Mitigating Position Bias Based on Causal Graph

Jiazheng Zhu, Linjuan Wu, Shaojuan Wu, Xiaowang Zhang(✉), Yuexian Hou, and Zhiyong Feng

College of Intelligence and Computing, Tianjin University, Peiyang Park Campus, Tianjin, China
{jiazhengzhu,linjuanwu,shaojuanwu,xiaowangzhang,yxhou,zyfeng}@tju.edu.cn

Abstract. Extractive Machine Reading Comprehension (MRC) requires models to obtain start and end positions of answers from a given passage. MRC models may tend to rely on position bias as a shortcut, and thus they fail to learn the multi-source knowledge from both passages and questions sufficiently. Recent debiasing methods proposed to exclude the position prior during inference. However, they cannot distinguish the good position context and bad position bias from the whole prior. In this paper, we propose a novel MRC framework CausalMRC based on causal graph to mitigate position bias. Motivated by causal inference, we design a causal graph for MRC to formulate the position bias as the direct causal effect of passages on answers. Specifically, we mitigate the position bias by subtracting the direct position effect from the total causal effect. Experiments demonstrate that our proposed CausalMRC achieves competitive performance on the biased SQuAD dataset while performing robustly on the original SQuAD.

Keywords: Machine reading comprehension · Causal graph · Position bias

1 Introduction

Question answering (QA) has become the cornerstone of many fundamental AI systems, such as dialog understanding [30], language inference [41] and coreference resolution [6]. MRC is one of the most significant question answering (QA) tasks, which requires machines to answer questions given a passage. Extractive MRC models are trained to predict the start and end positions as the answers based on the assumption that answers always lie in the passage. Simply applying such a task objective without much consideration causes the model to use spurious positional cues (*i.e.*, position bias) to locate the answer in the passage, as demonstrated in [21,25]. As shown in Fig. 1, BERT [7] is trained on the biased train set where answers are always in the first sentence. In this case, BERT tends only to consider the corresponding sentence to answer the question (*i.e.*, *2012* in the first sentence), even when tested on the dataset with different position distributions.

A. El Abbadi et al. (Eds.): DASFAA 2023 Workshops, LNCS 13922, pp. 251–266, 2023.
https://doi.org/10.1007/978-3-031-35415-1_18

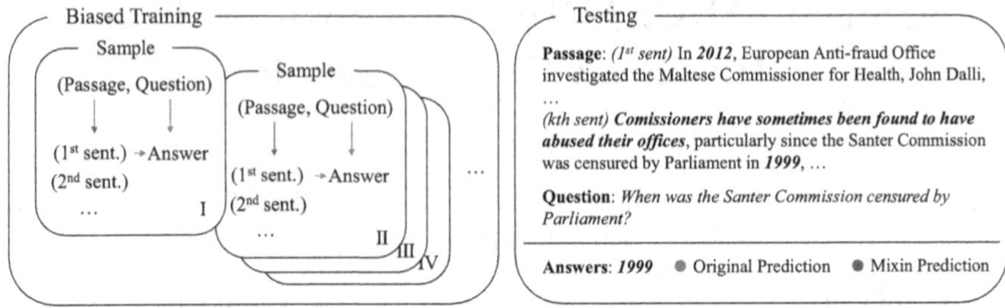

Fig. 1. An illustration of the position bias in MRC. Models are trained on the biased training set. The original prediction means BERT without extra methods, and the mixin prediction means BERT with the learned-mixin debiasing method.

In order to mitigate such position bias, existing methods mainly include data augmentation and model ensemble. The data augmentation methods balance the training data by human annotations [13] and counterfactual generation [9]. However, it is hard to manipulate the text for data augmentation in many cases [11]. Therefore, making unbiased inference under biased training remains a major challenge in MRC. The ensemble-based methods [3,35] employ a target model trained with a naive one that predicts answers exclusively based on dataset priors. The target model mitigates the prior during the test stage by excluding the naive model. However, the solution forces the target model to pay no attention to the position prior, which is not beneficial to predicting the answer position.

We argue that the position prior consists of both "bad" position bias (*e.g.*, excessive exploration of positional cues, and binding one sentence at the fixed position with answers) and "good" position context (*e.g.*, narrowing the answer position based on the question type "when" or "who"). Simply excluding the naive solution from the whole cannot utilize the good context. For example, Fig. 1 shows that the model using the learned-mixin debiasing method provided a wrong and irrelevant answer. The model made this prediction due to the absence of position information. Indeed, it is still challenging for recent debiasing MRC methods to distinguish the good and bad from the whole.

Motivated by causal inference and counterfactual explanations [26,36], we propose a novel MRC framework called CausalMRC based on causal graph to alleviate position bias in extractive MRC. Overall, we design a causal graph for MRC to formulate position bias as the direct causal effect of passages on answers and reduce the bias by subtracting the direct position effect from the total causal effect. Furthermore, we introduce two scenarios, traditional MRC and counterfactual MRC, to estimate the total causal effect and direct effect, respectively. The two scenarios are defined as follows:

Traditional MRC: *What will answer A be, if machine hears question Q, sees passage P, and extracts the multi-source knowledge S?*

Counterfactual MRC: *What would A be, if machine sees P, but had not extracted S or heard Q?*

In the case of traditional MRC, we can estimate the total causal effect of P and Q on A since both P and Q are available. However, traditional MRC cannot directly model the intended solution (*i.e.*, multi-source reasoning) because it is usually impossible to create a view of the data that excludes the bias. Therefore, we consider the situation where the machine sees P, but the multi-source knowledge S is intervened under the no-treatment condition, *i.e.*, the counterfactual question: "What would have happened if the machine had not performed multi-source reasoning?". Since Q and S are inaccessible, MRC models can only rely on the single-source impact. Thus, position bias can be identified by estimating the direct causal effect of P on A, i.e., pure position effect. The training stages follow ensemble-based methods [3,35] that train an ensemble model with a prevailing MRC model and a single-source branch. The debiased causal effect can be obtained by subtracting the pure position effect from the total effect. During the test stage, CausalMRC utilizes the debiased causal effect for inference.

Contributions to this paper are as follows.

- Our causal inference framework is the first to formulate the position bias in MRC as causal effects and mitigate the bias by subtracting the direct position effect from the total causal effect.
- We provide a novel counterfactual-based interpretation for ensemble-based MRC works, which can be unified into our causal inference framework.
- In order to prevent the competence of predicting correct answer spans from declining, we propose to utilize a weak model to learn the soft position bias instead of the fixed position prior.

Experimental results show that CausalMRC achieves competitive performance (from 14.29% to 81.79%) on the position-biased dataset, which almost recovers the performance of the model under the unbiased case, and remains stable on the relatively balanced SQuAD dataset. Furthermore, the debiased causal effect is general for different ensemble strategies.

2 Related Work

In this section, we mainly discuss the studies of extractive MRC and causal inference, which is related to our research object and technique.

2.1 Extractive Machine Reading Comprehension

Extractive MRC is a task to answer questions given a passage [8,19,29]. Many influential works rapidly progressed the development of effective MRC models [2,7,15,42]. However, these models predict the answer as positions without much consideration and unintentionally learn shortcuts to trick specific benchmarks [22,37,38], especially correlating answers with positions. We believe models should learn reasoning ability rather than superficial position correlation even in a skewed training set. Recently, a new variant of the reading comprehensive dataset, SQuAD$^k_{\text{train}}$, was proposed to evaluate the generalizability of MRC models on position bias [21]. Most of the recent solutions to reduce the position bias are ensemble-based methods that explicitly define and exclude the shortcut bias in the dataset [3]. They capture the shortcut bias by a bias-only model or statistical priors and ensemble with the bias to prevent models from converging to it [2,35]. The ensemble-based methods [3,4] can be unified into our proposed causal inference framework.

2.2 Causal Inference

[12,28] has been widely used in medicine, public policy, and epidemiology for years [1,24,31]. It not only is a framework for explaining data but also provides causal modeling tools [40] and solutions to overcome traditional impediments to achieving intended goals by estimating causal effects [27]. Recently, causal inference has also attracted increasing attention in natural language processing to mitigate the dataset bias [10]. We consider that data augmentation for debiasing can also be approximate to causal intervention [43]. These methods learn to generate additional counterfactual samples with human annotations [6,20] or generative models [14,23], which can be viewed as physical interventions on text input. Unlike these works that generate counterfactual examples or representations [5,11] for debiased training, our method focuses on counterfactual inference with even biased training data and offers a fundamental causal framework for mitigating position bias in MRC.

3 Causal Inference for MRC

Following the typical formulation, the extractive MRC models are required to extract the start and end position of an answer $A = a$ given a passage $P = p$ and a question $Q = q$. We represent a random variable as a capital letter and its observed value as a lowercase letter. We first introduce the causal graph for MRC, which is the core of our causal inference MRC framework CausalMRC.

3.1 Causal Graph

The causal graph for MRC is illustrated in Fig. 2(a). Causal graph reflects the causal relations between variables [12,33]. Here, X denotes the input variables $\{P, Q\}$, Y denotes the output variable A, and the mediator M denotes the multi-source knowledge variable. The effect of the input $\{P, Q\}$ on Y can be divided into the single-source impact and multi-source impact. The single-source impact captures the direct effect of $X = \{P\}$ on Y via $X = \{P\} \rightarrow Y$. This path reflects the position bias because only one source of inputs is utilized. The multi-source impact captures the indirect effect of $X = \{P, Q\}$ on Y via the multi-source knowledge M, *i.e.*, $X = \{P, Q\} \rightarrow M \rightarrow Y$. This path represents the comprehensive reasoning process, where MRC models make inferences based on multi-source knowledge.

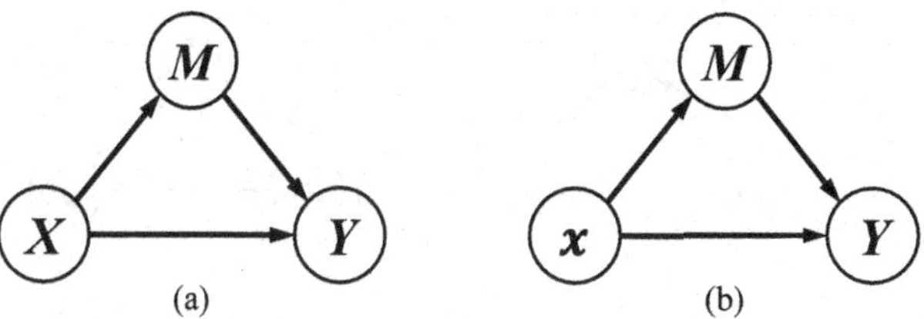

Fig. 2. (a) Causal graph for MRC. X: passage and question. M: multi-source knowledge. Y: output answer. (b) The intervention case of MRC. x: observed value of X.

Causal inference is used to translate causal assumptions from graphs to formulas. Figure 2(b) shows the intervention case in which the input X is set to x. The value to be obtained for the output Y, in this case, is defined as:

$$Y_x = P(Y|do(X = x)) \tag{1}$$

where M does not appear because M is naturally influenced by x without the intervention in this factual scenario.

3.2 Counterfactual Explanations

In the counterfactual scenario, X can be set as different values for M and Y. For example, Y_{x,M_x} describes the situation where X is set to x and M is set to the value when X had been x, *i.e.*, $Y_{x,M_x} = P(Y|X = x, M = P(M|X = x))$, which is similar to but different from the intervention situation. And $Y_{x,M_{x^*}}$ describes the situation where X is set to x and M is set to the value when X had been x^*, *i.e.*, $Y_{x,M_{x^*}} = P(Y|X = x, M = P(M|X = x^*))$. Note that X can be simultaneously set to different values x and x^* only in the counterfactual world. Figure 3(a) and Fig. 3(c) illustrates the two examples of counterfactual explanations [32].

Following the counterfactual explanations, we denote the answer that the output y would obtain if X is set to x (*i.e.*, $\{p, q\}$ or $\{p\}$) as

$$Y_{x,M_x}(y) = Y(y; X = x, M = m) \tag{2}$$

Without loss of generality, we omit y for simplicity and the M is denoted as $M_x = M(X = x)$. As shown in Fig. 3(a), there are two paths in total directly connected to Y, *i.e.*, $X \rightarrow Y$ and $M \rightarrow Y$. Therefore, Y_{x,M_x} can be rewritten as:

$$Y_{x,M_x} = Z_{x,m} = Z(X = x, M = m) \tag{3}$$

3.3 Causal Effects

The definition of causal effects is the comparison of two potential outcomes between the two states [34]. We suppose that $X = x$ represents "treatment

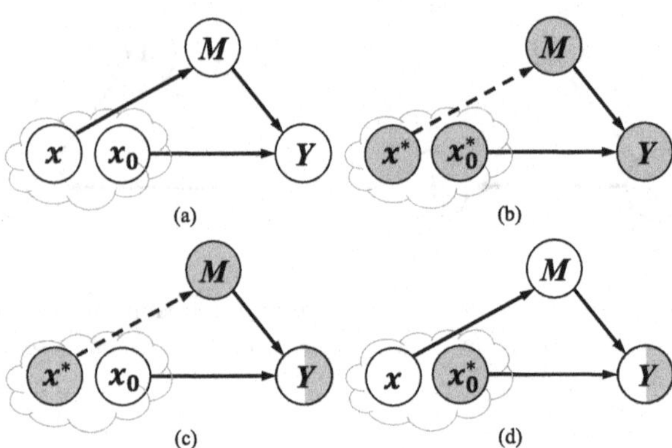

Fig. 3. The counterfactual cases for calculating Natural Direct Effect (NDE), Total Indirect Effect (TIE), Natural Indirect Effect (NIE) and Total Direct Effect (TDE). The four examples denote $Z_{x,m}$, Z_{x^*,m^*}, Z_{x,m^*} and $Z_{x^*,m}$. The white node is at the value $X = x_0 = p$ and $X = x = \{p, q\}$ while the gray node is at $X = x^*$ and $X = x_0^*$.

state" and $X = x^*$ represents "no-treatment state". The Total Effect (TE) of $X = x$ on $Y = y$ compares two counterfactual situations $X = x$ and $X = x^*$, which is denoted as:

$$TE = Y_{x,M_x} - Y_{x^*,M_{x^*}} = Z_{x,m} - Z_{x^*,m^*} \tag{4}$$

where $m^* = M_{x^*}$. Here x^* represents the counterfactual case where x is not given. The counterfactual situation of Z_{x^*,m^*} is shown in Fig. 3(b).

As we have discussed in Sect. 1, MRC models may suffer from the superficial correlation between passages and answers and thus fail to conduct effective multi-source reasoning. Therefore, we expect MRC models to exclude the direct impact of passages. To achieve this goal, we proposed CausalMRC to estimate the causal effect of position bias on $Y = y$ by blocking the effect of M.

Total effect can be decomposed into Natural Direct Effect (NDE) and Total Indirect Effect (TIE). As show in Fig. 3(c), the effect of M is blocked (*i.e.*, Z_{x,m^*}). NDE denotes the effect of X on Y with the mediator M blocked and the increase in Y with X changing from x^* to x. We obtain the NDE of P on Y by subtracting the no-treatment state from the counterfactual state illustrated in Fig. 3(c):

$$NDE = Z_{x,m^*} - Z_{x^*,m^*} \tag{5}$$

Since only the single-source path has an effect on Y, NDE explicitly captures the position bias. Thus, the effect of the multi-source path, *i.e.*, TIE, can be represented as:

$$TIE = TE - NDE = Z_{x,m} - Z_{x,m^*} \tag{6}$$

TE can also be decomposed into Natural Indirect Effect (NIE) and Total Direct Effect (TDE). Figure 3(d) shows that the effect of M is not blocked (*i.e.*,

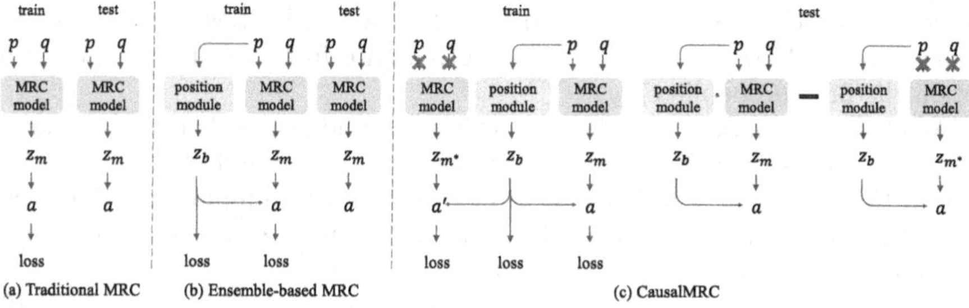

Fig. 4. Comparison between our CausalMRC and other MRC methods. (a) Traditional MRC methods use a single MRC model. (b) Ensemble-based methods [3,35] use an additional bias-only model to capture the bias during the training stage. The bias model is not used during testing. (c) CausalMRC maintains the bias-only model during the test stage and makes inferences based on the debiased causal effect.

$Z_{x^*,m}$). An alternative option to mitigate position bias is to subtract TDE of passages on answers from TE, which is formulated as:

$$TDE = Z_{x,m} - Z_{x^*,m} \tag{7}$$

$$NIE = TE - TDE = Z_{x^*,m} - Z_{x^*,m^*} \tag{8}$$

Intuitively, both TIE and NIE increase the confidence of the output answer y given the multi-source knowledge, *i.e.*, from m^* to m. The difference between TIE and NIE is the existence of $x = \{p\}$. The passage p is blocked to calculate NIE (*i.e.*, x^*), while $x = \{p\}$ is given to calculate TIE. We use TIE to reserve the position context. Similarly, the difference between TDE and NDE is the existence of the multi-source knowledge m. Note that we hope to exclude the effect directly caused by passages. Therefore, the mediator knowledge should be blocked when estimating the pure position effect, which is captured by NDE.

In Sect. 4.4, we will further discuss the differences between TIE and NIE. We select the answer distribution of TIE for predicting, which entirely differs from conventional solutions based on the posterior probability, *i.e.*, $P(y|x)$.

4 Causal MRC Model

Figure 4 shows the conceptual comparison between our CausalMRC, ensemble-based strategies [3,35], and traditional methods [7,42]. Our CausalMRC model makes inferences based on the debiased causal effect, which can mitigate the position bias and preserve the good context.

4.1 Parameterization

The calculation of the probabilities about all variables Z is parameterized by two neural functions and one ensemble function:

$$Z_b = \mathcal{F}_X(p), Z_m = \mathcal{F}_M(p, q), Z_{b,m} = h(Z_b, Z_m) \tag{9}$$

where $\mathcal{F}_X$ is the single-source branch (*i.e.*, $P \rightarrow Y$) and $\mathcal{F}_m$ is the multi-source branch (*i.e.*, $\{P, Q\} \rightarrow M \rightarrow Y$). The outputs are fused by the function h to obtain the final probability $Z_{b,m}$.

As illustrated in Sect. 3.2, the counterfactual condition is defined as ignoring the variable x. We represent this condition as $P = p^* = \phi$ and $Q = q^* = \phi$ while it is worth noting that MRC models cannot deal with void inputs. Therefore, we define the void inputs as a learnable parameter c. In this case, Z_b and Z_m can be defined as:

$$z_b^* = c, z_m^* = c \tag{10}$$

4.2 Ensemble Strategies

We expect that the ensemble probabilities $Z_{b,m}$ are a fusion of Z_b and Z_m. We utilize two ensemble variants, Product of Experts (PoE) [16] and Learned-Mixin (LM) [3]:

$$\text{(PoE)} \quad h(Z_m, Z_b) = \log(Z_m) * softmax(\log(Z_b)) \tag{11}$$

where $log(Z_m)$ is a log probability from the multi-source branch, $log(Z_b)$ is a log probability from the bias branch, and $Z_{b,m} = h(Z_m, Z_b)$.

$$\text{(LM)} \quad h(Z_m, Z_b) = \ softmax(\log(Z_m) + g(m)\log(Z_b)) \tag{12}$$

where g is a trainable function, and m is hidden representations before the softmax layer from the multi-source branch. $g(m)$ transforms the representations to obtain a scalar value.

Implementation of Biased Model. We employ two methods to calculate the bias log probability of the single-source branch.

We first utilize *answer prior* based on the location of each answer in a passage. First, we define the j-th passage with L sentences as a set $\{S_1^j, S_2^j, ..., S_L^j\}$. Then, for the j-th example, the answer prior distribution of the i-th word position is given by the number of times that answers appear in the l-th sentence:

$$\log(z_{b_i}^j) = \frac{1}{N}\sum_{k=1}^{N} \mathbf{1}[a^k \in S_l^k], \ i \in S_l^j \tag{13}$$

where we use the indicator function $\mathbf{1}[condition]$ and N is the number of training passages in the training set. For example, on $\text{SQuAD}_{\text{train}}^{1st}$, the answer prior distribution for $\text{SQuAD}_{\text{train}}^{1st}$ is 1 when $i \in S_1^j$, and 0 when $i \notin S_1^j$, which is a skewed case. The distributions are more even for general datasets.

Another solution is to employ *weak model* with a much smaller scale compared with a large-scale model such as BERT. Models with few computational resources and parameters can easily learn to answer questions with shortcut tricks like position bias rather than questions requiring challenging skills like co-reference [35]. The weak model performs relatively well on training data but is likely to perform very poorly on the challenging test conditions.

In the case of predicting the start and end positions of answers, the above sentence-level answer prior is too strong and excessively pays attention to the bias of the whole sentence. As a result, the competence to predict the correct answer span is weakened. Therefore, the weak model is utilized to learn the position bias instead of the fixed position prior as the biased model:

$$\log(z_b) = \mathcal{F}_w(p) \tag{14}$$

where $\mathcal{F}_w(\cdot)$ is the weak model function.

4.3 Training

The training strategy follows [21]. As illustrated in Fig. 4(c), given a triplet (p, q, a) where a is the ground-truth answer of the passage-question pair, the branches are optimized by minimizing the cross-entropy losses over the probabilities $Z_{b,m}$ and Z_b:

$$\mathcal{L}_{pos} = \mathcal{L}_{b,m}(p, q, a) + \mathcal{L}_b(p, a) \tag{15}$$

where $\mathcal{L}_{x,m}$ and $\mathcal{L}_x$ are over $Z_{x,m}$ and Z_x respectively. Note that we present a learnable parameter c in Sect. 4.1 which controls the influence of the probabilities distribution Z_{x,m^*} like the knowledge distillation [17]. We hypothesize that the influence of NDE should be similar to that of TE. Otherwise, an inappropriate c would result in TIE dominated by TE or NDE. Therefore, we use Kullback-Leibler divergence to estimate c:

$$\mathcal{L}_{kl} = \frac{1}{|\mathcal{A}|} \sum_{a \in \mathcal{A}} -p(a|b, m)\log\frac{p(a|b, m^*)}{p(a|b, m)} \tag{16}$$

where $p(a|b, m)$ denotes $softmax(Z_{b,m})$ and $p(a|b, m^*)$ denotes $softmax(Z_{b,m^*})$. Since c only refers to the loss $\mathcal{L}_{kl}$, c will only be updated when $\mathcal{L}_{kl}$ is minimized during back propagation. The total loss is the combination of $\mathcal{L}_{pos}$ and $\mathcal{L}_{kl}$:

$$\mathcal{L} = \mathcal{L}_{pos} + \mathcal{L}_{kl} \tag{17}$$

4.4 Inference

For inference, we use the debiased causal effect (*i.e.*, TIE in the counterfactual explanation) during the test stage, which is implemented as:

$$\begin{aligned} TIE = TE - NDE = Z_{b,m} - Z_{b,m^*} &= h(Z_b, Z_m) - h(Z_b, Z_{m^*}) \\ = (\log\sigma(z_m) + g(m)\log\sigma(Z_b)) &- (\log\sigma(c) + g(m^*)\log\sigma(Z_b)) \end{aligned} \tag{18}$$

Revisiting NIE and Learned-Mixin. The current ensemble-based methods MCE [4] and Learned-Mixin [3] use an ensemble model that consists of a robust multi-source branch $\mathcal{F}_M$ and a bias-only branch $\mathcal{F}_X$. However, they simply exclude $z_b = \mathcal{F}_X(p)$ and only use $z_m = \mathcal{F}_M(p, q)$ for inference during the test stage. These methods can be unified into our causal inference framework. They also follow the causal graph in Fig. 3, whereas NIE in Eq. (8) is used to predict answers. Using Learned-Mixin as an example, NIE is calculated as:

$$\begin{aligned} NIE &= Z_{x^*,m} - Z_{x^*,m^*} \\ &= (\log\sigma(z_m) + g(m)\log\sigma(c)) \\ &- (\log\sigma(c) + g(m^*)\log\sigma(c)) \propto z_m \end{aligned} \tag{19}$$

where $c, g(m)$ and $g(m^*)$ are constants for the same sample, and $\sigma(\cdot)$ is the softmax function. $NIE \propto z_m$ is equivalent to the output score of the multi-source branch $\mathcal{F}_M$. Therefore, these ensemble-based methods use NIE for inference in our causal effect perspective.

5 Experiments

5.1 Datasets and Settings

We mainly conduct the experiments on the SQuAD [29] dataset. SQuAD is a classic reading comprehension benchmark, requiring MRC models to extract answers from passages. We fine-tune our models on biased training sets and evaluate them on test sets. It is a well-known MRC dataset that is widely utilized in a large amount of studies. The biased training set is a subset divided from the full training set $\text{SQuAD}_{\text{train}}$ according to the position of the answer. For instance, $\text{SQuAD}_{\text{train}}^{1st}$ is the subset where all answers are in the first sentences. And the test set $\text{SQuAD}_{\text{dev}}$ is divided into $\text{SQuAD}_{\text{dev}}^{1st}$ for biased evaluation and $\text{SQuAD}_{\text{dev}}^{other}$ for debiased evaluation.

The standard evaluation metrics are Exact Match (EM) and F1 score. We use two pre-trained language models (*i.e.*, BERT and XLNet) as backbone models. We employ Product of Expert (PoE) and Learned-Mixin (LM) as ensemble strategies and adopt a significantly smaller pre-trained language model TinyBERT [39] as our weak model. For convenience, we abbreviate the answer prior as AP and weak model as WM.

We compare our main method CausalMRC with two bias ensemble methods: PoE and LM. We use the same hyperparameters and training procedures for all models, including batch sizes 12 and epochs 2. All our implementations are based on PyTorch and Transformers.

5.2 Effects of Debiasing Methods

We first fine-tune BERT and XLNet on $\text{SQuAD}_{\text{train}}^{1st}$ and evaluate them on $\text{SQuAD}_{\text{dev}}$, $\text{SQuAD}_{\text{dev}}^{1st}$, and $\text{SQuAD}_{\text{dev}}^{other}$.

Results with BERT. The results of using debiasing methods with BERT are in Table 1. Overall, compared to other debiasing methods, our proposed CausalMRC achieves competitive performance. The large gap between the performance of the baselines in $\text{SQuAD}_{\text{dev}}^{other}$ and $\text{SQuAD}_{\text{dev}}^{1st}$ reflects that their predictions are highly biased towards the first sentences. Especially, in the case of BERT, EM score on $\text{SQuAD}_{\text{dev}}^{1st}$ is 77.71% while on $\text{SQuAD}_{\text{dev}}^{other}$ is 9.12%.

Table 1. Results of applying various debiasing methods based on BERT and XLNet. The models are trained with $\mathbf{SQuAD}^{1st}_{\text{train}}$ and evaluated on $\mathbf{SQuAD}_{\text{dev}}$, $\mathbf{SQuAD}^{1st}_{\text{dev}}$, $\mathbf{SQuAD}^{other}_{\text{dev}}$. The best results of each ensemble strategy are highlighted in each column. AP denotes that the answer prior and WM indicates the weak model.

Test set		$\mathbf{SQuAD}^{1st}_{\text{dev}}$		$\mathbf{SQuAD}^{other}_{\text{dev}}$		$\mathbf{SQuAD}_{\text{dev}}$	
		EM	F1	EM	F1	EM	F1
Baseline	TinyBERT	37.09	51.29	4.13	8.11	15.46	22.97
	BERT	77.71	86.37	9.12	14.29	32.72	39.09
PoE	AP	78.16	**86.83**	15.33	20.94	36.95	43.61
	$\text{CausalMRC}_{\text{AP}}$	$78.02^{-0.14}$	$86.74^{-0.09}$	$\mathbf{41.99}^{+26.66}$	$\mathbf{50.49}^{+29.55}$	$53.93^{+16.98}$	$\mathbf{63.51}^{+19.90}$
	WM	**78.17**	86.80	12.60	18.50	35.13	41.91
	$\text{CausalMRC}_{\text{WM}}$	$77.87^{-0.3}$	$86.41^{-0.39}$	$41.38^{+28.78}$	$49.95^{+31.45}$	$\mathbf{54.47}^{+19.34}$	$63.04^{+21.13}$
LM	AP	77.16	85.25	71.33	79.68	73.35	81.59
	$\text{CausalMRC}_{\text{AP}}$	$77.54^{+0.38}$	$86.13^{+0.88}$	$72.68^{+1.35}$	$80.57^{+0.89}$	$74.69^{+1.34}$	$82.78^{+1.19}$
	WM	77.40	86.17	72.14	79.97	73.97	82.11
	$\text{CausalMRC}_{\text{WM}}$	$\mathbf{77.66}^{+0.26}$	$\mathbf{86.32}^{+0.15}$	$\mathbf{73.33}^{+1.19}$	$\mathbf{81.79}^{+1.82}$	$\mathbf{75.18}^{+1.21}$	$\mathbf{83.38}^{+1.27}$
Baseline	XLNet	79.32	87.36	20.85	26.34	42.90	49.47
LM	AP	68.76	82.16	64.51	77.76	66.23	79.23
	$\text{CausalMRC}_{\text{AP}}$	$74.81^{+6.05}$	$85.25^{+3.09}$	$65.32^{+0.81}$	$78.97^{+1.21}$	$68.71^{+2.48}$	$81.19^{+1.96}$
	WM	73.33	84.99	64.31	76.72	67.31	79.48
	$\text{CausalMRC}_{\text{WM}}$	$\mathbf{76.10}^{+2.27}$	$\mathbf{86.12}^{+1.13}$	$\mathbf{65.51}^{+1.2}$	$\mathbf{79.13}^{+2.41}$	$\mathbf{69.04}^{+1.73}$	$\mathbf{81.50}^{+2.02}$

With a deep look at the results of $\text{SQuAD}^{other}_{\text{dev}}$, our CausalMRC models with PoE or LM effectively improve the performance. The LM ensemble strategy works the best after applying the weak model (WM) as the biased module, while the score of CausalMRC with PoE is much lower than LM. However, CausalMRC with PoE obtains a significant gain (from ∼20% to ∼50%) compared with the ensemble-based methods, which indicates that the effect of position bias is effectively suppressed with the help of counterfactual explanations.

Furthermore, the sentence-level answer prior (AP) seems to provide strong position bias signals, and the weak model is softer than it. Therefore, the improvement of WM is greater than that of the answer prior in most cases.

Results with XLNet. The results of using debiasing methods with XLNet are in the bottom of Table 1. The overall performance is similar to BERT, especially on PoE, so we omit it. The results indicate that our causal framework is general to various MRC backbones. It is worth noting that the EM score of LM based on XLNet drops significantly by ∼10% on $\text{SQuAD}^{1st}_{\text{dev}}$, which reveals that the LM forces models to overcorrect the position bias. However, our CausalMRC consistently improves the performance on three test sets, demonstrating that our methods can prevent models from removing good position context.

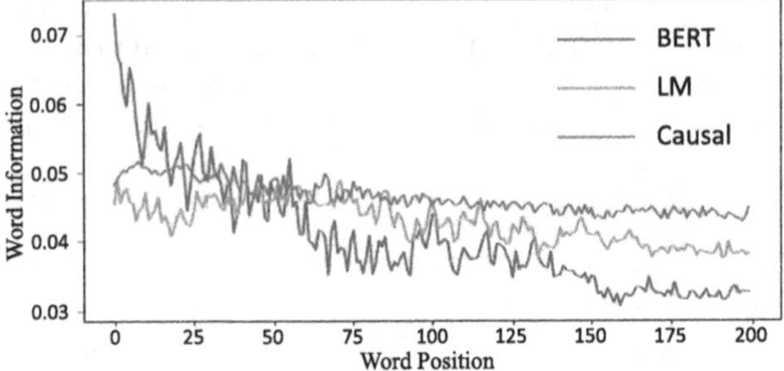

Fig. 5. Visualization of BERT models fine-tuned on $\text{SQuAD}^{1st}_{\text{train}}$ with LM_{WM} and $\text{CausalMRC}_{\text{WM}}$.

5.3 Visualization Analysis

To investigate the effect of debiasing methods, we visualize word information in the model. The word information for each word position is defined as the cosine similarity between the word embedding and its hidden state at the last layer. The similarities are averaged over $\text{SQuAD}_{\text{dev}}$. We visualize the three model fine-tuned on $\text{SQuAD}^{1st}_{\text{train}}$ in Fig. 5.

The BERT model (BERT) has higher similarities at the front of the passages. The similarity becomes smaller after the first few tokens, showing the position bias. The learned-mixin method (LM) seems to pass the word information safely across different positions. Our CausalMRC (Causal) carries more word information while delivering it more smoothly and stably across different positions.

5.4 Generalization Experiments

Next, we conduct generalization experiments on different sentence position biases, and the adversarial dataset such as the Adversarial SQuAD dataset [18].
Different Positions. We mainly fine-tune our models on $\text{SQuAD}^{1st}_{\text{train}}$ since the SQuAD training set has a large proportion of answers (~33%) in the first sentence. However, whether our method generalizes to different sentence positions is confusing. Depending on the positions of the answer sentences, we construct four additional biased training sets to validate this question, which is $\text{SQuAD}^{2nd}_{\text{train}}$, $\text{SQuAD}^{3rd}_{\text{train}}$, $\text{SQuAD}^{4nd}_{\text{train}}$, and $\text{SQuAD}^{others}_{\text{train}}$. We utilize BERT as the backbone and answer priors as the biased model to fine-tune models on four biased datasets. We evaluate them on $\text{SQuAD}_{\text{dev}}$ and use the F1 score as the metric.

The results are shown in Fig. 6. BERT suffers from position bias in every sentence position, and CausalMRC can promote ensemble-based methods with different answer position bias and improves performance. With the answer position of the sentence located more backward, the number of samples becomes less, and the sentence boundary becomes more blurred for machines. As a result, the answer priors are not as clear and accurate as $\text{SQuAD}^{1st}_{\text{train}}$. Therefore, we can see

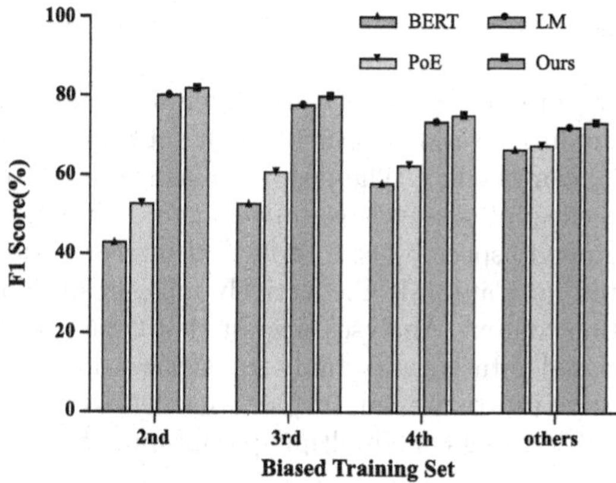

Fig. 6. Position bias in different positions. The debiasing methods (PoE, LM, Ours) utilize BERT as the backbone and AP as the biased model to fine-tune models on four biased datasets.

the performance of the models with different methods gets closer as the number (*i.e.*, 2nd, 3rd...) gets larger.

Table 2. F1 Score on the standard SQuAD dev set and Adversarial QA. * are the reported results, and ▲ represents our reproduced results of [35].

	$\text{SQuAD}_{\text{dev}}$	Adversarial QA	
		AddSent	AddOneSent
Bias Product [3]*	78.63	47.17	57.64
PoE [35]*	86.49	56.80	61.04
BERT-base	**88.49**	65.56	74.05
+ PoE▲	87.90	67.67	75.19
+ $\text{CausalMRC}_{\text{WM}}$	88.19	**68.41**	**76.04**

Adversarial Attack. In order to test the robustness of our CausalMRC, we fine-tune models on $\text{SQuAD}_{\text{train}}$ with the weak model as the biased model. We evaluate models on $\text{SQuAD}_{\text{dev}}$ and Adversarial SQuAD dataset [18] with F1 score. The Adversarial SQuAD is constructed by appending sentences to passages that would interfere with the prediction of models. As shown in Table 2, we compare our method CausalMRC with other debiasing methods based on PoE. BERT alone has low performance on adversarial sets but high performance on the original SQuAD. PoE training improves the adversarial performance (+2% on AddSent) while sacrificing a little performance of $\text{SQuAD}_{\text{dev}}$. Our CausalMRC reduces the cost and boosts adversarial robustness (+3% on AddSent and 2% on AddOneSent). Overall, our proposed CausalMRC can generalize to many cases.

6 Conclusion

In this paper, we have presented a novel causal inference framework, CausalMRC, that can effectively reduce position bias in MRC by subtracting the direct position effect from the total causal effect. The position bias is the first to be formulated as the direct causal effect of passages on answers and detached from the position prior in the fine-grained aspect by counterfactual interpretation. Experimental results demonstrate that CausalMRC effectively mitigates the position bias and preserves the position context. Analyses suggest that CausalMRC can generalize over the position-biased datasets and stably transfer word information across different positions. In the future, we will further consider extending our method to other extractive MRC datasets and multiple-paragraph settings.

References

1. Balke, A., Pearl, J.: Counterfactuals and policy analysis in structural models. arXiv preprint arXiv: 1302.4929 (2013)
2. Cheng, H., Chang, M., Lee, K., Toutanova, K.: Probabilistic assumptions matter: improved models for distantly-supervised document-level question answering. In: Proceedings of the 58th Annual Meeting of the Association for Computational Linguistics, pp. 5657–5667. Online (2020)
3. Clark, C., Yatskar, M., Zettlemoyer, L.: Don't take the easy way out: ensemble based methods for avoiding known dataset biases. In: Proceedings of the 2019 Conference on Empirical Methods in Natural Language Processing and the 9th International Joint Conference on Natural Language Processing, Hong Kong, China, pp. 4067–4080 (2019)
4. Clark, C., Yatskar, M., Zettlemoyer, L.: Learning to model and ignore dataset bias with mixed capacity ensembles. In: Findings of the Association for Computational Linguistics: EMNLP 2020, pp. 3031–3045. Online Event (2020)
5. Dalal, D., Arcan, M., Buitelaar, P.: Enhancing multiple-choice question answering with causal knowledge. In: Proceedings of Deep Learning Inside Out: The 2nd Workshop on Knowledge Extraction and Integration for Deep Learning Architectures, pp. 70–80. Online (2021)
6. Dasigi, P., Liu, N.F., Marasovic, A., Smith, N.A., Gardner, M.: Quoref: a reading comprehension dataset with questions requiring coreferential reasoning. In: Proceedings of the 2019 Conference on Empirical Methods in Natural Language Processing and the 9th International Joint Conference on Natural Language Processing, Hong Kong, China, pp. 5924–5931 (2019)
7. Devlin, J., Chang, M., Lee, K., Toutanova, K.: BERT: pre-training of deep bidirectional transformers for language understanding. In: Proceedings of the 2019 Conference of the North American Chapter of the Association for Computational Linguistics: Human Language Technologies, Minneapolis, MN, USA, pp. 4171–4186 (2019)
8. Dua, D., Wang, Y., Dasigi, P., Stanovsky, G., Singh, S., Gardner, M.: DROP: a reading comprehension benchmark requiring discrete reasoning over paragraphs. In: Proceedings of the 2019 Conference of the North American Chapter of the Association for Computational Linguistics: Human Language Technologies, Minneapolis, MN, USA, pp. 2368–2378 (2019)

9. Engstrom, L., Ilyas, A., Santurkar, S., Tsipras, D., Tran, B., Madry, A.: Learning perceptually-aligned representations via adversarial robustness. arXiv preprint arXiv: 1906.00945 (2019)
10. Feder, A., et al.: Causal inference in natural language processing: estimation, prediction, interpretation and beyond. arXiv preprint arXiv: 2109.00725 (2021)
11. Feder, A., Oved, N., Shalit, U., Reichart, R.: CausaLM: causal model explanation through counterfactual language models. Comput. Linguistics **47**(2), 333–386 (2021)
12. Fenton, N.E., Neil, M., Constantinou, A.C.: The book of why: the new science of cause and effect. Artif. Intell. **284**, 103–286 (2020)
13. Gardner, M., Artzi, Y., Basmova, V., Berant, J., Bogin, B., Chen, S.: Evaluating models' local decision boundaries via contrast sets. In: Findings of the Association for Computational Linguistics: EMNLP 2020, pp. 1307–1323. Online Event (2020)
14. Guo, J., Lu, S., Cai, H., Zhang, W., Yu, Y., Wang, J.: Long text generation via adversarial training with leaked information. In: Proceedings of the Thirty-Second AAAI Conference on Artificial Intelligence, New Orleans, Louisiana, USA, pp. 5141–5148 (2018)
15. He, P., Liu, X., Gao, J., Chen, W.: DeBERTa: decoding-enhanced BERT with disentangled attention. In: Proceedings of the 9th International Conference on Learning Representations, Austria. OpenReview.net (Online) (2021)
16. Hinton, G.E.: Training products of experts by minimizing contrastive divergence. Neural Comput. **14**(8), 1771–1800 (2002)
17. Hinton, G.E., Vinyals, O., Dean, J.: Distilling the knowledge in a neural network. arXiv preprint arXiv: 1503.02531 (2015)
18. Jia, R., Liang, P.: Adversarial examples for evaluating reading comprehension systems. In: Proceedings of the 2017 Conference on Empirical Methods in Natural Language Processing, Copenhagen, Denmark, pp. 2021–2031 (2017)
19. Joshi, M., Choi, E., Weld, D.S., Zettlemoyer, L.: TriviaQA: a large scale distantly supervised challenge dataset for reading comprehension. In: Proceedings of the 55th Annual Meeting of the Association for Computational Linguistics, Vancouver, Canada, pp. 1601–1611 (2017)
20. Kaushik, D., Hovy, E.H., Lipton, Z.C.: Learning the difference that makes a difference with counterfactually-augmented data. In: Proceedings of the 8th International Conference on Learning Representations, Addis Ababa, Ethiopia. OpenReview.net (2020)
21. Ko, M., Lee, J., Kim, H., Kim, G., Kang, J.: Look at the first sentence: position bias in question answering. In: Proceedings of the 2020 Conference on Empirical Methods in Natural Language Processing, pp. 1109–1121. Online (2020)
22. Lai, Y., Zhang, C., Feng, Y., Huang, Q., Zhao, D.: Why machine reading comprehension models learn shortcuts? In: Findings of the Association for Computational Linguistics: ACL/IJCNLP 2021, pp. 989–1002. Online Event (2021)
23. Lorberbom, G., Jaakkola, T.S., Gane, A., Hazan, T.: Direct optimization through arg max for discrete variational auto-encoder. In: Proceedings of the 2019 Annual Conference on Neural Information Processing Systems, Vancouver, BC, Canada, pp. 6200–6211 (2019)
24. MacKinnon, D.P., Fairchild, A.J., Fritz, M.S.: Mediation analysis. Annu. Rev. Psychol. **58**, 593–614 (2007)
25. Niu, Y., Zhang, H.: Introspective distillation for robust question answering. arXiv preprint arXiv: 2111.01026 (2021)

26. Pearl, J.: Direct and indirect effects. In: Proceedings of the 17th Conference in Uncertainty in Artificial Intelligence, Seattle, Washington, USA, pp. 411–420 (2001)
27. Pearl, J.: The seven tools of causal inference, with reflections on machine learning. Commun. ACM **62**(3), 54–60 (2019)
28. Pearl, J., Glymour, M., Jewell, N.P.: Causal Inference in Statistics: A Primer. Wiley, Chichester (2016)
29. Rajpurkar, P., Zhang, J., Lopyrev, K., Liang, P.: SQuAD: 100, 000+ questions for machine comprehension of text. In: Proceedings of the 2016 Conference on Empirical Methods in Natural Language Processing, Austin, Texas, USA, pp. 2383–2392 (2016)
30. Reddy, S., Chen, D., Manning, C.D.: CoQA: a conversational question answering challenge. Trans. Assoc. Comput. Linguistics **7**, 249–266 (2019)
31. Richiardi, L., Bellocco, R., Zugna, D.: Mediation analysis in epidemiology: methods, interpretation and bias. Int. J. Epidemiol. **42**(5), 1511–1519 (2013)
32. Robins, J.: A new approach to causal inference in mortality studies with a sustained exposure period-application to control of the healthy worker survivor effect. Math. Model. **7**(9), 1393–1512 (1986)
33. Robins, J.M.: Semantics of causal DAG models and the identification of direct and indirect effects. Oxford Statistical Science Series, pp. 70–82 (2020)
34. Rubin, D.B.: Bayesian inference for causal effects: the role of randomization. Ann. Stat., 34–58 (1978)
35. Sanh, V., Wolf, T., Belinkov, Y., Rush, A.M.: Learning from others' mistakes: avoiding dataset biases without modeling them. In: Proceedings of the 9th International Conference on Learning Representations. OpenReview.net, Virtual Event (2021)
36. Schölkopf, B.: Causality for machine learning. arXiv preprint arXiv: 1911.10500 (2019)
37. Shinoda, K., Sugawara, S., Aizawa, A.: Can question generation debias question answering models? A case study on question-context lexical overlap. arXiv preprint arXiv: 2109.11256 (2021)
38. Sugawara, S., Inui, K., Sekine, S., Aizawa, A.: What makes reading comprehension questions easier? In: Proceedings of the 2018 Conference on Empirical Methods in Natural Language Processing, Brussels, Belgium, pp. 4208–4219 (2018)
39. Turc, I., Chang, M., Lee, K., Toutanova, K.: Well-read students learn better: the impact of student initialization on knowledge distillation. arXiv preprint arXiv: 1908.08962 (2019)
40. Verma, T., Pearl, J.: An algorithm for deciding if a set of observed independencies has a causal explanation. In: Proceedings of the 8th Annual Conference on Uncertainty in Artificial Intelligence, Stanford, CA, USA, pp. 323–330 (1992)
41. Wang, A., et al.: SuperGLUE: a stickier benchmark for general-purpose language understanding systems. In: Proceedings of the 2019 Conference on Neural Information Processing Systems, Vancouver, BC, Canada, pp. 3261–3275 (2019)
42. Yang, Z., Dai, Z., Yang, Y., Carbonell, J.G., Salakhutdinov, R., Le, Q.V.: XLNet: generalized autoregressive pretraining for language understanding. In: Proceedings of the 2019 Conference on Neural Information Processing Systems, Vancouver, BC, Canada, pp. 5754–5764 (2019)
43. Yue, Z., Zhang, H., Sun, Q., Hua, X.: Interventional few-shot learning. arXiv preprint arXiv: 2009.13000 (2020)

Construct Fine-Grained Geospatial Knowledge Graph

Bo Wei[1,2], Xi Guo[1,2(✉)], Ziyan Wu[1,2], Jing Zhao[3,4(✉)], and Qiping Zou[5(✉)]

[1] School of Computer and Communication Engineering, University of Science and Technology, Beijing, China
xiguo@ustb.edu.cn
[2] Beijing Key Laboratory of Knowledge Engineering for Materials, Beijing, China
[3] State Key Laboratory of Precision Blasting, Jianghan University, Wuhan, China
jingzhao@jhun.edu.cn
[4] School of Artificial Intelligence, Jianghan University, Wuhan, China
[5] Key Laboratory of AI and Information Processing, Hechi University, Hechi, China
706611232@qq.com

Abstract. In this paper, we propose the fine-grained geospatial knowledge graph (FineGeoKG), which can capture the neighboring relations between geospatial objects. We call such neighboring relations strong geospatial relations (SGRs) and define six types of SGRs. In FineGeoKG, the vertices (or entities) are geospatial objects. The edges (or relations) can have "sgr" labels together with properties, which are used to quantify SGRs in both topological and directional aspects. FineGeoKG is different from WorldKG, Yago2Geo, and other existing geospatial knowledge graphs, since its edges can capture the spatial coherence among geospatial objects. To construct FineGeoKG efficiently, the crucial problem is to find out SGRs. We improve the existing geospatial interlinking algorithm in order to find out SGRs faster. We conduct experiments on the real datasets and the experimental results show that the proposed algorithm is more efficient than the baseline algorithms. We also demonstrate the usefulness of FineGeoKG by presenting the results of complicated spatial queries which focus on structural and semantic information. Such queries can help researchers (for example, ecologists) find groups of objects following specific spatial patterns.

Keywords: Geospatial knowledge graph · Geospatial Interlinking · Fine-grained · Strong geospatial relation

1 Introduction

Geospatial information is usually stored and managed in the form of vector data and raster data. However, such traditional data models cannot support complicated spatial queries efficiently if the queries contain structural and semantic

Supported by Beijing Key Laboratory of Knowledge Engineering for Materials, State Key Laboratory of Precision Blasting (Jianghan University), and Key Laboratory of AI and Information Processing (Hechi University).

A. El Abbadi et al. (Eds.): DASFAA 2023 Workshops, LNCS 13922, pp. 267–282, 2023.
https://doi.org/10.1007/978-3-031-35415-1_19

requirements. For example, "find a lake surrounded by a forest land and a desert" is a common task in spatial pattern analyses for ecologists, but they still do the work manually, because most GISs cannot answer such queries directly. In this paper, we propose the *fine-grained geospatial knowledge graph* (FineGeoKG) to capture structural and semantic information of geospatial objects.

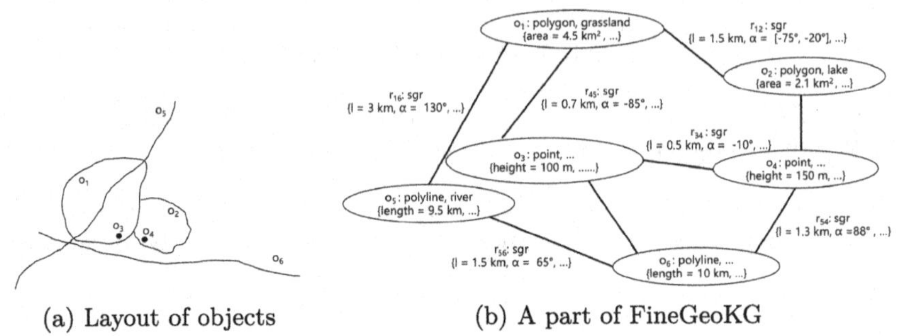

(a) Layout of objects (b) A part of FineGeoKG

Fig. 1. A toy example of FineGeoKG

As Fig. 1 shows, FineGeoKG focuses on describing the neighboring relations between objects. We define such relations as *strong geospatial relations* (SGRs). For example, Fig. 1(a) shows the natural layout of objects $\{o_1, o_2, \ldots, o_6\}$, where some objects are neighbors. For example, o_1 borders on o_2, o_5 intersects with o_1, o_3 and o_4 are close. We call they have SGRs. In FineGeoKG (Fig. 1(b)), the vertices (or entities) indicate objects, and an edge with the label "sgr" indicates two objects has SGR. For example, the edge r_{34} between o_3 and o_4 has a label "sgr". In addition, the properties of an "sgr" edge indicate the topological and directional characteristics of the relation. For example, the closeness of o_3 and o_4 is quantified by their distance $l = 0.5\,\text{km}$, and their relative direction is quantified by $-10°$. FineGeoKG is different from WorldKG [1], Yago2 [2,3], Yago2Geo [4,13], and other existing geospatial knowledge graphs, since FineGeoKG can capture fine-grained spatial information (i.e., neighboring relations among gound objects) while the other geospatial knowledge graphs focus on adding coarse-grained spatial information (for example, geographical coordinates) to a general purpose knowledge graph.

When constructing FineGeoKG, it is crucial to define SGRs and find out SGRs between objects. In this paper, we first define six types of SGRs considering there are three types of geospatial objects (i.e., point, polyline, polygon) in the perspective of geometry. The problem of finding out SGRs is in fact a geospatial interlinking problem [11] which aims at searching for object pairs having specific spatial relations. We improve the geospatial interlinking algorithm in [11] and our new algorithm can find SGRs faster. We conduct experiments on real datasets to evaluate the performances of our algorithms. We also demonstrate the usefulness of FineGeoKG by executing structural queries and semantic queries on FineGeoKG constructed. The two types of queries are subgraph matching queries [20] which contain structural and semantic requirements. The query results can help researchers analyze spatial patterns of ground objects and the layouts of facilities.

Our contributions are as follows:

- We define the fine-grained geospatial knowledge graph (FineGeoKG) which is characterized by SGRs. The SGRs can capture the spatial coherences among geospatial objects.
- We propose a fast geospatial interlinking algorithm which can find out SGRs between objects.
- We conduct experiments to evaluate the proposed algorithms and to demonstrate the usefulness of FineGeoKG.

2 Related Work

The mainstream knowledge graph techniques underemphasize geospatial semantics, which restricts the applications of knowledge graphs in GIS, mobile recommendation systems and location-based services (LBS). In recent years, researchers have made great efforts on spatiotemporal knowledge acquisitions and extractions [5,12,15]. To construct geospatial knowledge graph, the researchers extract triples from OpenStreetMap, build the alignment between WorldKG ontology and DBpedia ontology, and then integrate their spatiotemporal knowledge [1]. On the basis of GeoSPARQL ontology, some researchers expand the concept of geographical entities and attributes, and construct a spatiotemporal knowledge graph which contains geographical division entities and their spatial relations [8,9]. The common point of the techniques above is to add coarse-grained spatiotemporal information (for example, geographical coordinates) to a general purpose knowledge graph. These knowledge graphs lack fine-grained information which can describe the spatial coherences among ground objects. Some researchers study how to use a graph structure to describe the relations between ground targets in remote sensing images [6], but they focus on special scenes (for example, the airports) and small targets (for example, tarmac, runway, etc.). Papadakis et al. [11] use an intersection matrix [14] to determine whether there is a basic spatial topological relation between entities according to the intersection relation of two geographical entities. From the perspective of query processing techniques on geospatial knowledge graphs, researchers have propose a variety of indexing techniques and query algorithms, such as SS-Tree [19,21], Hilbert-Encoding [7], RisoTree [16–18]. In this paper, we focus on constructing knowledge graphs rather than processing spatial queries.

3 Problem Definitions

In this section, we formally define SGR and FineGeoKG.

Definition 1 (Strong Geospatial Relation (SGR)). *If two ground objects o_x and o_y are close to each other in the two-dimensional Euclidean space, they have a* ***strong geospatial relation*** *(SGR).*

In the definition, the criteria for measuring the closeness are defined as follows. Considering there are polygon objects, polyline objects, and point objects, (1) two polygons are close if they touches each other; (2) a polygon and a polyline are close if they intersect; (3) a point and a polygon are close if the point falls into the polygon; (4) two polylines are close if they intersect; (5) a point and a polyline are close if the minimum distance between the point and the polyline is less than ϵ, where ϵ is a very small; (6) two points are close if their distance is less than ϵ. Note that other criteria can be used according to different application scenarios.

Definition 2 (Fine-Grained Geospatial Knowledge Graph (Fine-GeoKG)). *A* ***fine-grained geospatial knowledge graph*** *(FineGeoKG) $KG = (E, R, L, P, V)$ consists of a set of entities E representing fine-grained ground objects, a set of relations R representing SGRs between two entities, a set of labels L indicating the classes of entities and relations, a set of properties P with their corresponding property value domains V.*

In a FineGeoKG, an entity can have multiple labels and a relation should have at least a "sgr" label to capture spatial coherences, as Fig. 1(b) shows. An entity or a relation can have mutiple properties and the values must be included in the corresponding domains.

Table 1 lists the properties used to quantify the six types of SGRs.

- **Polygon-Polygon** The length (l) of the shared border between o_x and o_y quantifies the relation topologically. The direction interval, namely, $\alpha = [\alpha_s, \alpha_e]$, quantifies the relation directionally[1].
- **Polygon-Polyline** The length (l) of the polyline (o_y) inside the polygon o_x quantifies the relation topologically. The clockwise angle (α) from the north direction to the intersection line quantifies the relation directionally[2].
- **Polygon-Point** The distance (l) from point o_y to the center of polygon o_x quantifies the relation topologically. The clockwise angle (α) from the north direction to the segment $o_x o_y$ quantifies the relation directionally.
- **Polyline-Polyline** The direction interval ($\alpha = [\alpha_s, \alpha_e]$) quantifies the relation directionally, where α_s and α_e are the clockwise angles from the north direction to the two segments which intersect[3]. Since it is difficult to quantify the relation topologically, we do not set the corresponding property.
- **Polyline-Point** The minimum distance (l) from point o_y to polyline o_x quantifies the relation topologically. The clockwise angle (α) from the north direction to the segment $o_y p$ quantifies the relation directionally, where p is the point on o_x nearest to o_y.

---

[1] The angles α_s and α_e indicate the directions of the endpoints on the shared border.

[2] The intersection line is the straight line passing through the intersections of o_x and o_y.

[3] A polyline consists of a group of consecutive segments. A segment of o_x and a segment of o_y intersect at a point.

Table 1. Properties used to quantify six types of SGRs

SGR Type	Natural Scene	FineGeoKG
polygon-polygon	north; $\alpha=[\alpha_s, \alpha_e]$: the direction interval; o_x; o_y; l : the length of the shared border	o_x: polygon ... — r_{xy}: sgr — o_y: polygon ... {l = 1.5km, α = [50°, 125°], ...}
polygon-polyline	o_y; l : the length of the polyline falling inside o_x; o_x; north; α : the slope	o_x: ploygon — r_{xy}: sgr — o_y: polyline {l = 3km, α = 130°, ...}
polygon-point	north; α : the direction; o_y; o_x; l: the distance	o_x: ploygon — r_{xy}: sgr — P_y: point {l = 0.7 km, α = 45°, ...}
polyline-polyline	$\alpha=[\alpha_s, \alpha_e]$: the direction interval; north; o_x; o_y; α_s; α_e	o_x: polyline — r_{xy}: sgr — o_y: polyline {α = [75°, 124°], ...}
polyline-point	north; o_x; α : the direction; o_y; l : the minimum distance from o_y to o_x	o_x: polyline — r_{xy}: sgr — o_y: point {l = 1.3 km, α = 130°, ...}
point-point	north; α : the direction; o_y; o_x; l : the distance between o_x and o_y	o_x: point — r_{xy}: sgr — o_y: point {l = 1 km, α = 80°, ...}

- **Point-Point** The distance (l) between o_x to o_y quantifies the relation topologically. The clockwise angle α from the north direction to the segment $o_x o_y$ quantifies the relation directionally.

Note that other definitions of properties can be adopted according to different application scenarios.

4 Constructing Fine-Grained Geospatial Knowledge Graph

There are two steps to construct a FineGeoKG. Firstly, we add the ground objects into KG as entities (or vertices). We can employ the image segmentation and target detection techniques to extract ground objects from remote sensing images. According to their geometries, vegetations and other characteristics, we add labels and property values to them. Secondly, we add the relations (or edges) to KG. If two vertices have SGR, we interlink them. The geospatial interlinking problem is defined as follows.

Definition 3 (Geospatial Interlinking Problem). *Given a set of polygon objects, polyline objects and point objects (O), the geospatial interlinking problem is to find out which objects have SGRs and to quantify the SGRs in topological aspect and directional aspect.*

4.1 Baseline Geospatial Interlinking Algorithms

R-tree Based Method. The geospatial interlinking problem is a kind of spatial join problems [10]. A straightforward way is as follows. At first, we organize the objects using an R*-tree. A polygon-object is abstracted as its minimum bounding rectangle (MBR). A polyline-object is abstracted as a set of MBRs and each MBR encloses a segment in the polyline. A point-object is abstracted as an MBR which corner points are the point itself. The objects are inserted into the R*-tree. Next, we find out the objects that have SGRs with each object $o_i \in O$ by using spatial range queries on R*-tree. If o_i is a polygon-object, the query range is its MBR. If o_i is a polyline-object, the query range is its MBR set. If o_i is a point, the query range is the MBR which is the circumscribed square of a circle with o_i as its center and ϵ as its radius. At last, we verify the SGRs between the query results and o_i, and return the objects passing the verifications. For each object found, we quantify the SGR according to Table 1 and add a relation r_{ij} with label "sgr" and the property values into the knowledge graph.

Grid Based Method. Another method is the grid based algorithm [11], which constructs a grid for the whole space and computes the cells covered by every object. In the grid, the cell length is Δ_x and the cell width is Δ_y, where Δ_x and Δ_y are determined by the average size of all objects. We give a number (n_x, n_y) to each cell as its identifier. The n_x and n_y are integers used to indicate the cells's position. The cells covered by a polygon-object o_i are the cells covered by o_i's MBR. Assuming the lower left corner point and the upper right corner point of an object's MBR are (x_1, y_1) and (x_2, y_2), the cells covered by the MBR are the cells with number $n_x \in \lfloor x_1/\Delta_x \rfloor..\lceil x_2/\Delta_x \rceil$ and $n_y \in \lfloor y_1/\Delta_y \rfloor..\lceil y_2/\Delta_y \rceil$. The cells covered by a polyline o_i are the cells crossed by o_i. The cells covered by a point o_i are the cells covered by the circle with o_i as its center and ϵ as its radius. A mapping table maintains the correspondences between each cell and the objects covering it. In the table, the keys are the identifiers of cells, the corresponding values are object identifier lists which record the objects covering the cells. The algorithm is similar to the R-tree based algorithm. The difference is that we retrieve objects using the mapping table rather than using the R-tree. For each object $o_i \in O$, we get its left corner point and its right corner point. Next, for each cell g_{uv} that o_i's MBR covers, we obtain its corresponding object list from the mapping table, and add the objects into the candidate set. For each candidate, like the R-tree based algorithm, we verify the SGR and insert a new relations into the knowledge graph.

4.2 Fast Geospatial Interlinking Algorithm

To find out SGRs faster, we propose Algorithm 1, which improves the grid based algorithm in three aspects. Firstly, we recognize the SGRs while constructing the mapping table rather than constructing the table beforehand. Secondly, we follow the "point→polyline→polygon" order to check the objects. Thirdly, to find the cells covered as accurately as possible, we propose a method which can find the cells passed through by a segment exactly.

Algorithm 1: Fast geospatial interlinking

Input: the geospatial object set $O = O_{point} \cup O_{polyline} \cup O_{polygon}$, the FineGeoKG KG
Output: the updated KG

```
Initialize (Δx, Δy) and the mapping table tbl;
for o_i ∈ O_point do
    (x1, x2, y1, y2) ← GetSearchRange(o_i, ε, Δx, Δy);
    C ← GetCandidObjs();
    Update KG.R;
    (u, v) ← GetCellPosOfPoint(o_i, Δx, Δy);
    L_uv ← L_uv ∪ {o_i};
for o_i ∈ O_polyline do
    R ← GetSearchRange(o_i, ε, Δx, Δy);
    for (a, b) ∈ R do
        C ← C ∪ GetObjList(g_ab, tbl);
    Update KG.R;
    UV ← GetCellPosOfPolyline(o_i, Δx, Δy);
    for (u, v) ∈ UV do
        L_uv ← L_uv ∪ {o_i};
for o_i ∈ O_polygon do
    R ← GetSearchRange(o_i, Δx, Δy);
    for (a, b) ∈ R do
        C ← C ∪ GetObjList(g_ab, tbl);
    Update KG.R;
    UV ← GetCellPosOfPolygon(o_i, Δx, Δy);
    for (u, v) ∈ UV do
        L_uv ← L_uv ∪ {o_i};
return KG;
```

Processing Point Objects. In Algorithm 1, after setting the unit sizes (Δ_x, Δ_y) and initializing the mapping table *tbl*, we check the point objects (i.e.,

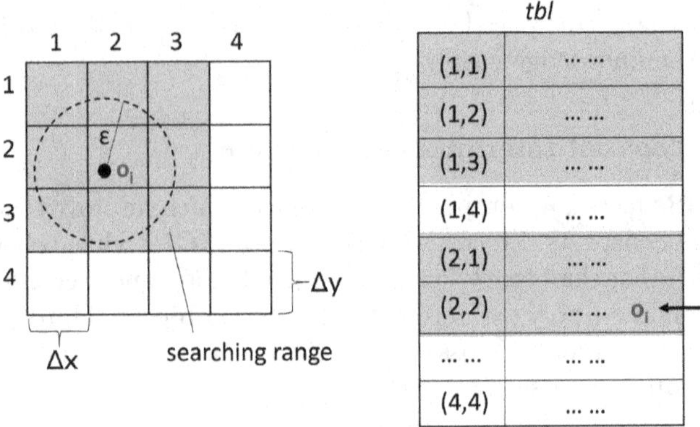

Fig. 2. Processing point objects

the for loop from line 2 to line 7) in order to find out the SGRs between two point objects (line 3 to line 5) and add the objects into *tbl* (line 6 to line 7). As Fig. 2 shows, the searching range of the point object o_i is the grey cells that are covered by the circle with center o_i and radius ϵ. In Algorithm 1, the function GetSearchRange() calculates the lower left identifier (x_1, y_1) and the upper right identifier (x_2, y_2) of the grey cells (line 3), the function GetCandidObjs() gets the candidate objects in the lists corresponding to the grey cells in *tbl* (line 4). See the grey tuples in Fig. 2. And then we determine the point objects that have SGRs with o_i and update edges in the knowledge graph KG (line 5). Next, we insert o_i into *tbl*. In Fig. 2, the arrow illustrates the inserting operation. The function GetCellPosOfPoint() finds the identifier (u, v) of the cell where o_i falls into (line 6) and then add o_i into the corresponding list L_{uv} (line 7).

Processing Polyline Objects. Firstly, we check each polyline object o_i in order to find out the SGRs between o_i and other point objects, as well as the SGRs between o_i and the polyline objects existing in *tbl*, i.e., the for loop from

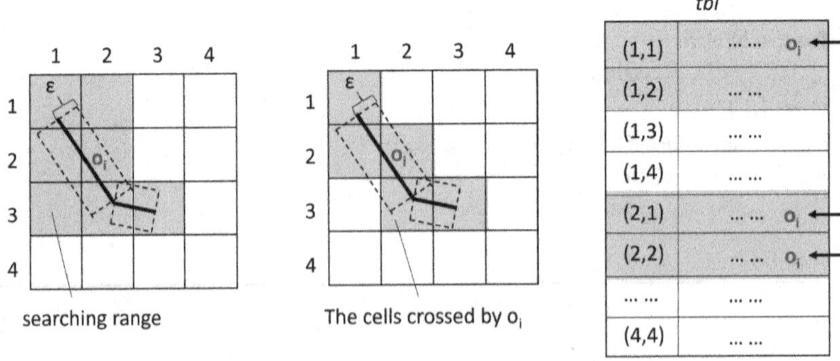

Fig. 3. Processing polyline objects

line 8 to line 15. On the one hand, the searching range for point objects is the cells covered by the rectangles extended from the segments of o_i. Since the closest distance between a SGR-neighbor object and o_i should be smaller than ϵ, we generate a rectangle by taking the segment as the central axis and extending ϵ widths to the left side and the right side of the segment. As the left figure of Fig. 3 shows, o_i consists of two segments and the rectangles are generated by using the extending method. The searching ranges are the gray cells covered by the rectangles. On the other hand, since the SGR-neighbor polyline should intersect with o_i, the searching range for polyline objects is the cells crossed by o_i (the red cells in the center figure), which are contained in the gray cells. Thus, GetSearchRange() function calculates the identifiers set R of the grey cells (line 9). The algorithm gets the candidate objects from *tbl* according to the identifiers (line 10 to line 11). Line 12 picks out the real SGR-neighbors from the candidates and adds edges to KG. Secondly, we insert the polyline object o_i into *tbl*. The function GetCellPosOfPolyline() calculates the identifier set UV of the cells crossed by o_i. See the red cells in the center figure. The algorithm inserts o_i to the object lists corresponding to the identifiers. The arrows in the right figure illustrate the inserting operations.

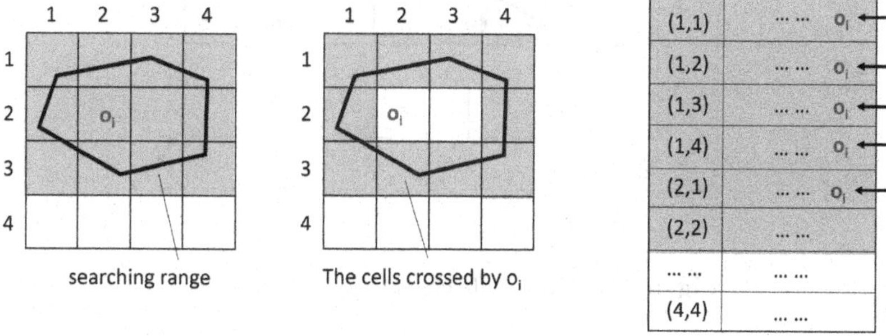

Fig. 4. Processing polygon objects

Processing Polygon Objects. Firstly, we check each polygon object o_i in order to find out its SGR-neighbors in the shape of points, polylines, and polygons, i.e., the for loop from line 16 to line 23. The searching range of point objects and polyline objects are the cells covered by o_i, since a point SGR-neighbor should fall inside the polygon and a polyline SGR-neighbor should fall inside the polygon partly or completely. See the grey cells in the left figure of Fig. 4. The searching range of polygon objects are the cells crossed by the boundary segments of o_i, since a polygon SGR-neighbor should touch o_i. See the red cells in the center figure, which are contained in the gray cell set. The function GetSearchRange() gets the identifiers R of the grey cells, and then the algorithm get the candidates C from *tbl* according to the identifiers (line 18 to line 19). After picking out the real SGR-neighbors, we add new edges to KG (line 12).

Secondly, we insert the polygon object o_i into *tbl*. The function GetCellPosOfPolygon() calculates the identifiers of the cells crossed by o_i's boundary (i.e., the red ones in the center figure), and insert o_i into the object lists corresponding to the identifiers (line 22 to line 23). We avoid inserting o_i to the object lists corresponding to the cells, which are covered by the interior area of o_i. For example, the cells with identifiers $(2, 2)$ and $(2, 3)$. There are two reasons. One reason is polygon objects often has large interior areas in the natural environments, and this strategy can help us reduce duplicate o_i's in *tbl*. Another reason is it is sufficient to store o_i's in the list of the cells crossed by the boundary. Two polygons might have SGR if some of their boundary segments fall into the same cell.

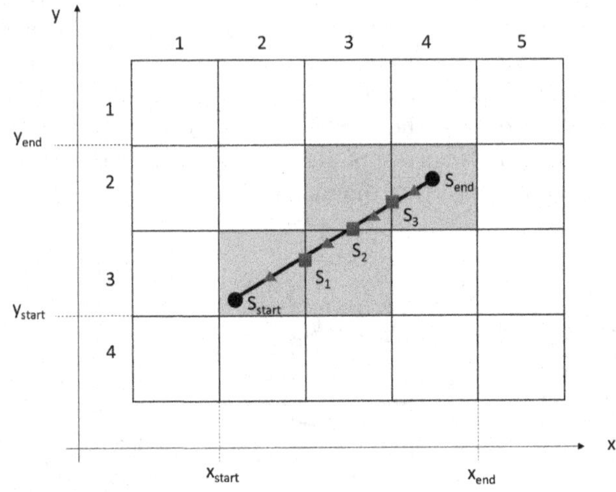

Fig. 5. Calculate cells crossed by a segment

Calculating Cells Crossed by a Segment. In Algorithm 1, a basic operation is to calculate the cells crossed by a segment. At line 9, line 13, line 17 and line 21, we have to do this basic operation. There are three steps to figure out the cells. Firstly, we delimit the spatial bounds of the cells, namely, x_{start}, x_{end}, y_{start} and y_{end}. Figure 5 shows the four bounds of the cells crossed by segment S whose endpoints are s_{start} and s_{end} (black points). Secondly, we derive the intersections of S with the vertical lines and the horizontal lines that are within the bounds. As the figure shows, the vertical lines are $x = x_{start} + \Delta_x$ and $x = x_{start} + 2 \times \Delta_x$. The horizontal line is $y = y_{start} + \Delta_y$. The blue squares illustrate the intersections, which are numbered as s_1, s_2 and s_3 according to the order of their x-coordinates. Thirdly, we calculate the midpoints of the consecutive sub-segments, which are obtained by dividing the segments by the intersections. In the figure, we have consecutive sub-segments $s_{start}s_1$, s_1s_2, s_2s_3, s_3s_{end}. The red triangles illustrate their midpoints. The cells are the ones we are looking for, if the midpoints fall into them. See the red cells in the figure.

Algorithm 2: Calculate cells crossed by a segment

Input: a segment S, the unit sizes (Δ_x, Δ_y)
Output: the identifiers UV of the cells crossed by S

1 $(x_{start}, x_{end}, y_{start}, y_{end}) \leftarrow \mathsf{GetBounds}(S, \Delta_x, \Delta_y)$;
2 $\{s_1, s_2, \ldots, s_k\} \leftarrow \mathsf{GetIntersections}(S, x_{start}, x_{end}, y_{start}, y_{end})$;
3 $p_{mid} \leftarrow \mathsf{getMidPoint}(s_{start}, s_1)$;
4 $(u, v) \leftarrow \mathsf{GetCellPosOfPoint}(p_{mid}, \Delta_x, \Delta_y)$;
5 $UV \leftarrow C \cup \{(u, v)\}$;
6 **for** $i \leftarrow 1$ *to* $k - 1$ **do**
7     $p_{mid} \leftarrow \mathsf{getMidPoint}(s_i, s_{i+1})$;
8     $(u, v) \leftarrow \mathsf{GetCellPosOfPoint}(p_{mid}, \Delta_x, \Delta_y)$;
9     $UV \leftarrow C \cup \{(u, v)\}$;
10 $p_{mid} \leftarrow \mathsf{getMidPoint}(s_k, s_{end})$;
11 $(u, v) \leftarrow \mathsf{GetCellPosOfPoint}(p_{mid}, \Delta_x, \Delta_y)$;
12 $UV \leftarrow C \cup \{(u, v)\}$;
13 **return** C;

Algorithm 2 summarizes the procedure of calculating the identifiers UV of the cells crossed by a segment S. Line 1 uses GetBounds() to calculate the four bounds of the cells. Line 2 uses GetIntersections() to calculate the intersections of S with the vertical lines and horizontal lines. Line 3 to line 5 calculate the midpoint p_{mid} of the first sub-segment $s_{start}s_1$ by using GetMidPoint(), get the identifier (u, v) of the cell where p_{mid} falls into, and then add (u, v) to the result set UV. The loop (line 6 to line 9) calculates the midpoints of the sub-segments from s_1s_2 to $s_{k-1}s_k$, and gets the identifiers of the cells where the midpoints fall into. Line 10 to line 12 calculate the midpoint of s_ks_{end} and add the corresponding identifier to the result set.

5 Experiments

We implement the proposed algorithms using Java and run the programs on a PC with an Intel(R) Xeon(R) W-2123 CPU(3.60 GHz), 32 GB of Memories, and a Windows 10 x64 operating system. To evaluate the algorithms, we use the geospatial objects in Shanghai which are collected from the OpenStreetMap (OSM). Table 2 summarizes the statistics of the four datasets (D_{1k}, D_{10k}, D_{100k}, D_{1m}) we used. Each dataset has polygon objects (for example, the buildings), polyline objects (for example, the highways) and point objects (for example, the restaurants). The subscript in the dataset name illustrates the approximate size of the dataset. The total number of the objects in the later dataset (for example, D_{10k}) is about 10 times the size of the previous one (for example, D_{1k}).

Table 2. Statistics of the datasets used in experiments

dataset	total	points	ratio	polygons	ratio	polylines	ratio
D_{1k}	980	203	21%	267	27%	510	52%
D_{10k}	10,426	1502	14%	3,542	34%	5,382	52%
D_{100k}	105,857	13,720	13%	32,531	31%	59,606	56%
D_{1m}	1,133,656	131,334	12%	272,154	24%	730,168	64%

5.1 Experiments of Geospatial Interlinking Algorithms

To construct a FineGeoKG, the key problem is to find out the SGRs between geospatial objects and interlink them. In this section, we compare the performances of the three algorithms for geospatial interlinking, namely, the R-tree based method (RBA), the grid-based method (GBA), and the method we proposed in Algorithm 1 (FSGR). Figure 6(a) shows the time consumptions of the three algorithms on different datasets. The time consumption of FSGR is the lowest, since FSGR can reduce the number of cells covered by polygon and polyline objects, and it can avoid repeating the computations of spatial relations between objects. Figure 6(b) shows that the memory consumptions of the three algorithms. The R*-tree index used by RBA is different from the mapping table used by GBA and FSGR. In the mapping table of FSGR, the object lists are shorter, so FSGR consumes less memory. Figure 6(c) shows the number of SGRs found in different datasets. On average, each object has relations with one or two objects in D_{1k} and D_{10k}, and each object has relations with three or four objects in D_{100k} and D_{1m}. Figure 6d shows the number of SGRs of different types in D_{1m}. The number of polyline-polyline SGRs is the largest, the number of polyline-polygon SGRs is the second largest, the number of point-polyline SGRs is the third largest, since in D_{1m} the polyline objects account for 64%, the polygon objects account for 24%, and the point objects account for 12% (Table 2).

To further evaluate the performances of the three algorithms in finding SGRs of different types, we divide each dataset into three parts and each part contains one type of objects. For example, D_{1k} is divided into D^1_{1k}, D^2_{1k}, and D^3_{1k}, where the superscript indicates the object type. The superscript 1 means "point", the superscript 2 means "polyline" and the superscript 3 means "polygon". Thus, D^1_{1k}, D^2_{1k}, and D^3_{1k} indicate the point objects, the polyline objects and the polygon objects from D_{1k}, respectively. We use such datasets to evaluate the performances of the algorithms in finding out the point-point SGRs, the polyline-polyline SGRs, and the polygon-polygon SGRs. To evaluate the other types of SGRs, for example, the point-polyline SGRs, we merge D^1_{1k} and D^2_{1k} and use D^{12}_{1k} indicate the object set. In the same, we create D^{13}_{1k} and D^{23}_{1k}.

Figure 7 shows the time consumptions of the three algorithms and the number of SGRs found in different datasets. Note that when we compute the SGRs

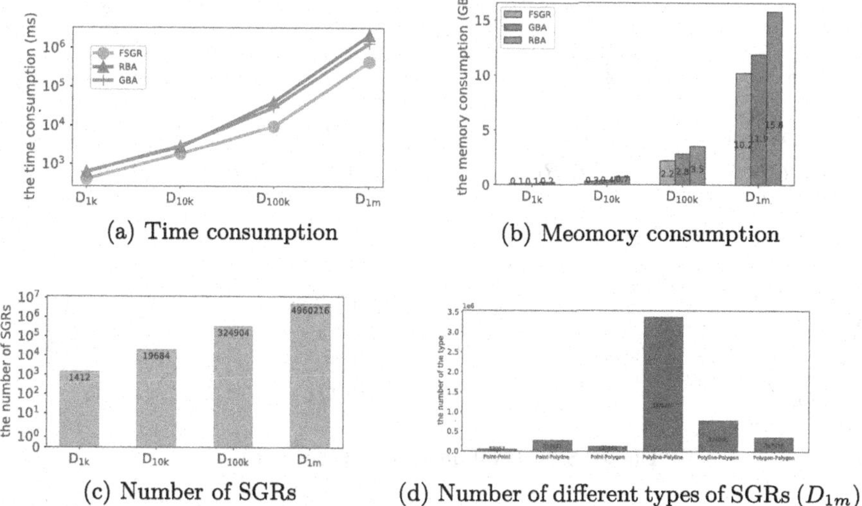

Fig. 6. Performances of geospatial interlinking algorithms on different datasets

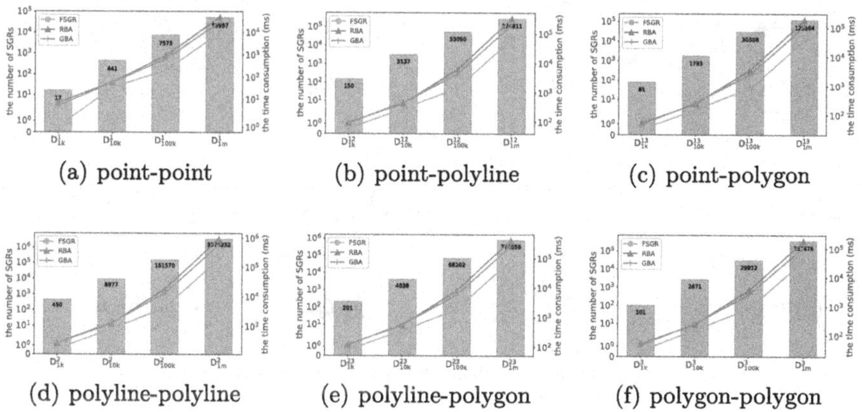

Fig. 7. Peformances of the algorithms in finding out different types of SGRs

of different types of objects (for example, point-polygon SGRs), we avoid finding out the SGRs of the same type of objects (for example, point-point SGRs). In the figures, the time consumption of FSGR is lower than RBA and GBA. Comparing the six types of SGRs, the time consumptions for point-point SGRs is the lowest, since the point objects is the fewest and the index structure taken fewest memories. The time consumptions for polyline-polyline SGRs is the highest, since the number of polyline objects is the largest and the index takes more memories. In addition, when computing the spatial relations and the SGR properties of polygons, the algorithms consume more time than processing the points

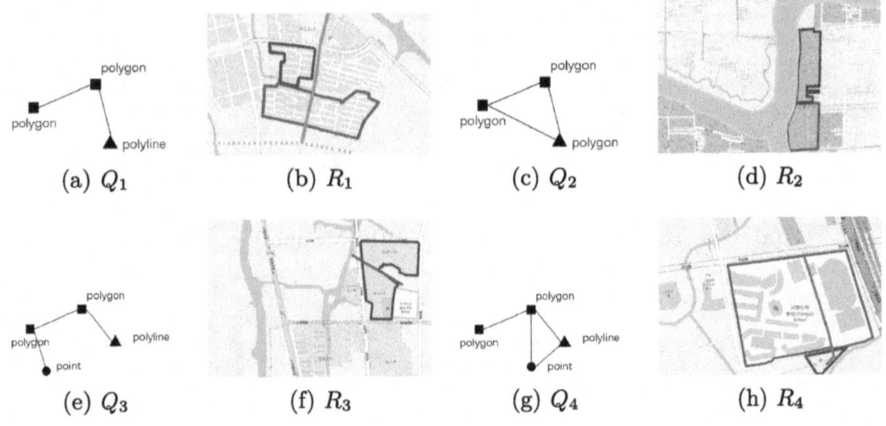

(a) Q_1 (b) R_1 (c) Q_2 (d) R_2

(e) Q_3 (f) R_3 (g) Q_4 (h) R_4

Fig. 8. Structural queries on FineGeoKG

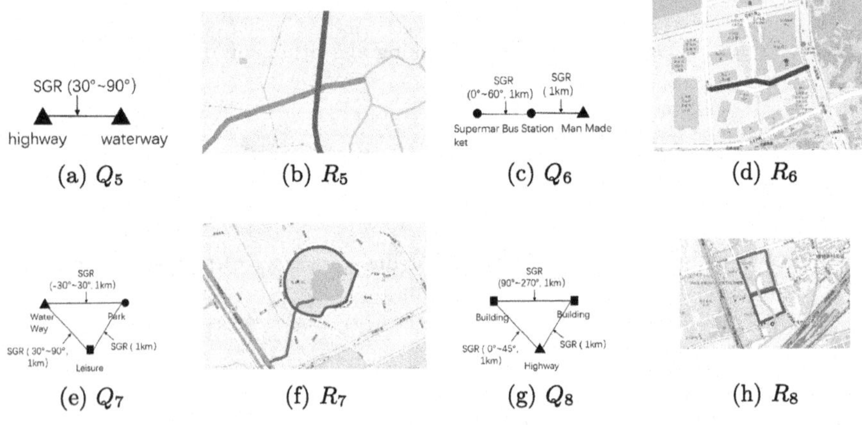

(a) Q_5 (b) R_5 (c) Q_6 (d) R_6

(e) Q_7 (f) R_7 (g) Q_8 (h) R_8

Fig. 9. Semantic queries on FineGeoKG

and the polylines. But in the datasets polygons are fewer than polylines. So the time consumption for polygon-polygon is not high.

5.2 Experiments of Spatial Queries on FineGeoKG

To demonstrate the usefulness of the FineGeoKG, this section shows the results of **structural queries** and **semantic queries** on the FineGeoKG. In the experiments, we use the FineGeoKG that is constructed based on the geospatial data of Shanghai. Structural queries and semantic queries are in fact the subgraph matching problem [20]. In the query graph of a structural query, the vertices indicate polygon objects, polyline objects, and point objects. The edges indicate the SGR relations. In the query graph of a semantic query, the vertices indicate the ground objects labelled by their functionalities, for example, the waterways, the bus stations, and so on. From the perspective of geometry, the objects may be polygons, polylines and points. The edges indicate the SGR relations and the query ranges of properties. The query results are groups of objects that can match the given query graphs.

Figure 8 shows the structural query graphs and the results found in the Fine-GeoKG of Shanghai. In the figures, there are four structural queries Q_1, Q_2, Q_3 and Q_4. The right figure of each query graph (for example, the figure (b)) shows an example of the results (R_1) w.r.t. the query (Q_1). The objects in the result are highlighted in different colors. The structural queries can help ecologists find out geospatial object groups following a given spatial pattern.

Figure 9 shows the semantic query graphs and the results. There are four queries Q_5, Q_6, Q_7 and Q_8 which query graphs are shown in the figures (a), (c), (e) and (g). In each query graph, the vertices have semantic labels, such as waterway, park, supermarket and so on. The edges have ranges of property values. The query graphs are more complicated than the query graphs in Fig. 8. The right figures of the query graphs show examples of the results. The semantic queries can help researchers in the urban planning field analyze the layouts of facilities.

6 Conclusions

In this paper, we propose FineGeoKG which is a geospatial knowledge graph characterized by SGRs. SGRs are used to capture the neighboring relations between ground objects, which are underemphasized in the existing knowledge graphs. Using FineGeoKG, ecologists can find out groups of ground objects matching the given spatial patterns. To find out SGRs fast, we improve the geospatial interlinking algorithm and evaluate the algorithm using real datasets. To illustrate the usefulness of FineGeoKG, we demonstrate the results of structural queries and semantic queries. It is an interesting future direction to design efficient algorithms to answer the queries fast considering characteristics of FineGeoKG.

Acknowledgments. This work was supported by Key Laboratory of AI and Information Processing (Hechi University) Education Department of Guangxi Zhuang Autonomous Region (2022GXZDSY007), National Natural Science Foundation of China (No. 61602031), Fundamental Research Funds for the Central Universities (FRF-IDRY-19-023).

References

1. Dsouza, A., Tempelmeier, N., Yu, R., Gottschalk, S., Demidova, E.: Worldkg: a world-scale geographic knowledge graph. In: CIKM, pp. 4475–4484 (2021)
2. Hoffart, J., Suchanek, F.M., Berberich, K., Lewis-Kelham, E., De Melo, G., Weikum, G.: Yago2: exploring and querying world knowledge in time, space, context, and many languages. In: WWW, pp. 229–232 (2011)

3. Hoffart, J., Suchanek, F.M., Berberich, K., Weikum, G.: Yago2: a spatially and temporally enhanced knowledge base from Wikipedia. Artif. Intell. **194**, 28–61 (2013)
4. Karalis, N., Mandilaras, G., Koubarakis, M.: Extending the YAGO2 knowledge graph with precise geospatial knowledge. In: Ghidini, C., et al. (eds.) ISWC 2019. LNCS, vol. 11779, pp. 181–197. Springer, Cham (2019). https://doi.org/10.1007/978-3-030-30796-7_12
5. Kyzirakos, K., et al.: Geotriples: transforming geospatial data into RDF graphs using R2RML and RML mappings. J. Web Semant. **52**, 16–32 (2018)
6. Li, Y., Zhang, Y.: A new paradigm of remote sensing image interpretation by coupling knowledge graph and deep learning. J. Wuhan Univ. **47**(8), 1176–1190 (2022)
7. Liagouris, J., Mamoulis, N., Bouros, P., Terrovitis, M.: An effective encoding scheme for spatial RDF data. PVLDB **7**(12), 1271–1282 (2014)
8. Liu, J., et al.: The construction of knowledge graph towards multi-source geospatial data. J. Earth Inf. Sci. **22**(7), 1476–1486 (2020)
9. Liu, J., Liu, H., Chen, X., Guo, X., Zhu, X.: Construction of knowledge graph based on geo-spatial data. Chin. J. Inf. **34**(11), 29–36 (2020)
10. Mamoulis, N., Papadias, D.: Multiway spatial joins. TODS **26**(4), 424–475 (2001)
11. Papadakis, G., Mandilaras, G., Mamoulis, N., Koubarakis, M.: Progressive, holistic geospatial interlinking. In: WWW, pp. 833–844 (2021)
12. Patroumpas, K., Skoutas, D., Mandilaras, G., Giannopoulos, G., Athanasiou, S.: Exposing points of interest as linked geospatial data. In: SSTD, pp. 21–30 (2019)
13. Punjani, D., et al.: Template-based question answering over linked geospatial data. In: Proceedings of the 12th Workshop on Geographic Information Retrieval, pp. 1–10 (2018)
14. Saveta, T., Fundulaki, I., Flouris, G., Ngonga-Ngomo, A.-C.: *SPgen*: a benchmark generator for spatial link discovery tools. In: Vrandečić, D., et al. (eds.) ISWC 2018. LNCS, vol. 11136, pp. 408–423. Springer, Cham (2018). https://doi.org/10.1007/978-3-030-00671-6_24
15. Sun, K., Hu, Y., Song, J., Zhu, Y.: Aligning geographic entities from historical maps for building knowledge graphs. Int. J. Geogr. Inf. Sci. **35**(10), 2078–2107 (2021)
16. Sun, Y., Sarwat, M.: A generic database indexing framework for large-scale geographic knowledge graphs. In: ACM SIGSPATIAL, pp. 289–298 (2018)
17. Sun, Y., Sarwat, M.: A spatially-pruned vertex expansion operator in the NEO4J graph database system. GeoInformatica **23**(3), 397–423 (2019)
18. Sun, Y., Yu, J., Sarwat, M.: Demonstrating spindra: a geographic knowledge graph management system. In: ICDE, pp. 2044–2047. IEEE (2019)
19. Wang, D., Zou, L., Feng, Y., Shen, X., Tian, J., Zhao, D.: S-store: an engine for large RDF graph integrating spatial information. In: Meng, W., Feng, L., Bressan, S., Winiwarter, W., Song, W. (eds.) DASFAA 2013. LNCS, vol. 7826, pp. 31–47. Springer, Heidelberg (2013). https://doi.org/10.1007/978-3-642-37450-0_3
20. Wang, X., Zou, L., Wang, C., Peng, P., Feng, Z.: Research on knowledge graph data management: a survey, pp. 2139–2174 (2019)
21. Zou, L., Özsu, M.T., Chen, L., Shen, X., Huang, R., Zhao, D.: gStore: a graph-based SPARQL query engine. VLDB J. **23**(4), 565–590 (2014)

A Multi-view Graph Learning Approach for Host-Based Malicious Behavior Detection

Chenfei Zhao(✉), Zhe Zhang(✉), Tiejun Wu, and Dunqiu Fan

Nsfocus Technologies Group Co., Ltd., Beijing, China
{zhaochenfei,zhangzhe3,wutiejun,fandunqiu}@nsfocus.com

Abstract. Due to the continuous evolution of malware, host-based malicious behavior detection is urgently needed in cyber security. An increasing amount of research is focused on developing techniques to detect host-based malicious behavior using rule engines and machine learning methods. However, current detection approaches concentrate on the specific or sequential features of malicious behaviors. Those methods ignore the internal information of an attack and structural information between multiple attacks. So it is difficult to detect malicious behaviors on the host comprehensively. To address the problem, we present MVGD, an approach for detecting malicious behavior using a multi-view graph learning method using audit logs. The multi-view information of the graph includes structural information, path information, and event-type information. Firstly, we construct a host-based behavior dependency graph, which describes the dependencies among processes, files, registries, etc. on the host. Then, in order to maintain the causality of a single attack path, a deepwalk technique is applied to capture sequential information. Next, the structural information of malicious behaviors is extracted using graph neural networks to describe the correlation between multiple attack paths. Finally, malicious behaviors can be detected by the combination of the structural, path, and event information. Experiments over three datasets indicate that MVGD achieves state-of-the-art performance.

Keywords: Host-behavior dependency graph · Multi-view information · Graph neural networks · Malicious behavior

1 Introduction

In recent years, numerous attacks have been against large enterprise networks, causing unimaginable damage. Therefore, malicious behavior detection on the host is an urgent need for cyber security. To defend against attacks, enterprises deploy traditional security defense devices, such as Firewalls and Intrusion Prevention Systems (IPS), on the network boundaries. However, these devices

C. Zhao and Z. Zhang—Equal contribution.

A. El Abbadi et al. (Eds.): DASFAA 2023 Workshops, LNCS 13922, pp. 283–299, 2023.
https://doi.org/10.1007/978-3-031-35415-1_20

cannot defend against Advanced Persistent Threats (APTs), which will break through the boundary and enter the intranet. After infecting the host, the attacker inevitably leaves traces of behavior on the host. Windows audit logs can record the actions of files, registries, processes, sockets, and other objects at the kernel level of the operating system (OS), providing the possibility for host-based threat detection.

The approaches of malicious behavior detection based on audit logs can be divided into two types. The first is the rule-based detection method, and the second is the machine-learning-based detection method. Traditional host-based malicious behavior detection methods usually rely on security rules. Rule-based methods [3] can accurately identify known attacks, but fail to detect continuously evolving or complex threats. In addition, rule-based methods acquire expensive costs for employing security experts to update and maintain the rules. With the development of machine learning, researchers attempt to apply machine learning algorithms in host-based malware detection [2,11], which can reduce experts' costs and have achieved certain performances. However, those methods usually leverage the attack paths, ignoring the structural correlation of multiple attack paths. Thus it is difficult to capture the attack pattern of malicious behaviors.

In this paper, we propose a graph learning approach for host-based malicious behavior detection based on Windows audit logs, called MVGD, which can capture multi-view information based on the dependency graphs containing attack structure information, attack path information, and event-type information. Compared with previous models, our model can extract more dimensional information to improve the performance of malicious behavior detection. In addition, the MVGD model is data-driven, which can be updated iteratively based on the data in a timely manner. It is more applicable to the field of malicious behavior detection because malware mutates quickly.

In the first stage, we build a host-based behavior dependency graph, which describes the relations among processes, files, and registries on the host. The behavior dependency graphs corresponding to different malicious behaviors are different. In the second stage, we apply the deepwalk [10] technique to capture sequential attack path information, which describes the causality of a single attack path. Then we employ graph neural networks [6,12] to extract structural information about malicious behaviors, which describes the correlation between multiple attack paths.

Our contributions are listed as follows:

- We propose a graph learning approach, which uses multi-view information containing attack path information, attack structure information, and event-type information. We are the first to fuse the path, event, and structure information in the cyber security field.
- We construct the host-based behavior dependency graph, which describes the dependency relationship among processes, files, registries, etc., and can reflect the attack pattern of malicious behavior.
- Our model is a data-driven approach that can be updated iteratively based on the data in a timely manner without costly maintenance of the rule base.

- Experiments are conducted on three real datasets. The experimental results validate the effectiveness of our model.

2 Related Work

MVGD is mainly related to two sub-topics about malicious behavior detection with audit logs, including rule-based methods and learning-based methods.

In previous works, many approaches focused on rule-based detection methods using audit logs to detect malicious behavior. Most of these approaches are heuristic and require expensive costs for employing security experts to update and maintain the rules. Milajerdi et al. [9] proposed HOLMES, a system for the detection of APTs that relied on a knowledge base of adversarial Tactics, Techniques, and Procedures (TTP). The central component of HOLMES is the TTP rules, which provide the mapping between audit events and APT steps. Hassan et al. [4] introduced a system, RapSheet, using a tactical provenance graph (TPG) that encodes related alerts into a provenance graph. And the alerts are produced by the MITRE ATT&CK behaviors rules matching on audit logs. Unlike rule-based approaches that require expensive expert experience, MVGD is a data-driven approach. Instead of requiring experts to develop rules based on different attack behavior patterns separately, threat detection can be performed by simply feeding audit logs to the model.

Another popular approach is the learning-based detection method using audit logs. Alsaheel et al. [1] developed ATLAS, a framework to reconstruct attack stories using audit logs. And the highlighted part is the sequence-based model that learns different patterns between attack sequences and non-attack sequences, which are extracted from the causal graphs. Wang et al. [13] formulated an approach, ProvDetector, that identifies stealthy malware by learning the sequences of benign paths of applications from a provenance graph. The most similar work to ours is given by Ring et al. [11] and Berlin et al. [2]. Berlin et al. [2] applied a logistic regression model to classify audit logs for a given-time window as malicious or benign. Ring et al. [11] focused on the LSTM model to capture the temporal dimension information by different subsets of features for malware detection. In contrast to these approaches, MVGD aims to fuse graph structure information and sequence information for host-based malicious behavior detection, which is a multi-view graph learning model that introduces additional information to improve accuracy and robustness.

3 Motivation and Definition

Motivation. We select a typical malicious family program to illustrate the motivation of the MVGD. The insight behind the MVGD approach has two key points.

First, although malicious programs conceal themselves, their actions will inevitably be monitored by the OS and recorded in the audit log. Therefore, it is necessary to construct a dependency graph based on audit logs, which can directly reflect the OS-level actions of the programs.

Second, malicious programs usually have meaningless attack behaviors to confuse security experts, however, through constructing a host-based dependency graph, the complete attack paths with event types can be visualized directly for subsequent analysis. These attack paths record the behavioral preferences of the malicious program and are excellent features to apply to identify malicious behavior. Hence, the relationship between multiple attack paths, the internal causality of a single attack path, and the event-type semantic information of the attack paths play an important role in malicious behavior detection using host-based audit logs. This will be explained in detail as follow.

Take Mydoom [14] as an example, Mydoom is a computer worm affecting Microsoft Windows, especially a fast-spreading e-mail worm. The dynamic malicious behavior of the Mydoom family is mainly reflected in sending spam through infected computers.

As depicted in Fig. 1, a part of the host-based dependency graph of the Mydoom family illustrates the correlation between multiple attack paths. Mydoom worm starts by creating files, including service.exe and java.exe. Then, the registry is modified to set services.exe and java.exe as self-starting items. Finally, the Mydoom Program process triggers network communications, reflected in the infected computer and multiple computers for network connection. Through the event-specific attribute information, it can be inferred that the infected computer sends spam.

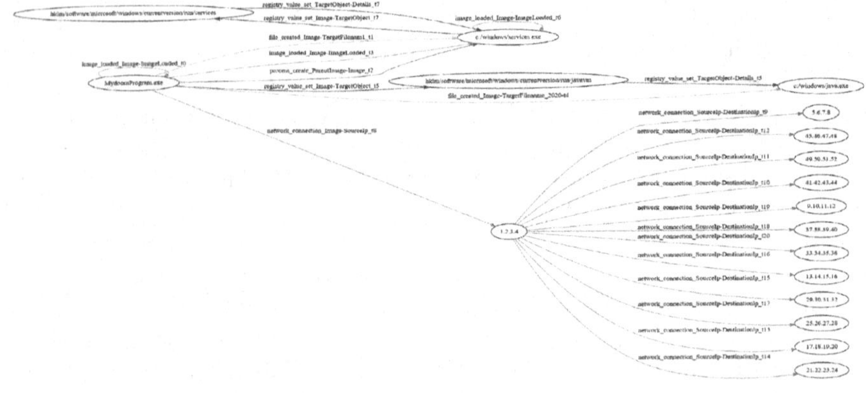

Fig. 1. A part of the host-based dependency graph of Mydoom family.

Figure 2 presents a time sequence of events triggered by a Mydoom program illustrating the causality of the attack path. The time sequences take the Mydoom Program process as the source entity, and every step means one event containing the event type and destination entity. A deepwalk technique is applied to an undirected graph, thus realistic attack paths can be captured by setting the appropriate walk length. This path can reflect the causal relationship of events.

It is worth noting that in the above analysis, the event-type information provides important semantic information, which represents the actions of the

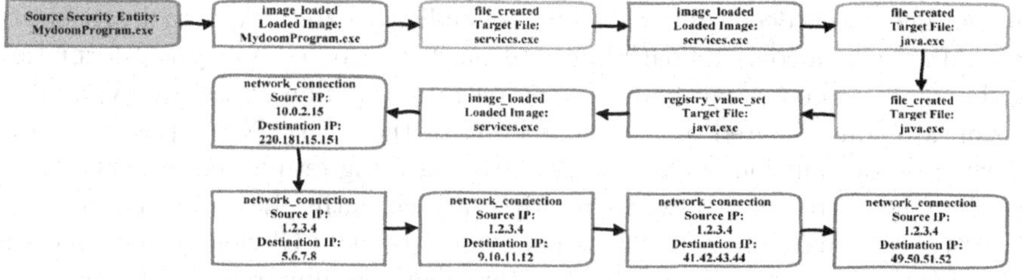

Fig. 2. Time series of the audit log events triggered by the Mydoom Program process.

malicious program. And the frequency of event type represents the malicious program's preference. So event-type information is extremely important when detecting malicious behavior. This part is also verified in the following experiments.

Definition. We formally define key concepts used in the rest of this paper.

- **Audit Logs.** Audit logs [7] record detailed activities of the operating system. In our experiments, audit logs are captured when a malicious program runs in the sandbox. They record the occurrence time of an event, the responsible program, the impacted entity, and detailed information.
- **Event.** An event $event = (src, dst, rel, time)$ models the interaction between two entities, where src means the source entity, dst means the destination entity, rel is the relation between src and dst that is the event type, $time$ records the timestamp when the event happened. For example, a process creates a file at a time on the file path. Among them, the process is src, the file is dst, create is rel, time is $time$.
- **Host-based Dependency Graph.** A host-based dependency graph $G(V, E, X)$ describes the dependencies among entities like processes, files, registries, etc. V are the set of entities, E are the set of event types, and X denotes the features of entities.
- **Dependency Graph based Malicious Behavior Detection.** Given a set of events, we construct a host-based dependency graph G. We aim to detect malicious behavior based on the graph G.

4 Model

4.1 Overall Architecture

The entire model is shown in Fig. 3. We aim to leverage the dependency graph containing wealth information for malicious behavior detection. First, we extract multiple events from unstructured audit logs. Second, we formulate multiple events offered by audit logs to a dependency graph G. Certain and meaningful entities and event types are designed to extract triplets (src, dst, rel) from events. Then, the deepwalk technique is applied to the graph to learn entity embedding, which contains path information. We encode event-type information by the applied one-hot mechanism. The path information and event-type information are concatenated as the embedding of the entity. Next, the structural information of malicious behaviors is extracted using graph neural networks to describe the correlation between multiple attack paths. Last, we utilize three aspects of information for malicious behavior detection. Obviously, our model is a data-driven approach. The details of our model are illustrated as follows.

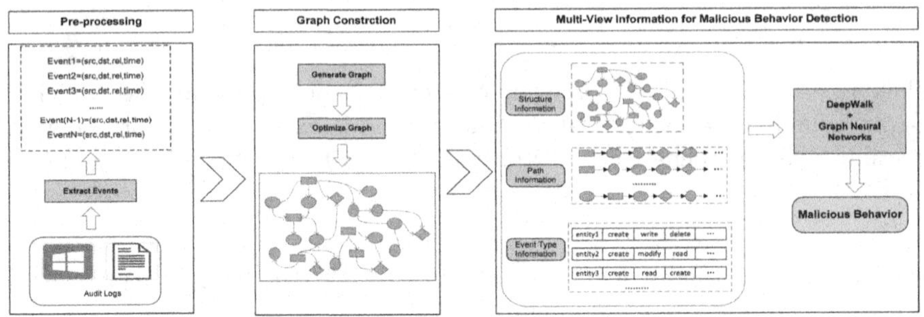

Fig. 3. The architecture of MVGD model.

4.2 Host-Based Behavior Dependency Graph

In the section, a host-based dependency graph is constructed and optimized for multi-view graph learning to detect malicious behavior. First, MVGD defines a general framework for mapping the data structures from audit logs. The framework defines the required fields for graph construction, involving *src*, *dst*, *rel*, and *time*. Second, graph optimization is applied to remove redundant information to detect malicious behaviors.

Table 1. The event entities (*src*, *dst*) and detailed security entities.

Event Entity	Security Entity	Detail
src, *dst*	process	system process, user process, programs process
dst	file	system file, user file, programs file
	socket	IP address, domain, url
	registry	hku registry, hklm registry, hkcr registry, hkcc registry, hkcu registry

Audit Log Pre-processing. To detect malicious behavior, MVGD provides a general framework to transform audit logs for constructing host-based behavior dependency graphs. Here, a general framework is used to extract standard event formats from multiple unstructured data formats. Similar to [1,5,8,13], the standard event format is defined as the event $event = (src, dst, rel, time)$ indicating a simplified audit log event. As defined in Sect. 3, src means the source entity, and dst means the destination entity, representing different entities in audit logs. Considering the characteristic of audit logs, we consider the following four types of entities: processes, files, sockets, and registries, detailed in Table 1. Besides, rel means the connection between src and dst, which is the event type. The relationship exists between the process and any other entities, indicating the event type in the audit log, which is detailed in Table 2. In addition, $time$ means the timestamp once the event occurred. $time$ is an indicator that records the number of events, thus the redundant information will exist in host-based dependency graphs.

Table 2. The relationship (rel) between security entities.

Source Security Entity	Destination Security Entity	Event Type (rel)
process	process	create, terminate, execute
	file	read, create, modify, delete
	socket	connect create, modify, delete
	registry	

Host-Based Dependency Graph Construction. Given the events $event = (src,\ dst,\ rel,\ time)$ in a certain duration, a host-based dependency graph $G(V, E, X)$ is built, which has directed and multiple edges between two nodes. The graph consists of nodes, which represent src and dst entities, and directed edges, which represent events from src to dst entities at certain timepoint $time$.

Host-Based Dependency Graph Optimization. As described above, we construct a host-based dependency graph using audit logs collected from hosts. To reduce log complexity and maintain the important information, graph optimization is needed. MVGD designs three techniques to remove redundant information.

- MVGD aggregates all edges which have the same elements, including src, dst, and rel, and then calculates the frequency of its occurrence as an attribute.
- MVGD removes the bias introduced by the sandbox in our experiments. Since the principle of sandbox operation, every software starts with certain processes to determine the type of inserted software and then execute

the inserted software. In our experiments, those certain processes can be python.exe, vboxservice.exe, and so on. Those specific processes can not provide effective information for detecting malicious behaviors and are inevitably collected due to the sandbox mechanism. We remove these nodes and connected edges introduced by sandbox bias.
- MVGD applies a data filtering strategy to remove audit logs that have limited behavior due to problems in the acquisition environment. Limited behavior specifically refers to audit logs with log data less than or equal to 2. According to our analysis, filtering audit logs is often a result of programs not running successfully or programs detecting a sandbox environment and terminating quickly on their own.

4.3 Multi-view Graph Learning Approach

After the host-based dependency graph is built, we will introduce the multi-view information for malicious behavior detection in this section.

Path Information. The attack path refers to the measures taken by malicious programs to infiltrate an enterprise's internal system. Therefore, security analysts often attempt to employ attack path information to analyze malicious behavior through a series of attack activities presented in audit logs. An attack path describes the causality of the path, which can be utilized as one of the key characteristics of malicious behavior detection. In this paper, we apply the deepwalk [10] technique to extract the path information as the characteristics of the entity.

We first utilize random walk to obtain paths based on dependency graph G. Given entity v_0, we can obtain many paths like $path = (v_0, v_1, ..., v_l)$, where l denotes the walk length. According to the idea of deepwalk, we learn the latent representations of entities based on paths. Obviously, the learned embedding of entities contains path information. The objective of deepwalk method is to estimate the likelihood:

$$Pr(v_i|(\Phi(v_1), \Phi(v_2), ..., \phi(v_{i-1}))) \tag{1}$$

where $\Phi : v \in V \mapsto R^{|V| \times d}$ is a mapping function, d means the dimension of entity embedding. In terms of entity representation modeling, this yields the optimization problem:

$$minimize_{\Phi} - logPr(v_{i-w}, ..., v_{i-1}, v_{i+1}, ..., v_{i+w}|\Phi(v_i)) \tag{2}$$

where w means the window size. Therefore, based on each dependency graph G, we can obtain the embeddings of entities $\mathbf{X}_p$ in G.

Structure Information. The dependency graph we constructed before, describes the dependencies among entities like processes, files, registries, etc. on the host. Therefore the structure of the graph can reflect the entire activities of malicious behavior. Also, we can model the correlation between multiple attack paths through the structure information. In this paper, graph neural network [12] is employed to extract structure information, specifically graph convolutional neural network. We adopt a multi-layer Graph Convolutional Network (GCN) [6] to model the structure information of the dependency graph. The formula is below:

$$\mathbf{H}^{l+1} = \sigma(\tilde{\mathbf{D}}^{-\frac{1}{2}}\tilde{\mathbf{A}}\tilde{\mathbf{D}}^{-\frac{1}{2}}\mathbf{H}^{(l)}\mathbf{W}^{(l)}) \tag{3}$$

where $\tilde{\mathbf{A}} = \mathbf{A} + \mathbf{I}_N$ is the adjacency matrix of the dependency graph G with added self-connections, $\mathbf{I}_N$ is the identity matrix, $\tilde{\mathbf{D}}_{ii} = \Sigma_j \tilde{\mathbf{A}}_{ij}$ and $\mathbf{W}^{(l)}$ is a layer-specific trainable weight matrix. $\sigma(\cdot)$ denotes an activation function, such as the $Relu(\cdot) = max(0, \cdot)$. $\mathbf{H}^{(l)} \in \mathbb{R}^{N \times D}$ is the matrix of the activations in the l_{th} layer; $\mathbf{H}^{(0)} = \mathbf{X}$, $\mathbf{X}$ means the embedding matrix of all entities that represent the features; and $\hat{\mathbf{A}} = \tilde{\mathbf{D}}^{-\frac{1}{2}}\tilde{\mathbf{A}}\tilde{\mathbf{D}}^{-\frac{1}{2}}$.

If we adopt three layers of GCN, the formula is below:

$$\mathbf{X}_s = Relu(\hat{\mathbf{A}}Relu(\hat{\mathbf{A}}Relu(\hat{\mathbf{A}}\mathbf{X}\mathbf{W}^{(0)})\mathbf{W}^{(1)})\mathbf{W}^{(2)}) \tag{4}$$

Event-Type Information. The event consists of $event = (src, dst, rel, time)$ introduced in Sect. 3. rel denotes the event type of events. Event-type information plays a very important role in malicious behavior detection. It reflects the actions or activities of malicious programs. And the frequency of event type represents the malicious program's preference and it is an indispensable feature. Researchers [2,11] employ event types as key features to carry out a number of combinations for malicious behavior detection. However, this will lead to a relatively large amount of feature engineering and will cause feature redundancy. The idea in this paper is to adopt event type as the edge information of the graph or as the entity information to be added to the entire model, which can avoid the huge feature engineering. The two methods are described below.

- **As edge feature.** We know that original graph neural networks do not consider edge information. But here event-type information can be leveraged as edge features in graph neural networks. The formula is below.

$$\mathbf{x}_i^{'} = \Theta\mathbf{x}_i + \Sigma_{j\in N(i)}\mathbf{x}_j \cdot h_\Theta(\mathbf{e}_{i,j}) \tag{5}$$

  where h_Θ denotes a neural network, i.e. MLP. $\mathbf{x}_i$ means the embedding of i_{th} entity. $\mathbf{x}_j$ means the embedding of j_{th} entity. $N(i)$ means the set of i_{th} entity's neighborhood. $\mathbf{e}_{i,j}$ means the edge information between i_{th} entity and j_{th} entity, which is event-type embedding. In this paper, the one-hot technique is applied to the event type for obtaining the event-type embedding.
- **As entity feature.** Event type models the interaction between entities. It means that each entity has more than one event type to which it belongs. That is to say, we can encode the frequency of event types to which an entity belongs as the embedding of that entity. Therefore, each entity contains rich information about event types and the frequency of event types. Incorporating

it into the model can improve the performance of our model. The formula is below.

$$\mathbf{x}_{i_{et}} = \mathbf{x}_i \oplus \mathbf{et}_i \tag{6}$$

where $\mathbf{et}_i$ denotes the event-type embedding of i_{th} entity, $\oplus$ denotes the concatenation.

Malicious Behavior Detection. To improve the malicious behavior detection performance, a multi-view model incorporates path information, structure information, and event information. We first concatenate the path information and event information as the input feature of entities as follows:

$$\mathbf{X}_{pet} = \mathbf{X}_p \oplus \mathbf{X}_{et} \tag{7}$$

where $\mathbf{X}_p$ denotes the entity feature matrix with path information, $\mathbf{X}_{et}$ denotes the entity feature matrix with event-type information.

Next, we extend Eq. 4 by incorporating three-view information as follows:

$$\mathbf{X}_{all} = Relu(\hat{\mathbf{A}}Relu(\hat{\mathbf{A}}Relu(\hat{\mathbf{A}}\mathbf{X}_{pet}\mathbf{W}^{(0)})\mathbf{W}^{(1)})\mathbf{W}^{(2)}) \tag{8}$$

Then we adopt the Linear component to predict the result of malicious behavior information.

$$y_{pred} = Linear(aggregate(\mathbf{X}_{all})) = \mathbf{W} \cdot aggregate(\mathbf{X}_{all}) + b \tag{9}$$

where if $y_{pred} = 1$, it indicates that the audit log contains malicious behaviors. Otherwise, the audit log does not contain malicious behaviors. $\mathbf{W}$ denotes the weight matrix, b denotes the bias. Here, *aggregate* means the integration operator like mean, sum, max, etc. In this paper, we use the mean aggregator. Our model is to optimize the following objective:

$$L = -y \cdot \log(y_{pred}) - (1 - y) \cdot \log(1 - y_{pred}) \tag{10}$$

5 Experiments

In this section, we introduce three datasets in detail. And experiments are conducted on three real datasets to demonstrate the effectiveness of our model. Moreover, a set of quantitative experiments are conducted to analyze the three-view information on the MVGD model. Analysis of graph construction and model training time will also be presented in this section.

5.1 Datasets

We describe the basic information about the three datasets used in the experiments regarding Windows audit logs. Table 3 shows the statistics information of the three datasets.

Table 3. The statistics information of dataset.

	Total	Malicious	Benign
CuckooBox Dataset	20,119	14,679	5,440
Sysmon Dataset	8,496	5,818	2,678
TAC Dataset	7,557	5,767	1,790

CuckooBox Dataset. The CuckooBox dataset is a public dataset about Windows audit logs given by Berlin et al. [2] which can be found on GitHub[1]. The malicious portion of the audit logs is generated by CuckooBox Sandbox (CuckooBox), an open-source automated malware analysis system. The benign part is collected from an internal enterprise network.

Sysmon Dataset. The Sysmon dataset is a self-constructed dataset about Windows audit logs collected from Windows virtual machines using sysmon monitors. Considering the characteristic of malicious behavior, the configuration of our virtual machine environment can be specified to filter captured events.

TAC Dataset. The TAC dataset is a self-constructed dataset about Windows audit logs generated by a commercial tool. The tool customizes the timing for each malware, so each file contains different times of audit log events.

5.2 Metric

In order to evaluate the performance of different methods of malicious behavior detection, we employ widely used precision, recall, and accuracy to measure the quality of detection performance.

5.3 Baseline

In this paper, we compare two baseline methods to verify the effectiveness of MVGD. To explore the effects of three-view information, we compare some variants of MVGD.

- **MDLR** [2]: MDLR uses domain knowledge to extract features out of audit logs. Then machine learning(Logistic Regression) technique is applied to compute a classifier that classifies audit logs based on the extracted features.
- **MDLSTM** [11]: MDLSTM extracts features out of the audit log events and uses LSTMs to capture sequential effects for malware detection.
- **DLR**: It is a variant of MVGD, which only uses deepwalk to extract features and use Logistic Regression to detect malicious behavior.
- **DSVM**: It is a variant of MVGD, which uses deepwalk to extract features and use SVM to detect malicious behavior.

[1] https://github.com/konstantinberlin/malware-Windows-audit-log-detection.

- **DMLP**: It is a variant of MVGD, which uses deepwalk to extract features and use MLP to detect malicious behavior.
- **MVGD$_{se}$**: It is a variant of MVGD, which only uses structure information and event information.
- **MVGD$_{s}$**: It is a variant of MVGD, which only uses structure information.

5.4 Experimental Results

Comparison Analysis. From Table 4, the accuracy of MDLR and MDLSTM is lower than that of MVGD. The feature engineering in the original MDLR is too large to run. Therefore, we optimize the feature engineering of the original MDLR and removed the redundant features, and then achieved good results. In addition, MDLSTM is higher than the accuracy of 0.9036 in [11]. The reason is that the aggregation strategy on the original audit logs in the graph optimization can also effectively removes the redundant information in the sequence. In conclusion, MVGD always achieves the best performance.

Table 4. Performance of MDLR, MDLSTM, and MVGD on three Datasets.

	MDLR			MDLSTM			MVGD		
	precision	recall	accuracy	precision	recall	accuracy	precision	recall	accuracy
CuckooBox	0.9875	0.9936	0.9861	0.9989	0.9957	0.9960	**0.9996**	**0.9991**	**0.9997**
Sysmon	0.9841	0.9936	0.9852	0.9759	0.9821	0.9714	**0.9906**	**0.9957**	**0.9906**
TAC	0.9523	0.9827	0.9491	0.7562	**1.0000**	0.7562	**0.9570**	0.9806	**0.9524**

Analysis of Three View Information. Table 5 presents the performance of methods with only path information on malicious behavior detection. We apply deepwalk technique to obtain entity embedding, then we aggregate them to the different classifiers to explore the effectiveness. From Table 5, only path information does not perform well. Among the different classifiers, DLR and DSVM perform inferior to DMLP. MVGD performs better than all methods with only path information. Although the recalls of DLR and DSVM achieve 1, those methods have a higher false alarm rate, which is intolerable in the security field.

Table 5. Performance of models with path information on three Datasets.

	DLR			DSVM			DMLP			MVGD		
	precision	recall	accuracy	precision	recall	accuracy	precision	recall	accuracy	precision	recall	accuracy
CuckooBox	0.8626	0.9691	0.8658	0.8598	0.9719	0.8648	0.8619	0.9718	0.8668	**0.9996**	**0.9991**	**0.9997**
Sysmon	0.6753	**1.0**	0.6753	0.6753	**1.0**	0.6753	0.8486	0.8089	0.7735	**0.9949**	0.9622	**0.9895**
TAC	0.7639	**1.0**	0.7639	0.7652	0.9669	0.7780	0.7599	**1.0**	0.7599	**0.9570**	0.9806	**0.9524**

Table 6. Performance of three-view information on three Datasets.

	$MVGD_s$			$MVGD_{se}$			MVGD		
	precision	recall	accuracy	precision	recall	accuracy	precision	recall	accuracy
CuckooBox	0.9337	0.9695	0.9272	0.9980	**0.9994**	0.9993	**0.9996**	0.9991	**0.9997**
Sysmon	0.6820	0.9636	0.6700	0.9500	**1.0**	0.9835	**0.9949**	0.9622	**0.9859**
TAC	0.7612	**1.0**	0.7612	0.9251	0.9914	0.9319	**0.9570**	0.9806	**0.9524**

Furthermore, we explore the effects of three-view information on our model. The experimental results are shown in Table 6. From Table 6, MVGD model with all information almost performs best on three datasets. Using only structural information is less effective than using structural information and event-type information. It's because event-type information can capture the attack actions and the frequency of actions, which reflects the attacker's intention. And path information also brings improvement, but not as much as event-type information. All in all, combining three views of information achieves the best malicious behavior detection performance.

To illustrate the learning capability of the model, we show the graph embedding of audit logs calculated by the MVGD and its variants $MVGD_s$ and $MVGD_{se}$ in Fig. 4. The benign audit logs are marked in blue points and the malicious audit logs are marked in orange points. Each point represents a set of audit log events generated from one program. As described in Fig. 4, the latent representation generated by MVGD with multi-view information can significantly distinguish between malicious and benign programs on three datasets.

Data-Driven Model. Figure 5 shows the flowchart of the data-driven model. For offline data with labels as well as misclassified online data, the MVGD model can be updated in time. Taking Sysmon dataset as an example, with 8,496 audit logs, its construction graph and training time does not exceed 1.96 h.

Efficiency Study. MVGD supports offline training and online prediction. The model's efficiency is illustrated below for the graph construction, and training phases respectively.

Table 7 shows the statistical values of the time overhead for graph construction on the three datasets. We can see that the time overhead required to build the dependency graph for each program is less than 0.6 s. Interestingly, for both the Sysmon and CuckooBox datasets, the maximum time is greater than the 20 s. This is due to the diversity of the programs. Still, the time to construct the dependency graph in our collection environment is consistent with the long-tail distribution, as shown in Fig. 6. So this part of the data is in the tail distribution and does not affect the overall efficiency of constructing the graph.

Tabel 8 shows the comparison of the training time of different models. Models based on path information, such as DLR, DSVM, and DMLP, cost relatively little training time, but are not performed very well. Therefore, using such models for malicious behavior detection in the online scenario is not recommended. As can

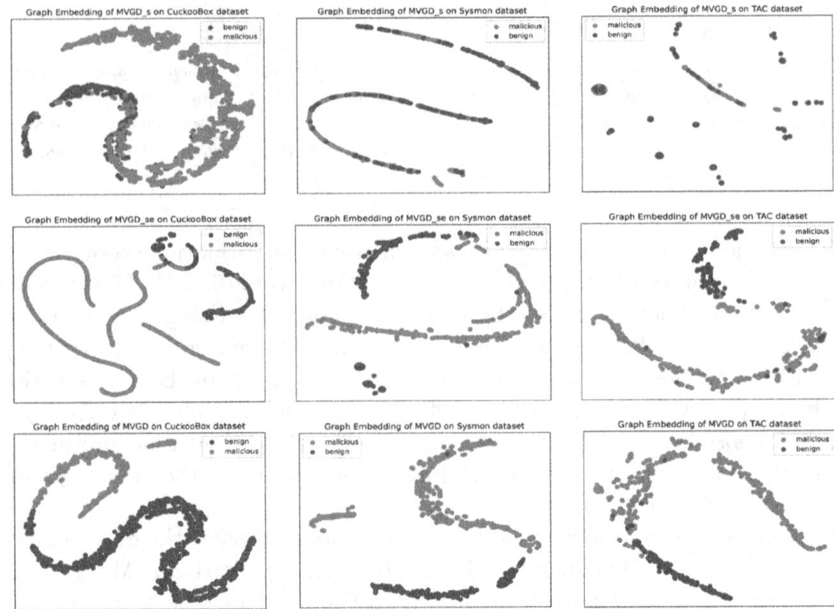

Fig. 4. Graph embedding of MVGD and its variants on three Datasets.

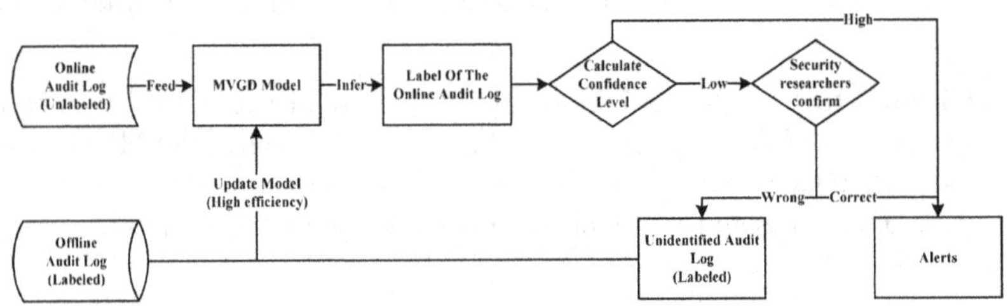

Fig. 5. The flowchart of the data-driven model.

Table 7. The time overhead (seconds) for constructing dependency graphs.

	CuckooBox	Sysmon	TAC
Mean Time	0.22 s	0.57 s	0.14 s
Minimum Time	0.05 s	0.09 s	0.07 s
Maximum Time	23.60 s	37.61	2.26 s
Lower Quartile Time	0.14 s	0.14 s	0.10 s
Median Time	0.18 s	0.19 s	0.11 s
Upper Quartile Time	0.22 s	0.28 s	0.14 s

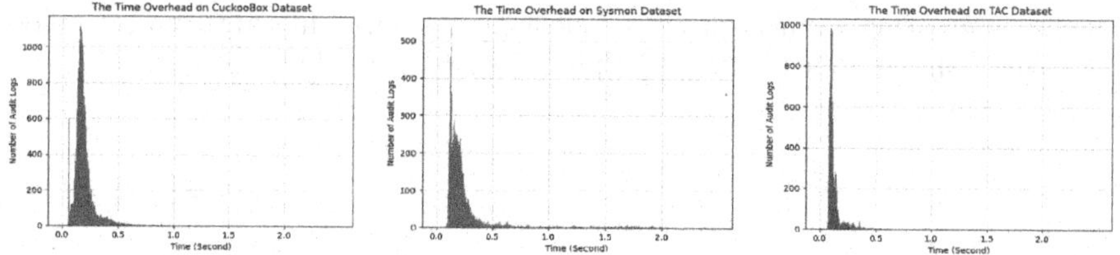

Fig. 6. The distribution histogram of time overhead for constructing dependency graph on three datasets.

be seen from the above, the detection results of MDLR are not bad, but its training time is significantly higher than other models.

Table 8. Comparison of training time of different Models.

	DLR	DSVM	DMLP	MDLR	MDLSTM	MVGD$_s$	MVGD$_{se}$	MVGD
CuckooBox	946.13 s	969.15 s	993.2 s	26588.13 s	5670.6 s	10523.04 s	12578.51 s	14418.3 s
Sysmon	731.31 s	711.63 s	746.78 s	3629.57 s	997.8 s	1856.92 s	1924.29 s	2212.10 s
TAC	440.85 s	458.43 s	443.69 s	2729.00 s	1274.69 s	1013.57 s	1414.14 s	1885.52 s

Case Study. The Nitol botnet is mostly involved in spreading malware and distributed denial-of-service attacks. A distinctive feature of the Nitol family is the use of lpk.dll hijacking. The lpk.dll hijacking enables the malicious program to be backed up horizontally across multiple disks on the victim device for the purpose of long-term persistence on the victim system and remote control.

Figure 7 depicts a partial host-based behavior dependency graph of the Nitol family. MVGD infers that there is a host threat in this audit log, and we analyze its maliciousness from its graph. Due to the acquisition environment, this graph does not reflect the communication process with C&C, but its attack technique is enough to indicate its maliciousness. We describe the attack process of the Nitol family in detail as follows.

- After execution, the Nitol Trojan starts copying itself to a random file name under "c:/windows/" called "c:/windows/imgkme.exe".
- it registers "c:/windows/imgkme.exe" as a service with the name "mnopqr tuvwxyab def".
- it writes the virus service information into the registry "HKLM/SYSTEM/CurrentControlSet/Services/" under the key-value pairs to achieve boot-up.
- The program, "c:/windows/imgkme.exe", releases a file "lpk.dll" in each path with ".exe" suffix for DLL injection.
- the malicious program can use the IPC to establish a connection, invade the user's host, send host information to the C&C server, and perform the corresponding operations through the instructions from the C&C host. This

includes creating a virus process, releasing the virus file anturun.inf in the system or mobile device disk, and planting and spreading the virus.

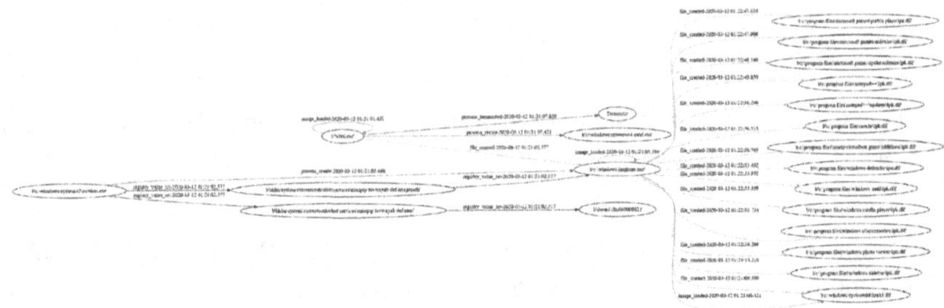

Fig. 7. A part of the host-based dependency graph of the Nitol family.

6 Conclusion

In this paper, we proposed MVGD, a threat detection approach to identify host-based malicious behaviors by Windows audit logs. MVGD employs novel multi-view graph learning techniques, combining structural information, path information, and event-type information from the host-based behavior dependency graphs. Experimental results over three datasets showed that MVGD achieves state-of-the-art performance, with higher precision and recall to detect host-based malicious behavior. We analyze the effects of structure, event type, and path information, and find event type information significantly improves the performance of MVGD. MVGD is more efficient in graph construction and can achieve a balance between training time and effectiveness. In addition, a case study about the Nitol family is also analyzed at the end of our experiments to illustrate the capability of MVGD.

Our model is currently able to identify malicious and benign behavior with high accuracy. In future work, we will make a detailed classification of malicious behaviors, such as identifying the family to which the malicious behavior belongs.

References

1. Alsaheel, A., et al.: {ATLAS}: a sequence-based learning approach for attack investigation. In: 30th USENIX Security Symposium (USENIX Security 2021), pp. 3005–3022 (2021)
2. Berlin, K., Slater, D., Saxe, J.: Malicious behavior detection using windows audit logs. In: Proceedings of the 8th ACM Workshop on Artificial Intelligence and Security, pp. 35–44 (2015)

3. Bridges, R.A., Glass-Vanderlan, T.R., Iannacone, M.D., Vincent, M.S., Chen, Q.: A survey of intrusion detection systems leveraging host data. ACM Comput. Surv. (CSUR) **52**(6), 1–35 (2019)
4. Hassan, W.U., Bates, A., Marino, D.: Tactical provenance analysis for endpoint detection and response systems. In: 2020 IEEE Symposium on Security and Privacy (SP), pp. 1172–1189. IEEE (2020)
5. Hassan, W.U., et al.: Nodoze: combatting threat alert fatigue with automated provenance triage. In: Network and Distributed Systems Security Symposium (2019)
6. Kipf, T.N., Welling, M.: Semi-supervised classification with graph convolutional networks. arXiv preprint arXiv:1609.02907 (2016)
7. Kwon, Y., et al.: MCI: modeling-based causality inference in audit logging for attack investigation. In: NDSS, vol. 2, p. 4 (2018)
8. Ma, S., Zhang, X., Xu, D., et al.: Protracer: towards practical provenance tracing by alternating between logging and tainting. In: NDSS, vol. 2, p. 4 (2016)
9. Milajerdi, S.M., Gjomemo, R., Eshete, B., Sekar, R., Venkatakrishnan, V.: Holmes: real-time apt detection through correlation of suspicious information flows. In: 2019 IEEE Symposium on Security and Privacy (SP), pp. 1137–1152. IEEE (2019)
10. Perozzi, B., Al-Rfou, R., Skiena, S.: Deepwalk: online learning of social representations. In: Proceedings of the 20th ACM SIGKDD International Conference on Knowledge Discovery and Data Mining, pp. 701–710 (2014)
11. Ring, M., Schlör, D., Wunderlich, S., Landes, D., Hotho, A.: Malware detection on windows audit logs using LSTMS. Comput. Secur. **109**, 102389 (2021)
12. Scarselli, F., Gori, M., Tsoi, A.C., Hagenbuchner, M., Monfardini, G.: The graph neural network model. IEEE Trans. Neural Networks **20**(1), 61–80 (2008)
13. Wang, Q., et al.: You are what you do: hunting stealthy malware via data provenance analysis. In: NDSS (2020)
14. Wong, C., Bielski, S., McCune, J.M., Wang, C.: A study of mass-mailing worms. In: Proceedings of the 2004 ACM Workshop on Rapid Malcode, pp. 1–10 (2004)

CECR: Collaborative Semantic Reasoning on the Cloud and Edge

Lei Sun[1], Tenglong Ren[1], Xiaowang Zhang[1,2(✉)], Zhiyong Feng[1,2], and Yuexian Hou[1]

[1] College of Intelligence and Computing, Tianjin University, Tianjin 300350, China
xiaowangzhang@tju.edu.cn

[2] Tianjin Key Laboratory of Cognitive Computing and Application, Tianjin, China

Abstract. Semantic reasoning could exploit the implicit information hidden in the graph and enrich the incomplete knowledge graph. Most existing research in semantic reasoning mainly focuses on the completeness of reasoning results and the efficiency of the reasoning algorithm, which neglects the practicality and scalability of reasoning systems. Especially in the Internet era of data explosion, traditional reasoning systems are gradually struggling to meet the demands of large-scale data reasoning. Therefore, scalability has become a focus for reasoning systems. In this paper, we combine cloud and edge computing for scalability with distributed storage based on data correlation and task scheduling. We specifically propose the query-driven backward reasoning optimization algorithm to improve efficiency and overcome the resource limitation of edge nodes. A reasoning system named CECR (Cloud and Edge Collaborative Reasoning System) is implemented to validate our approach. Experiments on three benchmarks demonstrate the soundness and completeness of reasoning results and scalability of CECR.

Keywords: Ontology reasoning · Knowledge graph · Edge computing

1 Introduction

With the prosperity of the Semantic Web, more and more applications with the help of the Semantic Web have greatly facilitated people's work and life [1,2]. Semantic Web plays an important role as the cornerstone of the semantic web. A large number of incomplete knowledge graphs is difficult to fully exploit their value because of missing semantics. Semantic reasoning could enrich and extend the original graph to discover implicit information hidden in the graph.

The traditional centralized single-node reasoning system improves the system performance at the algorithm level, which consists of two main directions: discovering the information implied in the knowledge graph based on materialization algorithms to enrich the original data [3–6]; rewriting the query with the help of ontology using relevant rules to obtain reliable and complete answers [7–9]. Steigmiller et al. [4] use the absorption technique to rewrite axioms efficiently. PAGOdA [5] combines a datalog reasoner and a full-fledged OWL 2 DL reasoner to obtain scalability. QuOnto [8] proposes a new framework to rewrite

A. El Abbadi et al. (Eds.): DASFAA 2023 Workshops, LNCS 13922, pp. 300–313, 2023.
https://doi.org/10.1007/978-3-031-35415-1_21

and approximate the OBDA ontology language specification. Ontop [9] takes a virtual approach to OBDA to avoid materializing and rewriting queries for ontology-based query processing. Traditional centralized reasoning systems are prone to single-point failures and communication bottlenecks, and their scalability is limited under large-scale data processing.

As the size of Semantic Web data and the number of network devices increase dramatically, scalability has become an important evaluation metric for reasoning systems. Distributed reasoning systems enhance scalability by collaborative computation of multiple nodes. DORS [11] performs data partitioning based on DHTs and reasoning by collaboration among distributed nodes. Mohamed et al. [12] combine an optimized execution strategy, pre-shuffling method and duplicate elimination strategy based on the Spark[1]-Hadoop[2]platform to achieve an effective distributed reasoning. However, distributed reasoning systems usually need more effective scheduling, and reasoning systems implemented based on the Spark-Hadoop computing platform would store data redundantly in multiple nodes, with high resource consumption and low practicality.

Although the scalability of distributed reasoning systems is improved, the lack of effective scheduling makes the system's stability questionable. Edge computing is a type of distributed computing that sinks computational tasks to the network's edge close to the data, enabling reduced network latency and easy scaling. Therefore, we combine cloud and edge computing paradigms, where the cloud center is responsible for large-scale data processing and node resource scheduling. The edge nodes are responsible for specific query processing and reasoning computation, thus overcoming their respective limitations. Considering the edge node's limited resources, we put forward a query-driven backward reasoning optimization algorithm to reduce reasoning computation effectively. Experimental results show that CECR can significantly reduce and avoid unnecessary computation to reason efficiently and is highly scalable on large-scale data processing.

The major contributions of our work are summarized as follows:

(1) We propose a cloud-edge collaborative reasoning mechanism that leverages the features of the cloud and edge computing to achieve high scalability.
(2) We propose a query-driven backward reasoning optimization algorithm that can materialize on all necessary data and rules, which effectively reduces the amount of materialized computation, thus overcoming the constraint of limited resources at the edge nodes.
(3) We implement a reasoning system named CECR and conduct experiments on three datasets, LUBM, UOBM, and DBpedia, which verifies the scalability, efficiency, and reliable completeness of the reasoning results of CECR.

[1] https://github.com/apache/spark.
[2] https://hadoop.apache.org/.

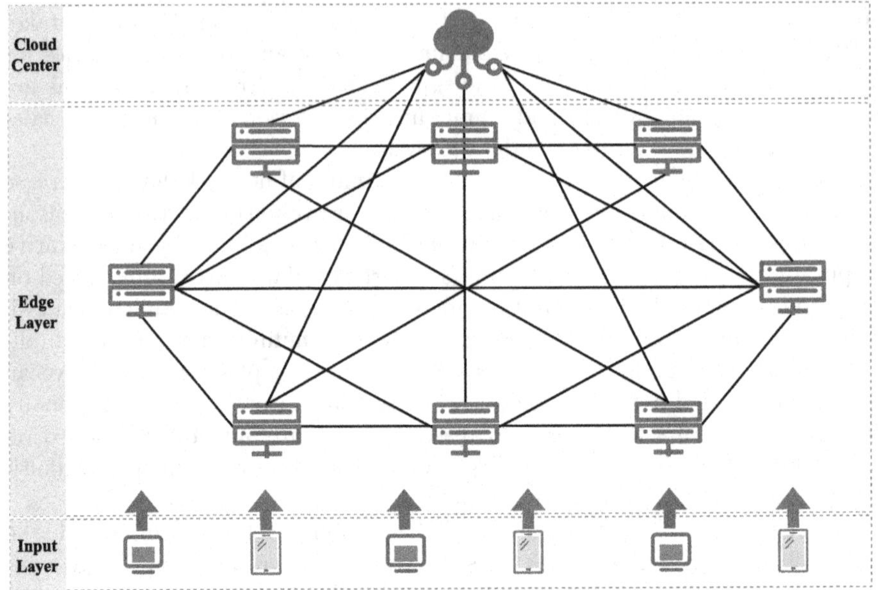

Fig. 1. System Overview

2 System Overview

2.1 System Architecture

Figure 1 depicts the system architecture of CECR. CECR is composed of two parts, a cloud center and edge nodes. In the start-up phase, the cloud center would partition and distribute the data and rules to each edge node for storage. The edge node exposes the API to the terminal and receives queries from multiple terminals to transmit the queries to the cloud center. The cloud center analyzes and schedules the queries in combination with the ontology and selects appropriate nodes for reasoning and querying according to the data storage conditions. When materializing for a specific query, the node may request data and rules required for materialization from other nodes. Figure 2 shows an example of the system running.

We assign the computing tasks to different nodes according to the characteristics of the computational tasks. Considering the performance difference between the cloud center and edge nodes, we assign tasks that require global data information and large-scale data processing to the cloud center and leave the specific query processing and reasoning tasks to the edge nodes near the data.

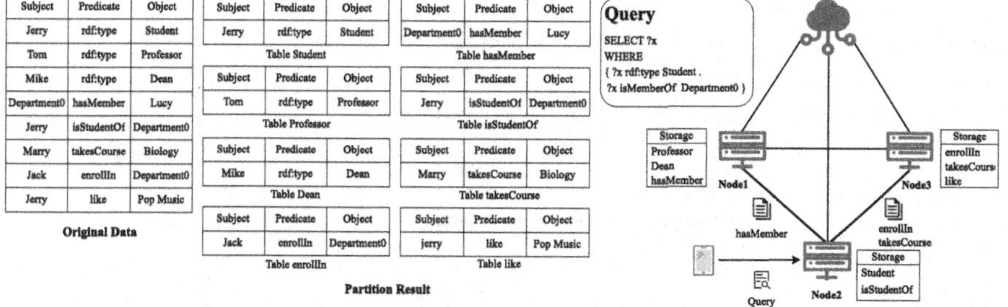

Fig. 2. A user sends a query to an edge node. Then the node sends the query to the cloud center. After analyzing, the cloud center selects the node (Node2) that stores the most matching data to process the query. Node2 requests relevant data and rules from other nodes. For example, **hasMember** is the *inverseProperty* of **isMemberOf**, and **enrollIn** is a *subProperty* of **isMemberOf**.

2.2 Cloud Center

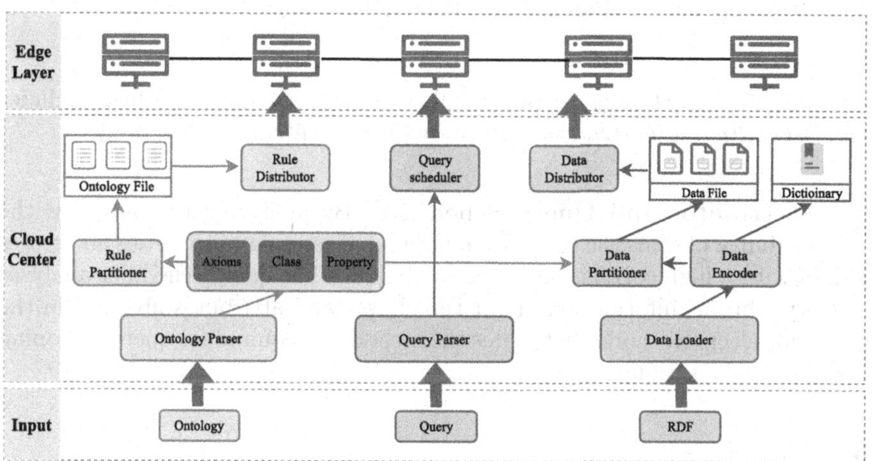

Fig. 3. Cloud Center Architecture

Data and Rule Processor. The data processor has two main submodules, a data encoder and a partitioner. The data encoder will read the input RDF data and encode every new element of each triple with an integer to form a dictionary to store the map between the element and its encoding. Data and rules are partitioned with the guidance of the ontology scheme for further optimization. To be specific, the whole ontology file would be partitioned into plenty of small ontology files relevant to the class or property, i.e., all axioms that

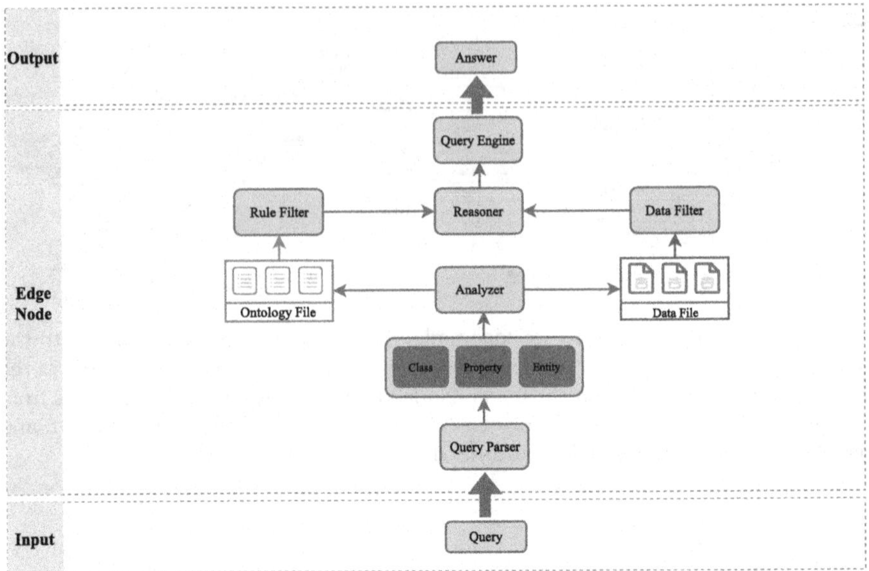

Fig. 4. Edge Node Architecture

are relevant to the class or property will be added into the file, which includes $SubclassOf$,$EquivalentClasses$, $DisjointClasses$, etc. (Fig. 3).

Data Distributor and Query Scheduler. By analyzing the ontology, the data distributor obtains the correlation between classes and properties and stores the data with a high correlation to the same node. The query scheduler analyzes the query, obtains information about the classes and attributes involved in the query, and selects the node that stores the most data required for query reasoning based on the data requirements.

2.3 Edge Node

Query Parser and Data Analyzer. The query parser obtains classes and properties information from the received SPARQL query. The data analyzer combines this information and ontology to obtain relevant data to ensure reliable completeness of the reasoning answers. By transforming SPARQL into a query graph, we obtain the smallest data graph containing the query graph that guarantees the completeness of the query answer with the help of ontology. The specific procedures is shown in Algorithm 1 (Fig. 4).

Data and Rule Filter. The data filter would filter out facts relevant to the entities in the query to ensure that the reasoner only materials about entity-related facts. The rule filter would filter out axioms relevant to these classes and properties and generate a new ontology with them to avoid the intervention of invalid rules.

Reasoner and Query Engine. Reasoner and query engine are not the focus of this work. Therefore, we use the open-source SUMA [6] as the reasoner and the open-source Apache Jena[3] as the query engine, and these modules can use other engines as alternatives.

2.4 Query-Driven Backward Reasoning Optimization Algorithm

The query-driven backward reasoning optimization algorithm for acquiring required data is shown in Algorithm 1. In the initialization phase, the query parser initializes *ClassSet* and *PropertySet* with the query. First, BGP in the query is obtained, and each triple in BGP is iterated to get *Class* and *property*. If its property is *type*, then get its object as *Class* and get *property* whose domain or range is *class*, otherwise get the domain and range of that property *Property* as *Class*. Then add *Class* and *Property* to the *ClassSet* and *PropertySet* respectively. The property's domain and range could reason the class information of the variables in the query to enrich the data and improve the reasoning results' completeness. Next, for each *EquivalentClassAxiom*, its *NamedClass* can be reasoned from the properties it contains. Therefore, it is necessary to add the *NamedClass* containing its *EquivalentClassAxiom* to the *ClassSet* of each property.

In the analysis phase, the analyzer gets as few relevant classes and properties data as possible while ensuring the soundness and completeness of reasoning results. First, iterate through each class in the *classSet*, get the equivalent class for each class, and add it to the *EquivalentClass* set. Also, for each *EquivalentClassAxiom* that contains properties, the properties involved in that axiom are added to the *PropertySet* set. Then, the *EquivalentClass* is added to the *ClassesSet*, and each class in the *ClassesSet* is iterated to obtain each class's subclasses, which will be added to the *SubClass*. Then, recurse the process until no new elements are generated in *EquivalentClass* and *SubClass*. All data relevant to the class is obtained. Next, iterate through each property in the *PropertySet*, get the equivalent and inverse properties for each property, and add them to the *EquivalentAndInverseProperties*. Then add *EquivalentAndInverseProperties* to the *PropertySet*, iterate through the *PropertySet*, get the sub-properties of each property, and add them to the *SubProperties*. Then add all the properties in the *SubProperties* to the *PropertySet*. Recursively get the relevant properties until no new *EquivalentAndInverseProperties* and *SubProperties* are generated. All relevant data for all classes and properties is obtained.

In the refinement phase, to further ensure the reliable completeness of the reasoning results, the analyzer would reason the possible derived equivalent classes from the existing data with *ClassSet* and *PropertySet*. The analyzer

[3] https://jena.apache.org/.

Algorithm 1. Query-Driven Backward Reasoning Optimization Algorithm

```
Input: Query Q Ontology O
Output: ClassSet, PropertySet
function InitializeClassSetAndPropertiesSet(Q)
    BGP ←GetBGP(Q)
    ClassSet ← ∅, PropertySet ← ∅
    for all tp ∈ BGP do
        if tp.pre = type then
            Class ← tp.obj
            Property ← GetPropertyFromDomainAndRange(Class)
        else
            Property ← tp.pre
            Class ←GetClassFromDomainAndRange(Property)
        ClassSet.add(Class)
        PropertySet.add(Property)
    for all Class ∈ ClassSet do
        for all Axiom ∈ EquivalentClassAxioms do
            if Axiom.contains(Class) then
                NamedClass ← axiom.getNamedClass()
                ClassSet.add(NamedClass)
    return ClassSet, PropertySet
function GetRelevantClass(ClassSet,PropertySet)
    EquivalentClass ← GetEquivalentClass(ClassSet,PropertySet);
    ClassSet.addAll(EquivalentClass);
    SubClass ←GetSubClass(ClassSet);
    SubClass.removeAll(ClassSet);
    ClassSet.addAll(SubClass);
    if !SubClass.isEmpty() then
        GetRelevantClasses(ClassSet,PropertySet)
function GetRelevantProperty(ClassSet,PropertySet)
    EquivalentAndInverseProperties ←
GetEquivalentAndInverseProperties(PropertySet)
    PropertySet.addAll(EquivalentAndInverseProperties)
    SubProperties ← GetSubProperties(PropertySet)
    SubProperties.removeAll(PropertySet)
    PropertySet.addAll(SubProperties)
    if !SubProperties.isEmpty() then
        GetInvolvedPropertiess(PropertySet,SubProperties)
function GetInverseEquivalentClass(ClassSet,PropertySet,Ontology)
    for all Axiom ∈ EquivalentClassAxioms do
        EquivalentClassAxioms ← GetEquivalentAxioms(Ontology)
        NamedClass ← Axiom.getNamedClass()
        Property ← Axiom.getProperty()
        Filter ← Axiom.getFilter()
        if ClassSet.contains(Filter)&&PropertySet.contains(Property) then
            ClassSet.add(NamedClass)
    return ClassSet, PropertySet
function Main(Query Q)
    ClassSet, PropertySet ← InitializeClassSetAndPropertiesSet(Q)
    ClassSet,PropertySet ← GetRelevantClass(ClassSet,PropertySet)
    ClassSet,PropertySet ← GetRelevantProperty(ClassSet,PropertySet)
    ClassSet,ProperySet ←
GetInverseEquivalentClass(ClassSet,PropertySet,O)
    return ClassSet, PropertySet
```

goes through the existing classes and properties to reason the possible equivalent classes deduced from these classes and properties. First, all the equivalent class axioms are obtained from the ontology. Then, iterate each equivalent class axiom to obtain the equivalent class, properties, and other classes in the axiom. Then, if $ClassSet$ and $PropertySet$ already contain the class and property data contained in that equivalent class, add that equivalent class to $ClassSet$, which means this equivalent class can be reasoned from these data. Finally, all relevant data are obtained and would be the input of the reasoner.

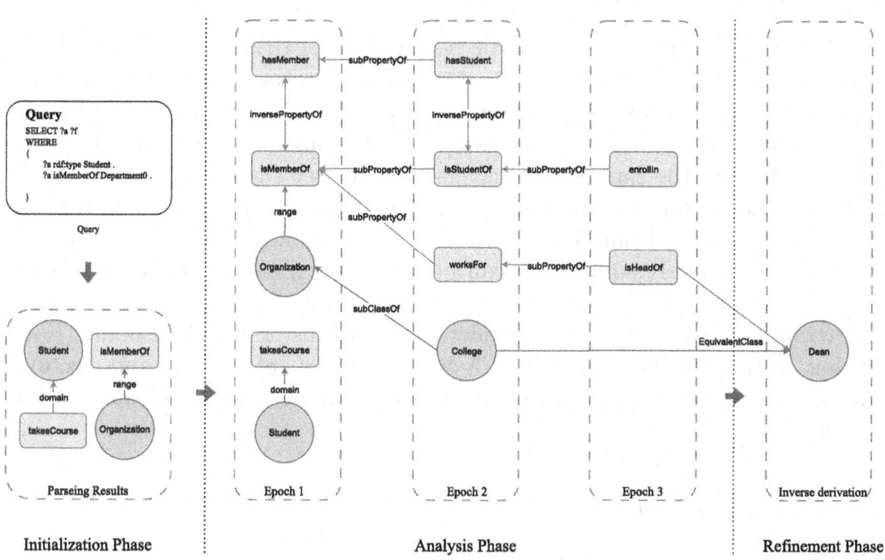

Fig. 5. An Example of Getting Relevant Data for the Query

Figure 5 gives an example of a query-driven backward reasoning optimization algorithm. In the initialization phase, $ClassSet$ and $PropertySet$ are obtained with query and ontology. The class information $Student$ and the property information $isMemberOf$ in the query are first obtained. Combined with the description of the domain and range of $isMemberOf$ in the ontology, $Organization$ is obtained and added into $ClassSet$, while the domain of $takseCourse$ is $Student$, so it is also added into $PropertySet$. In the analysis phase, iterate through $ClassSet$ and get the equivalent classes and subclasses of $Student$ and $Organization$, respectively, while $College$ is added to $ClassSet$ as a subclass of $Organization$. Then start recursively getting all the relevant data, get the inverse property $hasMember$ of $isMemberOf$ and add it to $PropertySet$, get the subclass $College$ of $Organization$ and add it to $ClassSet$. Recursively again, get the subclasses, equivalent classes, subproperties, equivalent properties, and inverse properties of each class and property, and add $College$, $isStudentOf$, $hasStudent$ and $worksFor$ into $ClassSet$

and *PropertySet* respectively. Again, repeat the above process recursively, adding *enrollIn* and *isHeadOf* to *PropertySet*. No new classes or properties can be added to *ClassSet* and *PropertySet*. In the refinement phase, according to the axiom $\exists isHeadOf \sqcap College \equiv Dean$ in the ontology, so *Dean* is inverted and added to the *ClassSet*.

3 Experiment

3.1 Experimental Setup

DataSets. Lehigh University synthetic benchmark[4] (LUBM) is used to evaluate the performance of repositories over a large-scale dataset with a realistic ontology. The University Ontology Benchmark[5] (UOBM) is an extension of LUBM with a more complex ontology. Dataset UOBM of arbitrary size can be generated by adjusting the number of universities. We set the data growth step of the dataset to 100 for the scalability test. The DBpedia[6] dataset is obtained by extracting knowledge from Wikipedia, which is quite large. However, the ontology rules are relatively simple and can be done by a reasoner supporting OWL 2 RL.

Setting. All experiments are conducted on the cluster with four nodes. Each node has an Intel(R) Xeon(R) CPU E5-4607 with 128G of memory and a configured capacity of 2T. The operating system of each node is CentOS Linux release 7.9.2009(Core).

3.2 Evaluation Metrics

We evaluate the system from three dimensions:

- **Completeness and soundness of answers**. The measure is the quality of the answer, i.e., the number of queries that can be answered correctly by the query answering system.
- **The scalability and efficiency of the reasoning system**. The evaluation of the scalability and efficiency of the system is to test the pre-processing time (pre-time), which includes data load time and materialization time (mat-time). The data load time includes all the data pre-processing steps before materialization, such as constructing a dictionary, generating an index, etc. The materialization time is the time taken by the reasoner to compute consequences.
- **The effectiveness of data and rule filters**. We measure the filter's effectiveness by comparing the number of facts after materialization with and without filters.

[4] http://swat.cse.lehigh.edu/projects/lubm/.
[5] https://www.cs.ox.ac.uk/isg/tools/UOBMGenerator/.
[6] https://www.dbpedia.org/.

3.3 Competing Models

We compare CECR with Pellet [3], PAGOdA [5] and SUMA [6]. All these systems, except Pellet, are tested for scalability and efficiency.

Pellet is a complete and sound OWL 2 DL reasoning system. It is the standard of completeness and soundness testing.

PAGOdA implements high-scalable reasoning with RDFox.

SUMA implements high-scalable reasoning performance with a partial materialization algorithm. It will compare with CECR and PAGOdA in the scalability and efficiency tests.

3.4 Experiment Results

The Soundness and Completeness Evaluation Table 1 shows that all systems obtain all answers on benchmarks, except for Pellet, who can not complete the DBpedia test in the effective time. CECR could obtain sound and complete reasoning results on benchmarks.

Table 1. Experiment Results on Quality of Answer.

System	LUBM	UOBM	DBPedia
CECR	24	16	1024
SUMA	24	16	1024
PAGOdA	24	16	1024
Pellet	24	16	*

The Scalability and Efficiency Test. As shown in Fig. 6, the data loading time of CECR increases linearly with the size of the dataset. The data loading time of SUMA also grows linearly with the size of the dataset, but its data loading time reaches seven times that of CECR. Because CECR can analyze the data necessary for reasoning about the query based on the class, attribute, and entity information contained in the query, combined with the ontology, CECR does not need to read all the data for reasoning, which effectively shortens the data loading time of the system. When the data size of LUBM reaches 700, the data reading time of SUMA can no longer grow linearly with the data size and reaches the bottleneck of system scalability. On the UOBM dataset, SUMA can no longer finish loading the dataset when the dataset size reaches 500. Even when the data size of the LUBM dataset reaches 700, the data loading time of CECR still grows steadily and linearly, which shows the good scalability of the CECR system.

As shown in Fig. 7, the materialization time of CECR grows linearly with the size of the dataset. When the size of the dataset is small, the materialization time of SUMA can grow linearly with the size of the dataset. However, when

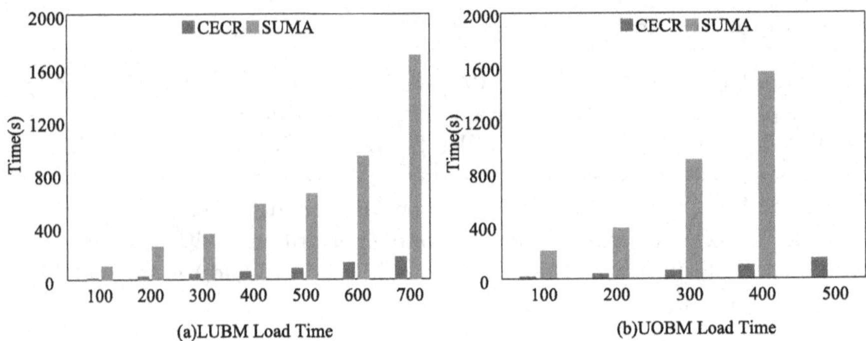

Fig. 6. Data Loading Experimental Results

the size of the dataset is large, the materialization time of SUMA grows steeply and lacks good scalability. On the LUBM dataset, the materialization time of SUMA is 2185 s on LUBM(700) and 981 s on LUBM(600), which is 2.23 times of the former. The materialization time of CECR is 284 s and 169 s on LUBM(700) and LUBM(600), respectively, which is 1.68 times of the former. CECR is based on a query to filter data and rules, not only data filtering reduces the amount of data for reasoning, but also rule filtering reduces the amount of computation for materialization, which improves the materialization efficiency of the system from two aspects. On the UOBM data, SUMA suffers memory overflow during materialization on UOBM(400) and cannot complete the reasoning task. At the same time, CECR is able to complete the reasoning task on UOBM(500), which shows the good scalability of the CECR system.

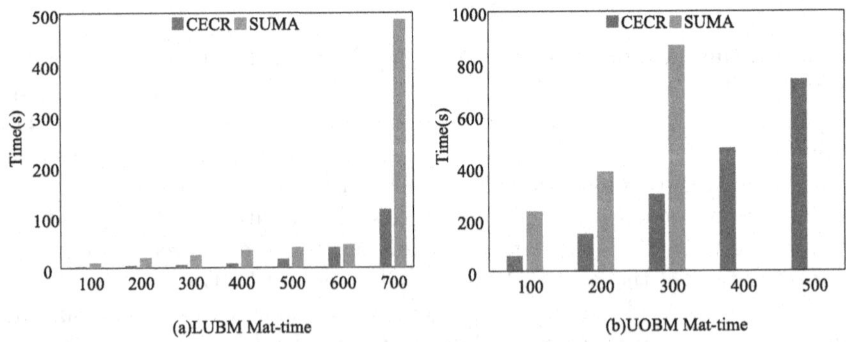

Fig. 7. Materialization Experimental Results

Because the data loading time, consistency checking time, and materialization time of the PAGOdA system could not be effectively counted, we composed the pre-processing time from the time the system reads the data to the completion of the materialization computation for comparing the reasoning performance of each system. As shown in Fig. 8, the pre-processing time of CECR is faster than that of SUMA and PAGOdA on either LUBM or UOBM datasets. On LUBM(700), the pre-processing time of CECR is 284 s, which is substantially faster than the pre-processing time of SUMA of 2185 s and that of PAGOdA of 2314 s. On the UOBM dataset, CECR achieves a two-order-of-magnitude performance lead over PAGOdA. The materialization phase will take more time because the UOBM ontology rules are more complex than LUBM. PAGOdA cannot complete the reasoning task in a limited time on UOBM(500), while CECR can still complete the reasoning task in 888 s, which shows that CECR has good performance compared to other systems.

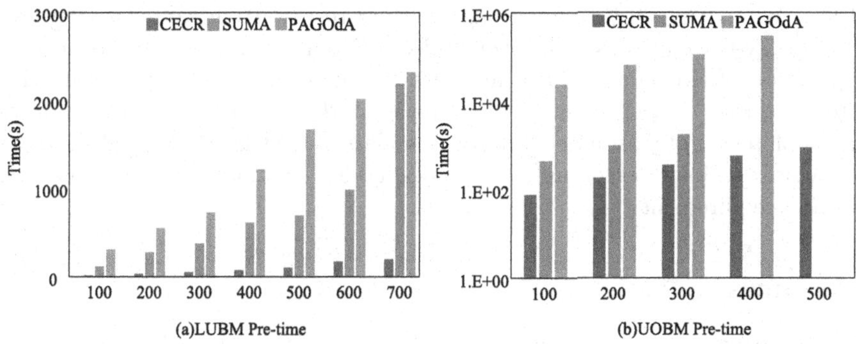

Fig. 8. Pre-processing Experimental Results

The Effectiveness Test of the Data Filter and Rule Filter. We tested filters on UOBM(100), and Fig. 9 shows that the data filter and rule filter can effectively reduce the amount of materialization computation. The amount of

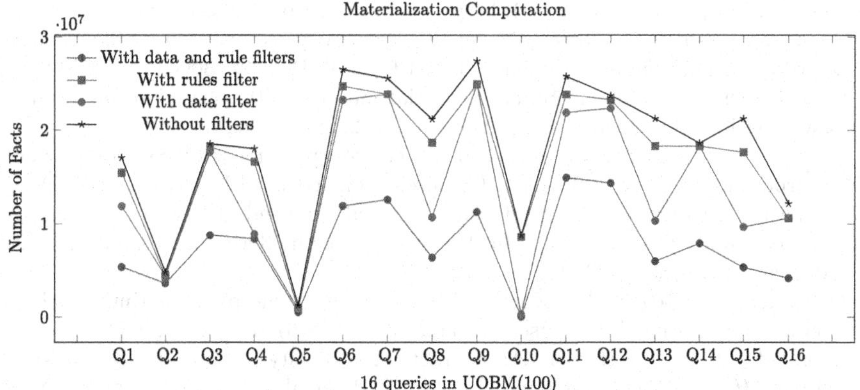

Fig. 9. Results of Data and Rule filters

materialized computation can be reduced by up to 50% on some queries with filters. The data filter efficiently reduces the input of materialization by analyzing the entities in the query to filter the data. The rule filter efficiently reduces the intermediate results of materialization by analyzing the axioms involved in the properties and the classes in the query, thus improving the efficiency of materialization.

4 Conclusion

In this paper, we propose a cloud-edge collaborative reasoning system that achieves scalability and efficiency improvement by combining cloud computing and edge computing to take advantage of their respective advantages. We propose a query-driven backward optimization reasoning algorithm to reduce the computation of reasoning, which improves the performance of the system with limited resources. Experiments on three benchmarks demonstrate the soundness and completeness of reasoning results and scalability of CECR. In the future, we will investigate joint reasoning among edge nodes for more complex reasoning tasks and requirements.

References

1. Bonte, P., Ongenae, F., De Turck, F.: Subset reasoning for event-based systems. IEEE Access **7**, 107533–107549 (2019)
2. Chien, Y., Lin, F.: Distributed semantic reasoning enabled by fog computing. In: Proceedings of 2019 International Conference on Internet of Things (iThings) and IEEE Green Computing and Communications (GreenCom) and IEEE Cyber, pp. 1033–1040. (2019)
3. Sirin, E., Parsia, B., Grau, B., et al.: Pellet: a practical OWL-DL reasoner. J. Web Semant. **5**(2), 51–53 (2007)
4. Steigmiller, A., Glimm, B.: Absorption-based query answering for expressive description logics. In: Ghidini, C., et al. (eds.) ISWC 2019. LNCS, vol. 11778, pp. 593–611. Springer, Cham (2019). https://doi.org/10.1007/978-3-030-30793-6_34
5. Zhou, Y., Grau, B.C., Nenov, Y., et al.: Pagoda: pay-as-you-go ontology query answering using a datalog reasoner. J. Artif. Intell. Res. **54**, 309–367 (2015)
6. Qin, X., Zhang, X., et al.: Suma: a partial materialization-based scalable query answering in OWL 2 DL. J. Data Sci. Eng. **6**, 229–245 (2021)
7. Eiter, T., Ortiz, M., Simkus, M., et al.: Query rewriting for Horn-SHIQ plus rules. In: Proceedings of the 26th AAAI Conference on Artificial Intelligence (2012)
8. Botoeva, E., Calvanese, D., Santarelli, V., et al.: Beyond OWL 2 QL in OBDA: rewritings and approximations. In: Proceedings of the 30th AAAI Conference on Artificial Intelligence, vol. 30, no. 1 (2016)
9. Calvanese, D., Cogrel, B., Komla-Ebri, S., et al.: Ontop: answering SPARQL queries over relational databases. Semant. Web J. **8**(6), 471–487 (2017)
10. Carral, D., Feier, C., Hitzler, P.: A practical acyclicity notion for query answering over *Horn-* $\mathcal{SRIQ}$ ontologies. In: Groth, P., et al. (eds.) ISWC 2016. LNCS, vol. 9981, pp. 70–85. Springer, Cham (2016). https://doi.org/10.1007/978-3-319-46523-4_5

11. Fang, Q., Zhao, Y., Yang, G., Zheng, W.: Scalable distributed ontology reasoning using DHT-based partitioning. In: Domingue, J., Anutariya, C. (eds.) ASWC 2008. LNCS, vol. 5367, pp. 91–105. Springer, Heidelberg (2008). https://doi.org/10.1007/978-3-540-89704-0_7
12. Mohamed, H., Fathalla, S., Lehmann, J., et al.: A scalable approach for distributed reasoning over large-scale OWL datasets. In: Proceedings of the 13th International Joint Conference on Knowledge Discovery, Knowledge Engineering and Knowledge Management, pp. 51–60. (2021)
13. Motik, B., Nenov, Y., Piro, R., et al.: Handling OWL: sameAs via rewriting. In: Proceedings of the 29th AAAI Conference on Artificial Intelligence, pp. 231–237. AAAI Press (2015)
14. Lutz, C., Przybylko, M.: Efficiently enumerating answers to ontology-mediated queries. In: Proceedings of the 41st ACM SIGMOD-SIGACT-SIGAI Symposium on Principles of Database Systems, pp. 277–289. ACM (2022)

Data-Augmented Counterfactual Learning for Bundle Recommendation

Shixuan Zhu, Qi Shen, Chuan Cui, Yu Ji, Yiming Zhang, Zhenwei Dong, and Zhihua Wei(✉)

Tongji University, Shanghai, China
{2130768,2130777,2033065,2230779, 2030796,1853155,zhihua_wei}@tongji.edu.cn

Abstract. Bundle Recommendation (BR) aims at recommending bundled items on online content or e-commerce platform, such as song lists or book lists. Several graph-based models have achieved state-of-the-art performance on BR task. But their performance is still sub-optimal, since the data sparsity problem tends to be more severe in real BR scenarios, which limits these models from more sufficient learning. In this paper, we propose a novel graph learning paradigm called Counterfactual Learning for Bundle Recommendation (CLBR) to mitigate the impact of data sparsity problem and improve BR by introducing counterfactual thinking. Our paradigm consists of two main parts: counterfactual data augmentation and counterfactual constraint. In counterfactual data augmentation, we design a heuristic sampler to generate counterfactual graph views for graph-based models to alleviate the data sparsity. We further propose counterfactual loss to constrain model learning for mitigating the effects of noise in augmented data and achieving more sufficient model optimization. Further theoretical analysis demonstrates the rationality of our design. Extensive experiments of BR models applied with our paradigm on two real-world datasets are conducted to verify the effectiveness of the paradigm.

Keywords: Bundle Recommendation · Data Augmentation · Graph Neural Network · Counterfactual Learning

1 Introduction

Bundled items are very common on nowadays online content or e-commerce platforms, which makes Bundle Recommendation (BR) [28] become an important task. The task aims to recommend a bundle with multiple items for users, which can improve users' experience for providing more variable options, as well as increase business profits for the expansion of order size. Typical BR scenarios include book lists recommendation on online reading websites [12], song lists recommendation on music streaming platforms, video collections recommendation on online video websites, etc.

S. Zhu and Q. Shen—Both authors contributed equally to this research.

A. El Abbadi et al. (Eds.): DASFAA 2023 Workshops, LNCS 13922, pp. 314–330, 2023.
https://doi.org/10.1007/978-3-031-35415-1_22

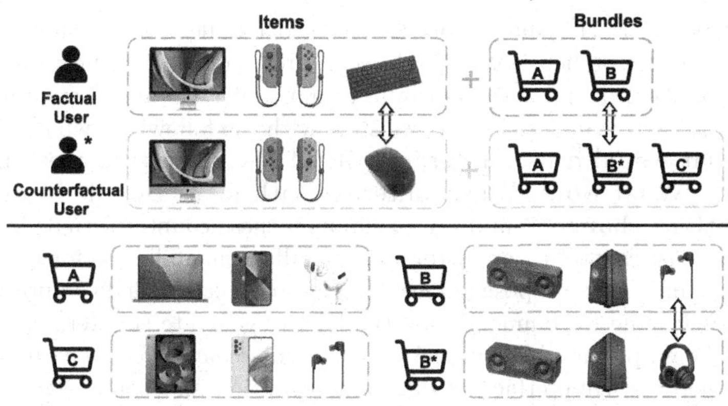

Fig. 1. A toy example of bundle recommendation in E-commerce, where the bundles are shopping carts that consist of multiple products.

On the basis of general recommendation methods, most existing works for BR introduce user-item interaction and bundle-item affiliation information as supplements for user-bundle interaction. For utilizing this key information, graph-based models are employed [2,17] to encode them into user and bundle embeddings [5,8,13], and some multi-task frameworks [3,6,8] are leveraged for model training.

However, the effectiveness of the aforementioned graph-based BR models usually depends on enough high-quality training data. Unfortunately, the data sparsity problem is more severe and complex in most realistic BR scenarios than in general recommendation scenarios, since there exists not only user-side sparsity but also creator-side sparsity problems. For the users of platform, similar to the ubiquitous user-item interaction sparsity [24,26], the user-bundle interactions are also very sparse. For the bundle creators, although there are massive possible item combinations for bundle generation, very few bundles are actually created and shared [7], which leads to sparse bundle-item affiliations. These prevent adequate training for the graph-based BR models and lead to sub-optimal recommendation results.

For confronting these data sparsity in BR, we focus on the counterfactual thinking in causal theory and resort to counterfactual data as a complement to the observed sparse data. In counterfactual thinking, the real data we observe in BR scenarios, whether interaction data or bundle-item affiliation relation, is only a small fraction of all possible data, or counterfactual data. For instance, Fig. 1 illustrates a toy example of BR in E-commerce, where a user can explore not only products but also shopping carts that consist of multiple products created by other users. We observe that a user purchased a keyboard instead of a mouse, and only chose *bundle A* and *bundle B*. But in the counterfactual world, due to possible different exposure mechanisms, the user may have bought the mouse, and additionally selected *bundle C* that has similar contents with *bundle A*. Or perhaps the contents of *bundle B* change due to the substitutability of some items, but the user still can accept the changed bundle. These counterfactual data can more comprehensively reveal users' preference, which naturally solve the data sparsity problem and contribute to more effective model training.

Motivated by the analysis above, we start with the following "what if" questions: **"What would a user interact with if his/her interaction history changes?" "What would a user interact with if the bundle-item affiliation relations change?"** and try to simulate these counterfactuals by intervening on the user's interaction history and bundle-item affiliation for improving BR model training. We propose a novel learning paradigm called Counterfactual Learning for Bundle Recommendation (CLBR) to generate counterfactual data and guide the graph-based BR model to utilize these data effectively. It consists of two main parts: counterfactual data augmentation and counterfactual constraint. For the data augmentation, we develop a simple yet efficient heuristic sampler to generate counterfactual graph views based on the original BR graph, which achieves effective noise control compared to the stochastic sampler. The counterfactual constraint helps models discard intractable noise in the generated data that deviates from the counterfactual distribution, and make efficient use of augmented data to learn accurate and robust representations for BR. We apply the CLBR paradigm to several state-of-the-art graph-based models. Extensive experiments on two real-world datasets demonstrate the effectiveness of the proposed paradigm.

Our main contributions in this work are summarized below:

- We introduce the data sparsity problem in bundle recommendation, and try to leverage counterfactual thinking into bundle recommendation for confronting the problem.
- We propose **C**ounterfactual **L**earning for **B**undle **R**ecommendation (CLBR), a novel learning paradigm for graph-based models, which includes counterfactual data augmentation with a heuristic sampler and counterfactual constraint for training, to achieve more sufficient model optimization.
- We theoretically analyze the noisy information in the augmented data, and interpret our proposed constraint term via information theory, to verify the rationality of our design.
- We apply CLBR on several state-of-the-art graph-based models and conduct extensive experiments based on real-world BR datasets, which verify the effectiveness of the paradigm.

2 Related Works

Bundle Recommendation. Recently, several efforts have been made in Bundle Recommendation, which recommends a bundle of items that users may have interest to interact together [4–6,10,11,13,16]. Sar Shalom et al. [16] introduced the list recommendation problem, and optimized a list's click probability based on collaborative filtering. Liu et al. [11] estimated the probability that a consumer would buy an item associated with already bought item to recommend product bundles. EFM [4] captured users' preferences over items and lists by factorization model. DAM [6] contributed a factorized attention network to gather item information into bundle representations in a multi-task way jointly for user-bundle and user-item interactions. More recently, graph-based recommendation models have become the state-of-the-art methods for BR. HFGN [10] proposed a Hierarchical Fashion Graph Neural Network to obtain more expressive representations for users and outfits, which incorporated both item-level and bundle-level semantics into the bundle representations. BGCN [5] utilized graph neural network (GNN) to learn representations from user-level and bundle-level graph views, and CrossCBR [13] further adopted contrastive learning to model the cooperative association between two views to achieve mutual enhancement.

However, previous researches, especially graph-based models rely on large amounts of reliable data. The severe data sparsity problems existed in BR scenarios limit the performance of graph-based models, which inspires the study of counterfactual data augmentation in this paper.

Counterfactual View in Recommendation. As a key theory of causal inference, counterfactual paradigm focuses on estimating causal effects with *What if* problems and has been widely adopted in many machine learning domains recently, including computer vision [1,9], natural language processing [29] and graph mining [19]. In the recommendation field, some works leveraged counterfactual thinking for debias problem. For example, Wang et al. [20] estimated the counterfactual click likelihood of the user for reducing the direct effect of exposure features and eliminating the clickbait issue. Zhang et al. [27] deconfounded the popularity bias in training phase, and adjusted the recommendation score with desired popularity information via causal intervention during inference phase. Besides, some researchers improved the recommendation models in causal view. Zhang et al. [25] empowered attention mechanism by the causal regularization, which uses individual treatment effect to measure the causal relation. In addition, some methods explored the counterfactual in data-augmentation manner [23]. For instance, Wang et al. [21] proposed data- and model-oriented sampling methods for generating better counterfactual sequences when training the anchor model. Xiong et al. [22] designed a learning-based method to discover more effective samples, actively intervened on the user preference, and predicted the user feedback based on a pre-trained recommender.

However, the counterfactual view in the heterogeneous graph field, e.g. the complex multi-relation in BR, is of great need but largely unexplored.

3 Preliminary

3.1 Task Definition

Given user set $U = \{u_1, u_2, \ldots, u_N\}$, item set $V = \{v_1, v_2, \ldots, v_L\}$, and bundle set $B = \{b_1, b_2, \ldots, b_K\}$, we formulate the user-item interaction data as $\mathcal{E}_{uv} = \{(u, v)|u \textit{ interacted } v, u \in U, v \in V\}$, user-bundle interaction data as $\mathcal{E}_{ub} = \{(u, b)|u \textit{ interacted } b, u \in U, b \in B\}$ and bundle-item affiliation relation as

$\mathcal{E}_{bv} = \{(b,v) \mid v \text{ belongs to } b, b \in B, v \in V\}$. Based on the above information, bundle recommender estimates the probability that user will interact with each candidate bundle.

3.2 A Unified Framework

Then we summarize previous bundle recommendation methods into a unified graph neural network framework for subsequent leveraging. Generally, we organize the user, item and bundle and their relation information as a heterogeneous user-bundle-item graph $\mathbf{G} = \{\mathcal{V} = \{U,V,B\}, \mathcal{E} = \{\mathcal{E}_{uv}, \mathcal{E}_{ub}, \mathcal{E}_{bv}\}\}$ following [5,10], with initialized node embeddings $\mathbf{X} \in \mathbb{R}^{(N+L+K)\times D}$ of $\mathcal{V}$, where D is the embedding dimension. Then we use the unified user-bundle-item relation graph defined above as the basis of graph-based BR models. The general structure of existing graph-based BR models can be summarized in the following two parts: (1) modeling the cross-semantic relation of user, item and bundle; (2) calculating the final matching score of user and bundle. Therefore, we formula the unified bundle recommender $M_\theta(\mathbf{X}, \mathbf{G})$ mainly as a graph-based encoder and an embedding-based predictor:

$$\mathbf{U}, \mathbf{B}, \mathbf{I} = \mathbf{Encoder}(\mathbf{X}, \mathbf{G})\ ;\quad s_{i,j} = \mathbf{Predictor}(\mathbf{U}_i, \mathbf{B}_j) \tag{1}$$

where $\mathbf{U}, \mathbf{B}, \mathbf{I}$ denote the learned embeddings of users, bundles and items respectively, and $s_{i,j}$ is the predicted probability of user i and bundle j based on their embeddings. The loss objective $L_{tasks}(M_\theta)$ can be designed flexibly for various operations, e.g. contrastive learning [13], hard negative sampling [5]. Through the minimization of loss, the bundle recommender parameters are optimized.

4 Methodology

In this section, we introduce the proposed counterfactual learning paradigm to effectively alleviate the sparsity problem in BR scenario. The complete process of our proposed Counterfactual Learning for Bundle Recommendation (CLBR) is presented in Algorithm 1, which consists of two main steps: counterfactual data augmentation with a heuristic node pairs sampler, and model training under counterfactual constraint.

4.1 Counterfactual Data Augmentation

In this part, we generate a set of counterfactual views based on the unified user-bundle-item relation graph as data augmentation. In detail, we use a sampler S to choose user-bundle, user-item and bundle-item pairs, and then change the structure of the original user-bundle-item graph by adding/dropping edges according to the sampling results. We set the augmentation ratios r_{ui}, r_{ub}, r_{bi} for three types of node pairs respectively, to control the disturbance ratio of counterfactual views. We denote that $\mathcal{E}_t^+$ is the set of node pairs selected to be added as edges to the counterfactual graph view, while $\mathcal{E}_t^-$ is the set of edges picked for dropping, where t denotes the type of node pairs which can be ui,ub and bi.

There are many feasible methods to implement the sampler. In the simplest stochastic sampling method, we randomly select node pairs for $\mathcal{E}_t^+$ from all unconnected node pairs in the original BR graph, as well as select node pairs for $\mathcal{E}_t^-$ from existing edges in BR graph, until the size of $\mathcal{E}_t^+$ and $\mathcal{E}_t^-$ for each type satisfy the ratio we set. Formally, it can be expressed as $|\mathcal{E}_t^+| \geq \alpha r_t|\mathcal{E}_t|, |\mathcal{E}_t^-| \geq (1-\alpha)r_t|\mathcal{E}_t|$ for $\forall t$, where $\alpha \in [0, 1]$ controls the proportion of adding and dropping. Then we can generate a counterfactual view of BR graph by applying these user-bundle, user-item and bundle-item edge adding and dropping operations to the original graph. Finally, we get a counterfactual view set through generating N_v counterfactual views, which is a naive simulation of the counterfactual world.

Algorithm 1: Learning Algorithm of CLBR

```
Data: A unified user-bundle-item relation graph G^f ; Number of counterfactual
      views N_v; Initialized embeddings X of users, items and bundles.
Result: Learned bundle recommendation model M_θ.
Pretrain the selection model M'_θ;
for i in [1, N_v] do                              // Counterfactual Data Aug
    for t in [ub, ui, bi] do
        E_t^+ = ∅,   E_t^- = ∅;
        while |E_t^+| + |E_t^-| < r_t|E_t| do
            Sample a batch of node pairs under t, and add them into E_t^+ and E_t^-
            following Eq. 2;
        end
    end
    Append the G^c = {V, E^c} into counterfactual view set 𝒢^c, where
    E^c = E ∪ Σ_t E_t^+ − Σ_t E_t^-;
end
for epoch in [0, T] do                            // Counterfactual Constraint
    Randomly choose a counterfactual view G^c ∈ 𝒢^c;
    Generate the user and bundle representations via bundle recommendation
    model: U^f, B^f = M_θ(X, G^f), U^c, B^c = M_θ(X, G^c) ;
    Normalize the embeddings of user and bundle ;
    Calculate the loss L through Eq. 4;
    Update θ by gradient descent;
end
Inference: Calculate the interaction probability score through M_θ(X, G^f) with
frozen parameters θ.
```

The stochastic mechanism is easy to implement, but it brings too much noise to the generated graph views and reduces the effectiveness of subsequent model learning. So we designed a simple yet efficient heuristic sampling method in Algorithm 1, which improve the reliability of sampling by considering noise control.

We pre-train a selection model $M_{\theta'}$ which has the same structure of the recommendation model M_θ, and utilize its graph encoder to generate all user, bundle and item embeddings U, B, I before we start the sampling phase. Considering the computational complexity, we randomly sample a batch of node pairs each time, query the embeddings of them, and calculate relevance scores for the node pairs in the batch. Taking user-bundle node pair (i,j) as an example, the relevance scores can be calculated as $rs_{i,j} = U_i^\top B_j$. Then we filter the noisy data by the following selection rule:

$$\begin{cases} \mathcal{E}_{ub}^{+} \leftarrow \mathcal{E}_{ub}^{+} \cup \{(i,j)\}, & if\; rs_{i,j} > \kappa^{+} and\, (i,j) \notin \mathcal{E}_{ub} \\ \mathcal{E}_{ub}^{-} \leftarrow \mathcal{E}_{ub}^{-} \cup \{(i,j)\}, & if\; rs_{i,j} \leq \kappa^{-} and\, (i,j) \in \mathcal{E}_{ub} \\ \text{No selection}, & else \end{cases} \tag{2}$$

where κ^+ and κ^- are batch-based thresholds for selection, which are defined as the maximum and minimum relevance scores of user-bundle pairs in the batch multiplied by fixed ratios α^+ and α^- respectively. This design simplifies the complex computation for full collection node pairs to the computation within every batch while ensuring noise filtering for effective augmentation. By repeating sampling and selection process, we keep filling the adding set $\mathcal{E}_t^+$ and dropping set $\mathcal{E}_t^-$ until they contain user-bundle, user-item and bundle-item node pairs with total numbers of $r_{ub}|\mathcal{E}_{ub}|$, $r_{ui}|\mathcal{E}_{ui}|$ and $r_{bi}|\mathcal{E}_{bi}|$ respectively. Then we merge $\mathcal{E}_t^+$ into $\mathcal{E}$ and remove $\mathcal{E}_t^-$ from $\mathcal{E}$ for all t to generate a counterfactual view, and further get a counterfactual view set by generating N_v counterfactual views.

4.2 Counterfactual Constraint

Then the question we would like to answer is: given the augmented data composed of the original factual graph and generated counterfactual views, how to achieve effective learning for BR model M_θ? Despite our noise control in the data augmentation process, some noise that deviates from the counterfactual distribution will inevitably remain in the generated data. In order to minimize the influence of residual noise, the model training needs to be additional constrained. Although the interaction data in the counterfactual world is different from that in the real world, user preferences, bundle and item features should be stable. While using GNN to perform semantic propagation on the counterfactual view and the original graph respectively, we expect that the embeddings of the same node learned on different graph views should be closer, while the distance between the node embedding distributions of the two views should vary within a reasonable range.

Therefore, for the normalized embeddings of user $\mathbf{U}$ and bundle $\mathbf{B}$, we have the following counterfactual constraints:

$$\begin{aligned} &\text{minimize} \quad L_{tasks}(M_\theta) \\ &\text{s.t.} \sum_{i=1}^{N} \mathcal{D}(\mathbf{U}_i^c, \mathbf{U}_i^f) \leq \lambda_u \sum_{i=1}^{N} \sum_{j=1, j\neq i}^{N} \mathcal{D}(\mathbf{U}_j^c, \mathbf{U}_i^f), \\ &\sum_{i=1}^{K} \mathcal{D}(\mathbf{B}_i^c, \mathbf{B}_i^f) \leq \lambda_b \sum_{i=1}^{N} \sum_{j=1, j\neq i}^{K} \mathcal{D}(\mathbf{B}_j^c, \mathbf{B}_i^f), \forall c \in \mathcal{G}^c \end{aligned} \tag{3}$$

where the function $\mathcal{D}(e_i, e_j)$ represents the distance between embeddings of node pair e_i and e_j in latent space, c denotes one potential counterfactual augmentation. The left consistency term in the two inequality reflects the consistency between the embeddings in the original graph and counterfactual view of the same user and bundle separately, and the right unrestraint term reflects the *average consistency* of every single node's embedding on the original graph with the embeddings of all other nodes in the augmented counterfactual view. Compared to the constant threshold λ_u, λ_b for the consistency between the real relation and counterfactual relation [23], we adopt the additional *average consistency* metrics for self-adaptive threshold settings. With lower *average consistency*, which can be explained as the user or bundle node being further from other nodes in latent space, we can make larger edge-level perturbations while ensuring that the node will not be confused with other nodes due to changes in embeddings caused by the perturbations. Through the above two constraints, the appropriate-intervention information is preserved and over-intervention information is discarded for exclusive user and bundle representations.

Directly solving the above constrained optimization problem is challenging for the constraints are not differentiable. Therefore we relax the constraints and convert the objective in Eq. 3 to the Lagrange optimization form as follows:

$$\text{minimize} \quad L = L_{tasks}(M_\theta) + \omega_u L_u + \omega_b L_b \tag{4}$$

where non-negative hyper-parameter ω_u, ω_b control the weight of the user- and bundle-side constraint loss L_u, L_b respectively, which are defined as follows:

$$L_u = \frac{1}{N}\sum_{i=1}^{N}\left(\mathcal{D}(\mathbf{U}_i^c, \mathbf{U}_i^f) - \lambda_u \sum_{j=1, j\neq i}^{N} \mathcal{D}(\mathbf{U}_j^c, \mathbf{U}_i^f)\right) \tag{5}$$

$$L_b = \frac{1}{K}\sum_{i=1}^{K}\left(\mathcal{D}(\mathbf{B}_i^c, \mathbf{B}_i^f) - \lambda_b \sum_{j=1, j\neq i}^{K} \mathcal{D}(\mathbf{B}_j^c, \mathbf{B}_i^f)\right) \tag{6}$$

In our experiments, the distance metric $\mathcal{D}(e_i, e_j)$ is computed as $-\exp(e_i^T e_j/\tau)$, where τ denotes the temperature hyper-parameter for tuning how concentrated the features are in the latent space. In practice, $\mathcal{D}(e_i, e_j)$ can be any other distance metric function. To reduce computational complexity, the average global representations are not explicitly sampled but generated from the other user/bundle nodes within the same minibatch. And the final loss is computed across all representations of users/bundles in the minibatch.

5 Further Discussion

Analysis About Noise Control. The data augmentation process in CLBR depends on sampler S. If S is not accurate, the augmented data can be noisy. To check the necessity of noise control in counterfactual data augmentation process, we would like to know how many data should be generated when given the noise level, if we want to ensure relatively good performance. So we theoretically analyze the relation between the amount of generated data and the noisy information introduced by S, under the pursuit of well performance based on

PAC learning framework [18]. We assume that our sampler can recover the true edges in the counterfactual graph views of the user-item-bundle graph with the probability of $1-\eta$, where $\eta \in (0, 0.5)$ indicates the noise level of S. Then we have the following theory:

Theorem 1. *Given a hypothesis class $\mathcal{H}$, for any $\epsilon, \delta \in (0,1)$ and $\eta \in (0, 0.5)$, if $h \in \mathcal{H}$ is the edge ranking model learned based on the empirical risk minimization (ERM), and the sample complexity (i.e., number of samples) is larger than $\frac{2\log(\frac{2|\mathcal{H}|}{\delta})}{\epsilon^2(1-2\eta)^2}$, then the error between the model estimated and true results is smaller than ϵ with probability larger than $1-\delta$.*

The proof of this theorem can refer to [21]. Assuming that the generated data is noisy, if the prediction error of h is larger than ϵ, we have the empirical mismatching rate of h is smaller than $\eta + \frac{\epsilon(1-2\eta)}{2}$, or the empirical mismatching rate of the optimal h^* is larger than $\eta + \frac{\epsilon(1-2\eta)}{2}$. In order to achieve good performance with given probability (i.e., ϵ and δ), at least $\frac{2\log(\frac{2|\mathcal{H}|}{\delta})}{\epsilon^2(1-2\eta)^2}$ samples are needed. That is to say, when the noise level of sampler η is larger, we need to choose much more edges for generating the counterfactual views, which is a huge computation consumption. This theory reveals the significance of our designed heuristic sampler with noise control. The adjustable threshold for the selection rule guarantees a trade-off between sample efficiency and noise level, so the heuristic sampler can generate counterfactual graph views with better reliability and outshone the stochastic one in the case of generating the same amount of data.

Table 1. Statistics of datasets used in experiments.

Statistic	NetEase	Youshu
No. of items	123,628	32,770
No. of users	18,528	8,039
No. of bundles	22,864	4,771
No. of user-bundle	302,303	51,377
No. of user-item	1,128,065	138,515
No. of bundle-item	1,778,838	176,667

6 Experiments

In this section, we conduct experiments on BR to evaluate the effectiveness of our method. We are aim to answer the following research questions: **RQ1:** Does our proposed counterfactual learning paradigm contribute to better performance on existing BR models? **RQ2:** How do different designs influence the final performance? **RQ3:** How do the hyper-parameters affect the effectiveness of our paradigm?

6.1 Experimental Setup

Dataset. We conduct extensive experiments on three public datasets: *NetEase* and *Youshu*, which are widely used in the BR research [5,6]. The statistics of all datasets after prepossessing are summarized in Table 1.

- *NetEase* is a dataset collected by Netease from its own music platform. Users can bind songs into a bundle or add bundles to their favorites on this platform.
- *Youshu* is collected by [4] from a book-review website Youshu[1], where every bundle is a list of books that users desired.

Baseline Models. To demonstrate the effectiveness of CLBR, we use following representative methods as baselines.

- **BPRMF** [15] uses a Bayesian Personalized Ranking pairwise learning framework to optimize the matrix factorization model.
- RGCN† [17] is a classic GNN-based model for heterogeneous graph, which consists of the multiple relations. Here we utilize it to model the user-item-bundle graph in BR.
- **DAM** [6] applies the attention mechanism multi-task learning to capture bundle-level association and collaborative signals.
- **BundleNet** [8] constructs a similar tripartite graph to well extract bundle representations from its included items' features.
- **BGCN** [5] is a state-of-the-art model for BR that constructs bundle-level and item-level graphs and explicitly models the user-bundle interaction, user-item interaction and bundle-item affiliation by GNN.
- **CrossCBR** [13] is also a state-of-the-art model which utilizes contrastive learning to model the cooperative association between bundle-level and item-level graphs to improve BR.

Among these methods, BPRMF and DAM are the representatives of traditional models without using GNN, while the rest are graph-based methods. To verify the effectiveness of our proposed counterfactual learning paradigm, we trained the four graph-based models with different paradigms, then compare the recommendation result of every model. It is worth noting that for a fair comparison, the graph-based models with additional side-information are not considered in our BR experiments.

As our attempt to apply counterfactual thinking in BR domain is new, there are few suitable methods for comparison. So we choose counterfactual method from other recommendation domains that can be adapted to BR as a reference. Each baseline model is trained with the following counterfactual approaches:

- **CASR** [21] is a counterfactual data augmentation framework for sequential recommendation, which proposed a model-orient method to choose items with larger recommendation loss and generate counterfactual sequences. Here we leverage this principle to sample user-bundle pairs that provide larger recommendation loss and generate counterfactual graph views for BR.

[1] http://www.youshu.com/.

- **CLBR** is our proposed counterfactual learning paradigm for bundle recommendation, which can refer to Sect. 4.

Evaluation Metrics. To evaluate the recommendation performance, we employ two widely used metrics: Recall (R@k) and Normalized Discounted Cumulative Gain (ndcg@k) following [5,8], where $k = \{20, 40\}$.

Implementation Details. We implemented the baseline models and our proposed learning paradigm based on Pytorch. We generate four counterfactual views as the counterfactual view set $\mathcal{G}^c$. The ratios r_{ui}, r_{ub}, r_{bi} for adding/dropping user-item, user-bundle and bundle-item edges are set as hyper-parameters, which can be adjustable for different models and datasets. The embedding dimension is set to 64. The temperature parameter τ is set with a default value of 1. The ratios α^+ and α^- are set to 0.8 and 1.2. All parameters are initialized through a Gaussian distribution with a mean of 0 and a standard deviation of 0.1. We employ the Adam optimizer to train the models with the mini-batch size 2048. We conduct the grid search over hyper-parameters as follows: learning rate in $\{1e\text{-}5, 3e\text{-}5, 1e\text{-}4, 3e\text{-}4, 1e\text{-}3, 3e\text{-}3\}$, learning rate decay in $\{0.01, 0.05, 0.1, 0.5\}$, learning rate decay step in $\{2, 3, 4\}$, augmentation ratios in $\{0.01, 0.02, 0.05, 0.1, 0.15, 0.2, 0.25, 0.3\}$ and controlling factors ω_u, ω_b in $\{0.01, 0.05, 0.1, 0.5, 1, 5, 10, 30\}$. For each baseline, we adopt BPR loss and set the negative sampling rate to 1. The temperature factor τ in counterfactual loss is set to 1. We tuned all the models through our best effort. Through 10 times experiments with different run-time seed settings, we recorded all model's average results.

Table 2. Experimental results (%) of models with different training paradigms in R@{20, 40}, and ndcg@{20, 40} on two datasets. For all datasets, we perform 10-times to evaluate the performance. The bold number indicates the improvements over the related baseline are statistically significant ($p < 0.01$) with paired t-tests.

Models	NetEase				Youshu			
	R@20	ndcg@20	R@40	ndcg@40	R@20	ndcg@20	R@40	ndcg@40
BPRMF	3.512	2.010	6.217	2.734	19.63	11.17	27.35	13.27
DAM	4.721	2.404	7.732	3.210	21.02	12.05	28.92	14.10
RGCN†	4.652	2.370	7.714	3.170	20.52	11.52	28.68	13.67
RGCN†+CASR	4.782	2.437	7.998	3.192	20.70	11.63	29.01	13.82
RGCN†+CLBR	**5.210**	**2.643**	**8.494**	**3.499**	**22.17**	**11.88**	**31.05**	**14.25**
BundleNet	4.776	2.545	8.120	3.225	22.45	12.00	30.59	14.19
BundleNet+CASR	4.921	2.583	8.320	3.402	22.45	12.03	30.79	14.34
BundleNet+CLBR	**5.332**	**2.711**	**8.730**	**3.648**	**22.53**	**12.31**	**32.11**	**14.60**
BGCN	5.760	3.060	9.020	3.918	22.69	12.98	31.07	15.28
BGCN+CASR	5.911	3.092	9.315	4.007	22.70	13.16	31.87	15.68
BGCN+CLBR	**6.330**	**3.298**	**9.885**	**4.255**	**24.74**	**14.10**	**34.29**	**16.73**
CrossCBR	8.418	4.565	12.62	5.689	28.11	16.68	37.82	19.37
CrossCBR+CASR	8.423	4.569	12.66	5.699	28.16	16.70	37.85	19.40
CrossCBR+CLBR	**8.721**	**4.683**	**12.85**	**5.782**	**28.48**	**16.98**	**38.30**	**19.52**

6.2 Overall Comparison (RQ1)

We can obtain the following significant observations from the comparison results shown in Table 2.

Comparison of Different Baselines. While training the baselines under their original methods, we can observe that graph-based models generally achieve superior results, which proves the effectiveness of graph-based models for BR task. CrossCBR outperforms all baselines, demonstrating the effectiveness of applying graph contrastive learning in bundle recommendation. In other graph-based models, it should be noted that BGCN significantly performs better than BundleNet and RGCN. We attribute the reason to BGCN's bundle-level and item-level graph modeling, which is better at differentiating users' behavioral similarity and bundles' content relatedness than BundleNet and RGCN's user-bundle-item tripartite graph, and this design is also inherited by CrossCBR.

Table 3. Experimental results (%) of BGCN with different designs in Recall@20 and ndcg@20 on two datasets. Here we omit "@20" on the evaluation metrics for simplicity due to space constraints.

Different Designs	NetEase		Youshu	
	Recall	ndcg	Recall	ndcg
stochastic data sampler	6.266	3.241	24.60	13.99
w/o Counterfactual loss	5.922	3.094	23.36	13.21
Constant threshold	6.146	3.138	24.16	13.59
Ours	**6.330**	**3.298**	**24.74**	**14.10**

Observation of Paradigm Effectiveness. Compared with original models, the extended models with CLBR paradigm achieve significant performance improvement on all metrics consistently, while the improvement brought by the adjusted CASR is very limited, demonstrating the superiority of our method. For CLBR, the average improvements of each model on NetEase dataset are respectively 11.0%, 9.7%, 8.95%, 2.4% for RGCN†, BundleNet, BGCN and CrossCBR while on Youshu dataset the improvements are 5.9%, 2.6%, 9.3%, 1.3%. The CLBR paradigm brings the underperforming model RGCN close to SOTA graph-based models. BGCN and CrossCBR have already obtained superior performance without augmentation, but still can achieve vast performance improvement through our CLBR. These results demonstrate the effectiveness and generality of our learning paradigm on graph-based models for BR.

We attribute these performance improvements to the combination of data augmentation and counterfactual constraint. After the data augmentation of CLBR, the data used for training has the potential to better represent the integral data distribution in the counterfactual world. Combined with the specially designed counterfactual loss, CLBR also achieves effective noise control during the model training step, so the models could learn more accurate and BR robust embeddings for users and bundles under limited training data. Therefore, the data sparsity problem in BR is alleviated, and the recommendations based on these embeddings show to be superior.

6.3 Ablation Study (RQ2)

In this part, we aim to investigate the impact of different designs. We substitute sampling methods with a stochastic data sampler on the SOTA model BGCN, and compare the performance under CLBR paradigm with different counterfactual loss settings, including our proposed counterfactual loss with a self-adaptive threshold, constraint with a constant threshold, and no counterfactual loss. As shown in Table 3, we can find that both the stochastic and heuristic sampler improve the models' performance on most evaluation metrics, but the heuristic method is obviously more effective than the stochastic one. These results verify our conjecture about the disadvantage of stochastic mechanism in Sect. 4.2 for introducing too much noise in the process of generating enough counterfactual data, and prove the validness of heuristic sampler. We can observe that the performances of counterfactual loss versions are higher than the no counterfactual version, and our paradigm with self-adaptive threshold loss outperforms the constant threshold loss version consistently. These could be due to that introducing the counterfactual loss can effectively filter out the remaining noise after aug-

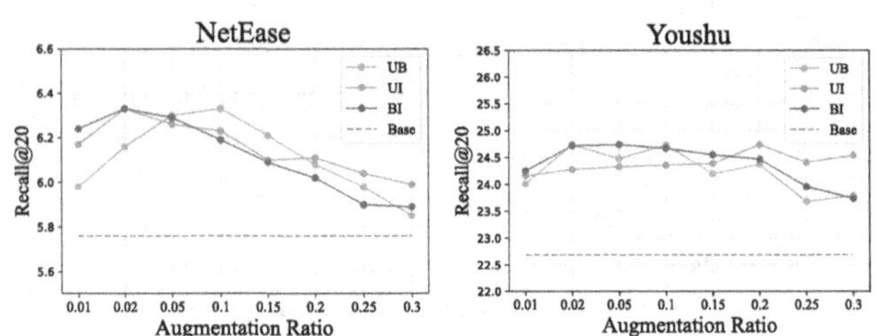

Fig. 2. Influence of Augmentation ratios.

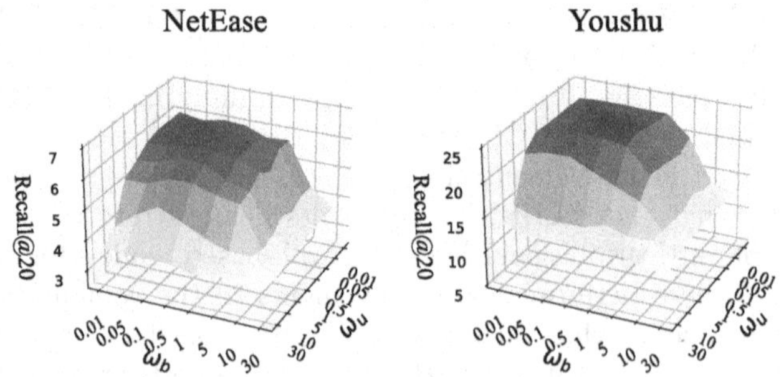

Fig. 3. Influence of Controlling Factors.

mentation, and the self-adaptive threshold additionally considers the dynamic penalty term based on the distance between the local and global user/bundle representation distributions.

6.4 Hyper-Parameters Study (RQ3)

Impact of Augmentation Ratios. For each augmentation ratio, we set the other two ratios and all remaining hyper-parameters to the optimal value that we've tuned in the above experiments. Then we conduct experiments with varying the ratio from 0.01 to 0.3. We present the BGCN's performance of Recall@20 on NetEase and Youshu datasets, respectively in Fig. 2. The results of other models are omitted here for their similarity to BGCN's and the space limitation. The polyline "UB" shows the experiment results of different r_{ub} while r_{ui} and r_{bi} are fixed, and "UI" and "BI" likewise denote the results of various r_{ui} and r_{bi}. The dotted line in the figure presents the basic performance of BGCN without any augmentation. From Fig. 2 we can discover that the performances continue to rise until reaching the optimal augmentation ratios, then drop slowly as the ratios go up. We notice that the results remain stable when the augmentation ratios are around the optimal value, and even sub-optimal results still have considerable improvements compared to the base model. For example, on Youshu dataset, the results stabilize above 24.65 when tuning r_{bi} in $\{0.02, 0.05, 0.1\}$, which are very close to the optimum. It reveals that the effectiveness of our CLBR is not sensitive to the augmentation ratios. In other words, we just need to adjust the augmentation ratios to a relatively loose range and then get nearly optimal results, which means the CLBR could dramatically reduce the workload of parameter adjustment.

Impact of Controlling Factors. In Sect. 6.3, we have verified that counterfactual constraint will help improve graph-based models' performance for BR. Here we want to further explore how the weights of constraint affect the performance of CLBR. For the controlling factors ω_u and ω_b of our proposed counterfactual loss, we fix all remaining hyper-parameters to the optimal value and conduct experiments. We show BGCN's results using different factor combinations as 3D Surface in Fig. 3. It can be observed that on NetEase and Youshu datasets, CLBR paradigm works well when both ω_u and ω_b are in a certain range from 0.01 to 1. The performance of BGCN in terms of Recall@20 is relatively stable when the controlling factors are not too large, as shown in the plateau of the figures. In other words, the CLBR paradigm is insensitive to ω_u and ω_b in a reasonable range, which demonstrates the robustness of it. But if we set either factor too large (e.g., > 5), the performance of the model is severely degraded, even much worse than the original version. This reveals that overlarge weight makes the counterfactual constraint dominate the total loss, thus the total loss deviates too much from the BR task, which leads to a drastic performance drop. So it is important to balance the weight of counterfactual loss by adjusting the parameters for different models and datasets.

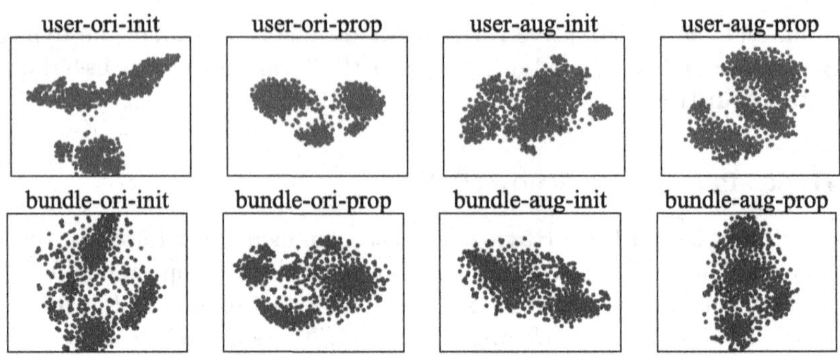

Fig. 4. Embedding visualizations of users & bundles from Youshu dataset.

6.5 Visualization and Analysis

As shown in Fig. 4, we utilize T-SNE [14] to visualize the embedding distribution of users and bundles learned by BGCN on Youshu dataset, to find out how does CLBR paradigm affect representation learning in BR models. We use the "ori" to indicate the embeddings learned under original training paradigm, while "aug" means the embeddings are learned with augmented data under CLBR paradigm. The "init" or "prop" suffix indicates that the embeddings are obtained before or after the graph propagation. The nodes of user and bundle embeddings are colored green and purple respectively. We get the following observations: First, from the initialization case, compared with using original data, the embeddings initialized using augmented data evenly dispersed in representation space. Second, the graph propagation under the counterfactual constraint of CLBR paradigm can make the embeddings distribution become multiple uniform clusters, which is aligned with our expectation for the optimal situation. These reveal that the augmented data has more potential to cover the counterfactual space, and the residual noise in the augmented data is effectively suppressed under our counterfactual constraints. Especially for users, the embeddings' distribution of original data is in an extreme disequilibrium after initialization, even graph propagation does not alleviate this situation well. But training with augmented data effectively solves this problem.

7 Conclusion

In this paper, we pay special attention to the causal view on bundle recommendation, to find a solution for the data sparsity problem and address insufficient learning of representations to improve recommendation performance. Inspired by counterfactual thinking, we propose a novel graph-based counterfactual learning algorithm, including counterfactual data augmentation and counterfactual constraint. We apply the CLBR paradigm to several SOTA graph-based BR models, and conduct extensive experiments on real-world datasets to verify the effectiveness of it. As to future work, we would like to explore more interesting methods for counterfactual data augmentation. It's also meaningful to generalize this paradigm into more recommendation fields.

Acknowledgments. The work is partially supported by the National Nature Science Foundation of China (No. 61976160, 61906137, 61976158, 62076184, 62076182), the Natural Science Foundation of Shanghai (Grant No. 22ZR1466700), Shanghai Science and Technology Plan Project (No. 21DZ1204800) and Technology research plan project of Ministry of Public and Security (No. 2020JSYJD01).

References

1. Abbasnejad, E., Teney, D., Parvaneh, A., Shi, J., Hengel, A.V.D.: Counterfactual vision and language learning. In: CVPR, pp. 10044–10054 (2020)
2. Berg, R.V.D., Kipf, T.N., Welling, M.: Graph convolutional matrix completion. arXiv preprint arXiv:1706.02263 (2017)
3. Cao, D., Nie, L., He, X., Wei, X., Zhu, S., Chua, T.S.: Embedding factorization models for jointly recommending items and user generated lists. In: Proceedings of the 40th International ACM SIGIR Conference on Research and Development in Information Retrieval, pp. 585–594 (2017)
4. Cao, D., Nie, L., He, X., Wei, X., Zhu, S., Chua, T.S.: Embedding factorization models for jointly recommending items and user generated lists. In: SIGIR (2017)
5. Chang, J., Gao, C., He, X., Jin, D., Li, Y.: Bundle recommendation with graph convolutional networks. In: SIGIR, pp. 1673–1676 (2020)
6. Chen, L., Liu, Y., He, X., Gao, L., Zheng, Z.: Matching user with item set: Collaborative bundle recommendation with deep attention network. In: IJCAI (2019)
7. Deng, Q., et al.: Build your own bundle - a neural combinatorial optimization method (2021)
8. Deng, Q., et al.: Personalized bundle recommendation in online games. In: Proceedings of the 29th ACM International Conference on Information & Knowledge Management (2020)
9. Goyal, Y., Wu, Z., Ernst, J., Batra, D., Parikh, D., Lee, S.: Counterfactual visual explanations. In: ICML, vol. 97, pp. 2376–2384 (2019)
10. Li, X., Wang, X., He, X., Chen, L., Xiao, J., Chua, T.S.: Hierarchical fashion graph network for personalized outfit recommendation. In: SIGIR, pp. 159–168 (2020)
11. Liu, G., Fu, Y., Chen, G., Xiong, H., Chen, C.: Modeling buying motives for personalized product bundle recommendation. TKDD **11**(3), 1–26 (2017)
12. Liu, Y., Xie, M., Lakshmanan, L.V.: Recommending user generated item lists. In: RecSys, pp. 185–192 (2014)
13. Ma, Y., He, Y., Zhang, A., Wang, X., Chua, T.S.: Crosscbr: cross-view contrastive learning for bundle recommendation. arXiv preprint arXiv:2206.00242 (2022)
14. Van der Maaten, L., Hinton, G.: Visualizing data using t-SNE. J. Mach. Learn. Res. **9**(11) (2008)
15. Rendle, S., Freudenthaler, C., Gantner, Z., Schmidt-Thieme, L.: BPR: Bayesian personalized ranking from implicit feedback. arXiv preprint arXiv:1205.2618 (2012)
16. Sar Shalom, O., Koenigstein, N., Paquet, U., Vanchinathan, H.P.: Beyond collaborative filtering: the list recommendation problem. In: WWW, pp. 63–72 (2016)

17. Schlichtkrull, M., Kipf, T.N., Bloem, P., van den Berg, R., Titov, I., Welling, M.: Modeling relational data with graph convolutional networks. In: Gangemi, A., et al. (eds.) ESWC 2018. LNCS, vol. 10843, pp. 593–607. Springer, Cham (2018). https://doi.org/10.1007/978-3-319-93417-4_38
18. Shalev-Shwartz, S., Ben-David, S.: Understanding Machine Learning: From Theory to Algorithms. Cambridge University Press, Cambridge (2014)
19. Sui, Y., Wang, X., Wu, J., He, X., Chua, T.S.: Deconfounded training for graph neural networks. arXiv preprint arXiv:2112.15089 (2021)
20. Wang, W., Feng, F., He, X., Zhang, H., Chua, T.S.: Clicks can be cheating: counterfactual recommendation for mitigating clickbait issue. In: SIGIR (2021)
21. Wang, Z., Zhang, J., Xu, H., Chen, X., Zhang, Y., et al.: Counterfactual data-augmented sequential recommendation. In: SIGIR, pp. 347–356 (2021)
22. Xiong, K., et al.: Counterfactual review-based recommendation. In: CIKM, pp. 2231–2240 (2021)
23. Xu, S., Ge, Y., Li, Y., Fu, Z., Chen, X., Zhang, Y.: Causal collaborative filtering. arXiv preprint arXiv:2102.01868 (2021)
24. Yuan, X., Chen, H., Song, Y., Zhao, X., Ding, Z., et al.: Improving sequential recommendation consistency with self-supervised imitation. In: IJCAI (2021)
25. Zhang, J., Chen, X., Zhao, W.X.: Causally attentive collaborative filtering. In: CIKM, pp. 3622–3626 (2021)
26. Zhang, S., Yao, D., Zhao, Z., Chua, T.S., Wu, F.: Causerec: counterfactual user sequence synthesis for sequential recommendation. In: SIGIR, pp. 367–377 (2021)
27. Zhang, Y., et al.: Causal intervention for leveraging popularity bias in recommendation. In: SIGIR (2021)
28. Zhu, T., Harrington, P., Li, J., Tang, L.: Bundle recommendation in ecommerce. In: SIGIR, pp. 657–666 (2014)
29. Zmigrod, R., Mielke, S.J., Wallach, H., Cotterell, R.: Counterfactual data augmentation for mitigating gender stereotypes in languages with rich morphology. arXiv preprint arXiv:1906.04571 (2019)

Multi-domain Fake News Detection with Fuzzy Labels

Zhenghan Chen[1(✉)], Changzeng Fu[2(✉)], and Xunzhu Tang[3]

[1] Peking University, Beijing, China
1979282882@pku.edu.cn
[2] Osaka University, Suita, Japan
changzeng.fu@irl.sys.es.osaka-u.ac.jp
[3] University of Luxembourg, Esch-sur-Alzette, Luxembourg
xunzhu.tang@uni.lu

Abstract. Fake news commonly exists in various domains (*e.g.*, education, health, finance), especially on the Internet, which cost people much time and money to distinguish. Recently, previous researchers focused on fake new detection with the help of a single domain label because fake news has different features in different domains. However, one problem is still solved: A piece of news may have semantics even in one domain source and these meanings have some interactions with other domains. Therefore, detecting fake news with only one domain may lose the contextual semantics of global sources (*e.g.*, more domains). To address this, we propose a novel model, `FuzzyNet`, which addresses the limitations above by introducing the fuzzy mechanism. Specially, we use BERT and mixture-of-expert networks to extract various features of input news sentences; Then, we use domain-wise attention to make the sentence embedding more domain-aware; Next, we employ attention gate to extract the domain embedding to affect the weight of corresponding expert's result; Moreover, we design a fuzzy mechanism to generate pseudo domains. Finally, the discriminator module uses the total feature representation to discriminate whether the news item is fake news. We conduct our experiment on the Weibo21 dataset and the experimental results show that our model outperforms the baselines. The code is open at https://anonymous.4open.science/r/fakenewsdetection-D2F4.

Keywords: Fuzzy Labels · Fake News Detection · Social Media

1 Introduction

With the rapid popularization of the Internet, social media such as Weibo [2], Twitter [4], and Reddit [1], has become an important platform to acquire news. However, fake news also disseminate on these platforms in various types, such as texts, pictures, voices, and videos. Unfortunately, the fake news incites people's emotions and stimulates users to forward, so they can be widely spread. More important, Tavernise [34] showed that important social events were affected due

A. El Abbadi et al. (Eds.): DASFAA 2023 Workshops, LNCS 13922, pp. 331–343, 2023.
https://doi.org/10.1007/978-3-031-35415-1_23

to moderated fake news campaigns. In statics analysis, there are over 70 thousand pieces of fake news spread on Weibo all year round in year 2020. Since the spread of fake news may result in people's panic and social dislocation and bring great loss to social medias [31]. Thus, fake news detection is a critical problem to address.

By now, there are many previous approaches have been proposed and most of them focus on single domain methods for fake news detection [6,20,21,23, 25]. There are two methods of social media fake news detection. One is social background detection method, and the other is content-based detection method [17]. Social background method aims to study users' social network structure, users' personal information, microblog forwarding and reply relationship and rumor's propagation patterns. The content-based research method aims to detect the text, voice and video carried by microblog news. Our work focuses on the content information extraction.

On the Internet, fake news arises in various domains day by day many previous works employ Multi-domain Fake News Detection (*MFND*) method to detect fake news with multiple domain information. There are two main challenges in *MFND*. First, the performance of techniques generally drops if news records are coming from different domains (e.g., politics, entertainment), and a possible explanation of this could be the rather unique content and style of each domain [5].

For example, during the 2020 US election, political fake news and Covid19 fake news were widespread simultaneously. It is difficult to detect both political fake news and Covid19 fake news at the same time, because of the different or even conflicting features between political fake news and Covid19 news. Thus, a variety of previous works [6,9,11] focused on rumor detection in a single domain. But herein lies another problem. In a single-domain, there may be too little data to train a good model. Therefore, Nan et al. [15] proposed a Multi-domain Fake News Detection Model (*MDFEND*) which use single domain to solve the above two problems. Single domain label can be used to describe that a piece of news belongs to a certain domain, such as science and technology, education, health and so on. The model receives news content and single domain label as input. Utilizing these data, the model can extract the common features of fake news in all different domains, and can distinguish the specific features of certain domain by domain label.

In this work, we focus on addressing one main problem: A piece of news may have features of several domains, case of which as listed above. Therefore, we propose a novel model, Multi-Domain Fake News Detection with Fuzzy labels (`FuzzyNet`). `FuzzyNet` introduces a fuzzy inference mechanism, a domain-wise attention module, a attention gate, and mixture-of-expert networks. Fuzzy mechanism is used to generate pseudo domains for multi-semantic sentence detection (addressing the key problem); Domain-wise attention is used to make the sentence embedding more domain-aware; Attention gate is used to employ the domain embedding to affect the weight of corresponding expert's result; BERT and mixture-of-expert networks are used to extract various features of input news sentences.

The fuzzy inference mechanism can solve the above two problems. The fuzzy inference mechanism constructs a fuzzy domain label for each news item. Compared with single domain label, fuzzy domain label can better describe the domain features of news, so that help the model better extract the multi-domain features of news. We demonstrated this by the experiment on the weibo21.

Our Contributions are:

1. We propose a novel model, Fuzzy-domain Fake News Detection Model (`FuzzyNet`), which can extract the multi-domain features of news content by the fuzzy domain part.
2. To the best of our knowledge, we are the first to employ domain-wise attention module to make model more domain-aware.
3. In order to describe the multi-domain features of news, we introduced fuzzy inference mechanism to the multi-domain fake news detection task. The fuzzy domain label constructed by fuzzy inference mechanism can more accurately describe the multi-domain features that news has.
4. Experimental results show that `FuzzyNet` outperforms all baselines and fuzzy part play the most important role in `FuzzyNet`.

2 Related Work

2.1 Fake News Detection

Many approaches have been proposed to tackle the challenges of fake news detection, especially with hand-craft features [5,6,14]. In [16], the authors proposed an ensemble classification model for fake news detection. Their model obtains relevant features from a fake news dataset and then uses an ensemble model to classify the extracted features, but these works are based only on news texts. Some recent research works use propagation patterns for structural modeling [15,25], others jointly used both textual and visual features for multi-modal modeling [19,29]. In [28], the authors study and analyze whether fake news can be distinguished from mainstream news by text writing style and whether fake news can be detected only by writing style by building a model. Rawat et al. [31] proposed a method to automatically collect fake news detection tasks online. For each piece of news data, they collect evidence and generate their summaries as another input to the model to help detect news text.

2.2 Multi-domain (Multi-task) Learning

The thought of multi-domain (multi-task) learning is to jointly learn a group of domains (tasks), which has been proved effective in many applications [24,30]. These researches focus on capturing relationships of different tasks with multiple representations. And each task is reinforced by the interconnections, including inter-task relevance difference. Expired by these, In [29], author propose a Multi-domain Visual Neural Network (MVNN) framework consisting of three main parts to mix all the features together. and In [27], authors propose a domain gate and use mixture-of-expert networks with multi-domain strategy for fake news detection.

However, [26] demonstrates that a set of similar words of one noun are much more precious to represent the meaning of the noun than the noun itself. Thus, it is necessary to design a fuzzy labels generating network for detecting fake news with multi-domain information.

3 Approach

Figure 1 shows an overview of the FuzzyNet. Firstly, to capture the inter information between domains and texts, we learn the domain-wise text representations via BERT [12] and domain-wise attention part. Secondly, to make the model robuster, we exploit GRUs [8] to generate the Pseudo domain embeddings for later expert calculation (fuzzy part). Thirdly, we take the advantage of Mixture-of-Expert [18, 24, 41] to extract various representations of news. Fourthly, we use attention-gate part to get effective weights of domain name to T experts. Finally, we combine the results from attention gate and fuzzy part to multiply T weights from corresponding expert and conduct a classifier to generate 0-1 probability for judging if the input news is true or not.

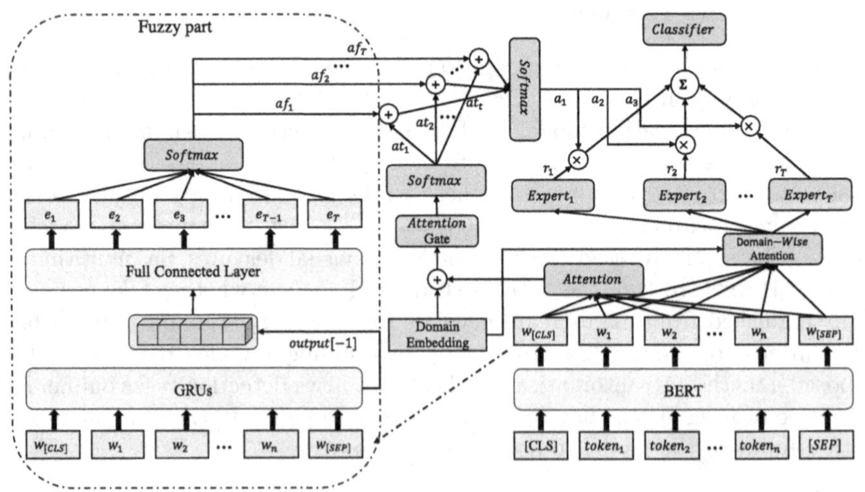

Fig. 1. The architecture FuzzyNet.

3.1 BERT Encoder

For a piece of news, we use BertTokenizer [12] to tokenize its contents and add stating tag: [CLS] and separation tag: [SEP] in the token list. Then we obtain a list of tokens $[[CLS], x_1, \ldots, x_N, [SEP]]$, where n is the length of tokens. Next, we feed these tokens into BERT [13,36,37] and map them into a low dimensional word embedding space, which can denoted as $W = \{w_{[CLS]}, w_1, w_2, \ldots, w_{[SEP]}\}$.

$$W = BERT(X) \in R^{N*d} \tag{1}$$

where d = 768 because we use the uncased version of BERT-base.

Domain	**Science**	**Military**	**Education**	**Accidents**	**Politics**
Probability (%)	$3.86*10^{-24}$	$3286*10^{-24}$	$2.92*10^{-24}$	$3.46*10^{-24}$	$3.49*10^{-24}$
Domain	Health	Finance	Entertainment	Society	-
Probability (%)	$9.32*10^{-24}$	$3.16*10^{-24}$	$9.9998*10^{-4}$	$2.08*10^{-5}$	-

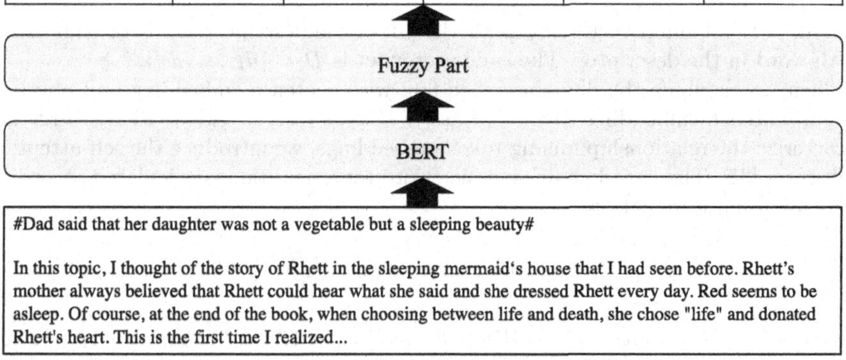

Fuzzy Part

BERT

#Dad said that her daughter was not a vegetable but a sleeping beauty#

In this topic, I thought of the story of Rhett in the sleeping mermaid's house that I had seen before. Rhett's mother always believed that Rhett could hear what she said and she dressed Rhett every day. Red seems to be asleep. Of course, at the end of the book, when choosing between life and death, she chose "life" and donated Rhett's heart. This is the first time I realized...

Fig. 2. Running Example of Fuzzy Part

3.2 Fuzzy Mechanism

A fuzzy set descriptions are much more precious than a single definition for describing one thing [26,39]. In LotClass [26], authors use generated candidates from BERT instead of masked noun itself to represent its meaning in a sentence achieve great performance in weak-supervised text classification tasks. For example, sport is a label of classification task, it is more robust to use a set of {basketball, football, etc. } to represent sport label instead of sport itself. Thus, in this section, as shown in Fig. 1, we employ fuzzy part to generate potential tags for given a piece of news. Fuzzy part consists of GRU for generating a hidden state embedding, full connected layer with multiple sigmoid functions for generating multiple 0-1 probability distribution for each domain tag and softmax for generating last scores for each pseudo domain tag. The whole calculation of score for T pseudo domain tags is shown as follows:

$$\mathbf{AF} = \mathbf{softmax}(\mathbf{MLP}(\mathbf{GRU}(\mathbf{W}; \theta_1); \theta_2)) \tag{2}$$

where θ_1 and θ_2 are parameters of GRU network and MLP network, respectively. T equals 9. *softmax* is the normalized exponential function. $AF = [af_1, \ldots, af_T]$

Figure 2 is a running example of fuzzy part in **FuzzyNet**, the example is a English-version translated from a piece of Chinese news. The input news is a description about the story of Rhett's donations. Fuzzy part predicted pseudo domain names for this text, and of these names, entertainment gets the highest score and society also gets a higher score. Obviously, the story has both entertainment and society information.

3.3 Domain-Wise Attention

After encoding by BERT, we obtain the sentence representation $W = \{w_{[CLS]}, w_1, w_2, \ldots, w_{[SEP]}\}$. Since we need to assign domain names for each sentence, we employ domain-wise attention to learn relevant sentence representations. Firstly, we generate the domain vector for each domain via averaging the word embeddings of descriptor from *wikipedia*[1]:

$$d_i = \frac{1}{N_d} \Sigma_{j=1}^{N_d} w_j, i = 1, \ldots, T \tag{3}$$

where d_i is the domain vector, N_d is the length of the descriptor, w_j is th embedding of j-th word in the descriptor. The code vector set is $D = \{d_1, \ldots, d_T\}$.

Then, we combine the domain embedding with sentence embedding and obtain a domain-sentence embedding: $Wv = \{d_i, w_{[CLS]}, w_1, w_2, \ldots, w_{[SEP]}\}$. Next, to further characterize the relationship among token embeddings, we introduce the self-attention mechanism [35]. The new domain-sentence word representation is calculated according to the following formulation:

$$Attention = softmax(\frac{Q_v * K^t}{\sqrt{d}}) \tag{4}$$

where $Q_v = W^Q * Embed$, $K_v = W^K * Embed$, $V_v = W^V * Embed$. Q_v, K_v, V_v are learnable parameter matrices. With *Attention*, we can further obtain new presentations for the given sentence. The new domain-sentence embedding is calculated by following equation:

$$Wv_{atten} = \Sigma_{j=1}^{L} V_j * Attention_j \tag{5}$$

where L is the length of domain-sentence, and domain-wise sentence embedding can be represented as $W = Wv_{atten}[1:]$.

3.4 Mixture-of-Expert Part

Expired by previous works, to obtain various representations of given piece of news, we employ mixture-of-expert to extract features of news' content. Each expert network is a TextCNN [22]. Each expert has its own area of expertise and is set to extract certain features of corresponding domain. The calculation process of mixture-of-expert is shown as follows:

$$r_i = TextCNN(W; \phi_i) \tag{6}$$

[1] https://www.wikipedia.org/.

3.5 Attention Gate

Attention gate is used to generate high-quality news representations that can represent news from different domains appropriately. Intuitively, we can average representations by all experts. However, the simple average operation will the domain-specific information [27]. Thus, the synthetic representation may not be good for **FuzzyNet**. Moreover, different expert specialize in different area. Thus, we use attention gate to process the domain embedding as well as sentence embedding. Domain embedding is calculated in Eq. 3. We donate the attention gate as $G(\cdot;\phi)$, the calculation process is as follows:

$$AT = softmax(G(d_i \oplus e^s);\phi) \tag{7}$$

where $G(\cdot;\phi)$ is a feed-forward network. e^s is a sentence attention embedding. We use *softmax* function to normalize the output of $G(\cdot)$. $AT = [at_1, \ldots, at_T]$

3.6 Combination

We combine AT and AF and them conduct a *softmax* function to get final parameters $A = [a_1, \ldots, a_T] = softmax(AT + AF)$. Finally, final feature vectors is obtained by following equation:

$$fv = \Sigma_{i=1}^{T} a_i * r_i \tag{8}$$

3.7 Prediction

The new's final vector is fed into the classifier, which is a multi-layer perception network with a softmax output layer, for detecting fake news:

$$\hat{y} = softmax(MLP(fv, \phi)). \tag{9}$$

The purpose of fake news detection is to identify whether the news is fake or not. Thus, fake news detection can be seen as a binary prediction task. We use y^i to represent ground truth and $\hat{y}^i$ to represent the predicted label. We employ binary cross-entropy loss function for this classification:

$$\mathcal{L} = y^i \log \hat{y}^i + (1 - h^i) \log(1 - \hat{y}^i) \tag{10}$$

4 Experiment Design

4.1 Dataset

Different from the single or two domain provided in previous works [3,9,11,25,32,40], We investigate the feasibility of our approach open dataset with multi domains: Weibo21 [27]. Weibo consists of 4,488 pieces of fake news and 4640 pieces of true news on 9 different domains: Science, Military, Education, Disasters, Politics, Health, Finance, Entertainment, Society. Data distribution and frequent words distribution for different domains are shown in Table 1 and Fig. 3, respectively.

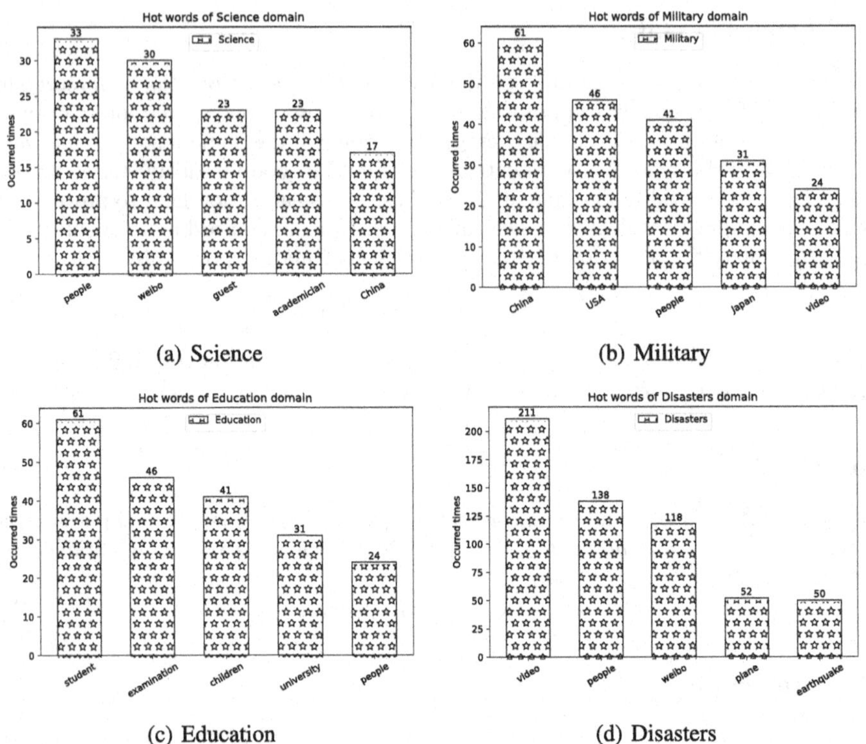

(a) Science (b) Military (c) Education (d) Disasters

Fig. 3. Distribution of Top_5 words in 4 category domains: Science, Military, Education, and Disasters. Due to the page limitation, we list all statics in data folder of our open source. All the examples in the paper are translated from Chinese.

4.2 Metric

Following previous works [27,32,40], our objective in fake news identification is to ensure that the proposed approaches can get high effectiveness of both precision and recalling true news. As F1 score metric is a comprehensive evaluation method of precision and recall, we take it as the metric in evaluating our approach and baselines. Recall, Precision, and F1 equations are shown as follows:

$$F1 = \frac{2 * Precision * Recall}{Precision + Recall} \tag{11}$$

4.3 Experiment Setting

In this section, we will introduce details of settings of our approach. The hidden dimension of BERT is set as 768. The max length of the sentence is 170 and we pad it to 256. We employ Adam [10] as the optimizer of our model. Learning rate is set to 5e-4. To make the result more credible, we conduct our experiment with 10 cross-fold and use the average F1 score to evaluate the performance of detection.

Table 1. Data Statistics of Weibo21 Dataset.

domain	Science	Military	Education	Disasters	Politics	Health	Finance	Entertainment	Society	All
real	143	121	243	185	306	485	959	1000	1198	4640
fake	93	222	248	591	546	515	362	440	1471	4488
all	236	343	491	776	852	1000	1321	1440	2669	9128

4.4 Baselines

For Single-Domain and Mixed-domain:

- TextCNN_single and TextCNN_all [7]: Using two Parallel CNN and Deep CNN to solve the multi-label classification problem.
- BiGRU_single and BiGRU_all [25]: Using recurrent neural networks (RNN) for learning the hidden representations that capture the variation of contextual information of relevant posts over time.
- BERT_single and BERT_all [13]: Using a large pre-trained natural language model for single domain fake news detection

For Multi-domain:

- EANN [38]: Using an end-to-end framework named Event Adversarial Neural Network (EANN) to derive event-invariant features and thus benefit the detection of fake news on newly arrived events.
- MMOE [24]: Introducing a novel multi-task learning approach, Multi-gate Mixture-of-Experts (MMoE) to model task relationships from data.
- MOSE [30]: Introducing a novel framework, Mixture of Sequential Experts (MoSE) for user activity streams
- EDDFN [33]: Introducing a novel framework that jointly preserves domain-specific and cross-domain knowledge in news records to detect fake news from different domains.
- MDFEND [27]: Introducing a domain gate to aggregate multiple representations extracted by a mixture of experts for fake news detection.

4.5 Implementation and Availability

For easy reproduction, we keep our codes open and anonymous. At the same time, spitted data is included. The dataset and the replication package of `FuzzyNet` are publicly available at:

https://anonymous.4open.science/r/fakenewsdetection-68BC

5 Experimental Result

5.1 Compared with Baselines

Nan et al. [27] have showed the effectiveness of their multi-domain fake news detection model (MDFEND). Further, the results in Table 4 support that our idea is right. The fake news detection performance of `FuzzyNet` compared with MDFEND proves that

Table 2. Multi-domain with Fuzzy labels Fake News Detection Performance (F1-score)

Model	Science	Military	Education	Disasters	Politics	Health	Finance	Entertainment	Society	All
TextCNN_single	0.7470	0.778	0.8882	0.8310	0.8694	0.9053	0.7909	0.8591	0.8727	0.8380
BiGRU_single	0.4876	0.7169	0.7067	0.7625	0.8477	0.8378	0.8109	0.8308	0.6067	0.7342
BERT_single	0.8192	0.7795	0.8136	0.7885	0.8188	0.8909	0.8464	0.8638	0.8242	0.8272
TextCNN_all	0.7254	0.8839	0.8362	0.8222	0.8561	0.8768	0.8638	0.8456	0.8540	0.8686
BiGRU_all	0.7269	0.8724	0.8138	0.7935	0.8356	0.8868	0.8291	0.8629	0.8485	0.8595
BRET_all	0.7777	0.9072	0.8331	0.8512	0.8366	0.9090	0.8735	0.8769	0.8577	0.8795
EANN	0.8225	0.9274	0.8624	0.8666	0.8705	0.9150	0.8710	0.8957	0.8877	0.8975
MMOE	0.8755	0.9112	0.8706	0.8770	0.8620	0.9364	0.8567	0.8886	0.8750	0.8947
MOSE	0.8502	0.8858	0.8815	0.8672	0.8808	0.9179	0.8672	0.8913	0.8729	0.8939
EDDFN	0.8186	0.9137	0.8676	0.8786	0.8478	0.9379	0.8636	0.8832	0.8689	0.8919
MDFEND	0.8301	0.9389	0.8917	0.9003	0.8865	0.9400	0.8951	0.9066	0.8980	0.9137
FuzzyNet	**0.8765**	**0.9412**	**0.9191**	**0.9173**	**0.8902**	**0.9501**	**0.9120**	**0.9076**	**0.9055**	**0.9249**

the fuzzy domain label generated by fuzzy inference mechanism can better help the model extract the features for fake news detection task.

We compare our `FuzzyNet` with single-domain, mixed-domain, and multi-domain methods to testify the effectiveness. The experimental results are shown in Table 2. Experimental results reveal several insightful observations. (1) Generally, Mixed-domain models and multi-domain models work much better than single-domain models and multi-domain models perform better than mixed-domain models, which demonstrates that (i) additional data makes sense; (ii) multi-domain is of great important. (2) Not every mixed-domain model performs better than single-domain, such as TextCNN_single and TextCNN_all on health domain. This case shows that using mixed-domain information may bring noisy data into the model, which may lead to negative effect. (3) `FuzzyNet` performs the best compared with all baselines, which indicates the performance of combination of fuzzy mechanism and multi-domain. To see more details of which part works, we conduct ablation study on `FuzzyNet`.

5.2 Ablation Study

In this section, we conduct ablation investigation to examine the effectiveness of each part in our `FuzzyNet`. To evaluate a module, we remove it (denoted as without, w/o) and perform the rest part on the datasets. The experimental results of ablation study are shown in Table 3.

Table 3. Ablation study by removing the main components.

Model	Science	Military	Education	Disasters	Politics	Health	Finance	Entertainment	Society	All
FuzzyNet	**0.8765**	**0.9412**	**0.9191**	**0.9173**	**0.8902**	**0.9501**	**0.9120**	**0.9076**	**0.9055**	**0.9249**
w/o attention gate	0.8501	0.9335	0.9188	0.9114	0.8887	0.9480	0.9037	0.8978	0.8955	0.9140
w/o fuzzy part	0.8641	0.9333	0.9057	0.8706	0.8860	0.9410	0.8992	0.8907	0.9003	0.9137
w/o domain-wise attention	0.8695	0.9302	0.9117	0.9100	0.8854	0.9443	0.9087	0.8990	0.8975	0.9169

Impact of *attention gate*. We remove the *attention gate* from the full model. As shown in Table 3, `FuzzyNet` without attention gate achieves lower scores of F1.

Impact of *fuzzy part*. Compared with the w/o fuzzy part, the full **FuzzyNet** improves the score on all F1 from 0.9137 to 9249 (1.2% improvement). Table 3 shows that **FuzzyNet** without fuzzy part achieves the lowest part, which demonstrates that fuzzy part plays the most important part in **FuzzyNet**.

Impact of *domain-wise attention*. The F1 score that **FuzzyNet** w/o domain-wise attention achieves is between **FuzzyNet** w/o attention gate and **FuzzyNet** w/o fuzzy part on four domains: Education, Disasters, Health, Society.

Table 3 shows that **FuzzyNet** without fuzzy-part performs the worst on weibo21 datasets. The results from Table 3 demonstrates the effectiveness of different modules in **FuzzyNet**. In addition, *Fuzzy* module plays a more important role in **FuzzyNet** compared with other modules.

6 Conclusion

In this work, we introduce a novel architecture with fuzzy labels for detecting fake news. Specially, we use BERT and mixture-of-expert networks to extract various features of input news sentences; use domain-wise attention to make the sentence embedding more domain-aware; use attention gate to employ the domain embedding to affect the weight of corresponding expert's result; use fuzzy mechanism to generate pseudo domains for multi-semantic sentence detection.

We conduct our experiment on a open dataset: weibo21 and outperforms all the baselines, including three types: single-domain, mixed-domain, and multi-domain models. Additionally, we do ablation study and demonstrate that fuzzy mechanism play the most important role in our model **FuzzyNet**.

References

1. Reddit. https://reddit.com/
2. Sina weibo. http://www.weibo.com
3. Tencent rumor government report. https://tech.qq.com/a/20171220/026316.html
4. Twitter. http://www.twitter.com
5. Ajao, O., Bhowmik, D., Zargari, S.: Sentiment aware fake news detection on online social networks. In: ICASSP 2019-2019 IEEE International Conference on Acoustics, Speech and Signal Processing (ICASSP), pp. 2507–2511. IEEE (2019)
6. Castillo, C., Mendoza, M., Poblete, B.: Information credibility on twitter. In: Proceedings of the 20th International Conference on World Wide Web, pp. 675–684 (2011)
7. Chen, Y.: Convolutional neural network for sentence classification. Master's thesis, University of Waterloo (2015)
8. Cho, K., et al.: Learning phrase representations using RNN encoder-decoder for statistical machine translation. arXiv preprint arXiv:1406.1078 (2014)
9. Cui, L., Lee, D.: Coaid: Covid-19 healthcare misinformation dataset. arXiv preprint arXiv:2006.00885 (2020)
10. Da, K.: A method for stochastic optimization. arXiv preprint arXiv:1412.6980 (2014)
11. Dai, E., Sun, Y., Wang, S.: Ginger cannot cure cancer: battling fake health news with a comprehensive data repository. In: Proceedings of the International AAAI Conference on Web and Social Media, vol. 14, pp. 853–862 (2020)

12. Devlin, J., Chang, M.W., Lee, K., Toutanova, K.: BERT: pre-training of deep bidirectional transformers for language understanding. In: NAACL (2019)
13. Devlin, J., Chang, M.W., Lee, K., Toutanova, K.: BERT: pre-training of deep bidirectional transformers for language understanding (2019)
14. Giachanou, A., Rosso, P., Crestani, F.: Leveraging emotional signals for credibility detection. In: Proceedings of the 42nd International ACM SIGIR Conference on Research and Development in Information Retrieval, pp. 877–880 (2019)
15. Guo, H., Cao, J., Zhang, Y., Guo, J., Li, J.: Rumor detection with hierarchical social attention network. In: Proceedings of the 27th ACM International Conference on Information and Knowledge Management, pp. 943–951 (2018)
16. Hakak, S., Alazab, M., Khan, S., Gadekallu, T.R., Maddikunta, P.K.R., Khan, W.Z.: An ensemble machine learning approach through effective feature extraction to classify fake news. Futur. Gener. Comput. Syst. **117**, 47–58 (2021)
17. Hangloo, S., Arora, B.: Fake news detection tools and methods-a review. arXiv preprint arXiv:2112.11185 (2021)
18. Jacobs, R.A., Jordan, M.I., Nowlan, S.J., Hinton, G.E.: Adaptive mixtures of local experts. Neural Comput. **3**(1), 79–87 (1991)
19. Jin, Z., Cao, J., Guo, H., Zhang, Y., Luo, J.: Multimodal fusion with recurrent neural networks for rumor detection on microblogs. In: Proceedings of the 25th ACM International Conference on Multimedia, pp. 795–816 (2017)
20. Jin, Z., Cao, J., Guo, H., Zhang, Y., Wang, Yu., Luo, J.: Detection and analysis of 2016 US presidential election related rumors on Twitter. In: Lee, D., Lin, Y.-R., Osgood, N., Thomson, R. (eds.) SBP-BRiMS 2017. LNCS, vol. 10354, pp. 14–24. Springer, Cham (2017). https://doi.org/10.1007/978-3-319-60240-0_2
21. Kwon, S., Cha, M., Jung, K., Chen, W., Wang, Y.: Prominent features of rumor propagation in online social media. In: 2013 IEEE 13th International Conference on Data Mining, pp. 1103–1108. IEEE (2013)
22. Lai, S., Xu, L., Liu, K., Zhao, J.: Recurrent convolutional neural networks for text classification. In: Twenty-Ninth AAAI Conference on Artificial Intelligence (2015)
23. Ma, B., Lin, D., Cao, D.: Content representation for microblog rumor detection. In: Angelov, P., Gegov, A., Jayne, C., Shen, Q. (eds.) Advances in Computational Intelligence Systems. AISC, vol. 513, pp. 245–251. Springer, Cham (2017). https://doi.org/10.1007/978-3-319-46562-3_16
24. Ma, J., Zhao, Z., Yi, X., Chen, J., Hong, L., Chi, E.H.: Modeling task relationships in multi-task learning with multi-gate mixture-of-experts. In: Proceedings of the 24th ACM SIGKDD International Conference on Knowledge Discovery & Data Mining, pp. 1930–1939 (2018)
25. Ma, J., et al.: Detecting rumors from microblogs with recurrent neural networks (2016)
26. Meng, Y., et al.: Text classification using label names only: a language model self-training approach. arXiv preprint arXiv:2010.07245 (2020)
27. Nan, Q., Cao, J., Zhu, Y., Wang, Y., Li, J.: MDFEND: multi-domain fake news detection. In: Proceedings of the 30th ACM International Conference on Information & Knowledge Management, pp. 3343–3347 (2021)
28. Potthast, M., Kiesel, J., Reinartz, K., Bevendorff, J., Stein, B.: A stylometric inquiry into hyperpartisan and fake news. arXiv preprint arXiv:1702.05638 (2017)
29. Qi, P., Cao, J., Yang, T., Guo, J., Li, J.: Exploiting multi-domain visual information for fake news detection. In: 2019 IEEE International Conference on Data Mining (ICDM), pp. 518–527. IEEE (2019)

30. Qin, Z., Cheng, Y., Zhao, Z., Chen, Z., Metzler, D., Qin, J.: Multitask mixture of sequential experts for user activity streams. In: Proceedings of the 26th ACM SIGKDD International Conference on Knowledge Discovery & Data Mining, pp. 3083–3091 (2020)
31. Rawat, M., Kanojia, D.: Automated evidence collection for fake news detection. arXiv preprint arXiv:2112.06507 (2021)
32. Shu, K., Mahudeswaran, D., Wang, S., Lee, D., Liu, H.: Fakenewsnet: a data repository with news content, social context, and spatiotemporal information for studying fake news on social media. Big Data **8**(3), 171–188 (2020)
33. Silva, A., Luo, L., Karunasekera, S., Leckie, C.: Embracing domain differences in fake news: cross-domain fake news detection using multi-modal data. In: Proceedings of the AAAI Conference on Artificial Intelligence, vol. 35, pp. 557–565 (2021)
34. Tavernise, S.: As fake news spreads lies, more readers shrug at the truth. The New York Times, vol. 6 (2016)
35. Vaswani, A., et al.: Attention is all you need. In: Advances in Neural Information Processing Systems, vol. 30 (2017)
36. Wang, S., Tang, D., Zhang, L.: A large-scale hierarchical structure knowledge enhanced pre-training framework for automatic ICD coding. In: Mantoro, T., Lee, M., Ayu, M.A., Wong, K.W., Hidayanto, A.N. (eds.) ICONIP 2021. CCIS, vol. 1517, pp. 494–502. Springer, Cham (2021). https://doi.org/10.1007/978-3-030-92310-5_57
37. Wang, S., Tang, D., Zhang, L., Li, H., Han, D.: HieNet: bidirectional hierarchy framework for automated ICD coding. In: Bhattacharya, A., et al. (eds.) DASFAA 2022. LNCS, vol. 13246, pp. 523–539. Springer, Cham (2022). https://doi.org/10.1007/978-3-031-00126-0_38
38. Wang, Y., et al.: EANN: event adversarial neural networks for multi-modal fake news detection. In: Proceedings of the 24th ACM SIGKDD International Conference on Knowledge Discovery & Data Mining, pp. 849–857 (2018)
39. Zadeh, L.A.: Fuzzy sets. In: Fuzzy sets, fuzzy logic, and fuzzy systems: selected papers by Lotfi A Zadeh, pp. 394–432. World Scientific (1996)
40. Zhang, X., Cao, J., Li, X., Sheng, Q., Zhong, L., Shu, K.: Mining dual emotion for fake news detection. In: Proceedings of the Web Conference 2021, pp. 3465–3476 (2021)
41. Zhu, Y., et al.: Learning to expand audience via meta hybrid experts and critics for recommendation and advertising. In: Proceedings of the 27th ACM SIGKDD Conference on Knowledge Discovery & Data Mining, pp. 4005–4013 (2021)

Correction to: Auto-TSA: An Automatic Time Series Analysis System Based on Meta-learning

Tianyu Mu, Zhenli Sheng, Lekui Zhou, and Hongzhi Wang

Correction to:
Chapter "Auto-TSA: An Automatic Time Series Analysis System Based on Meta-learning" in: A. El Abbadi et al. (Eds.): *Database Systems for Advanced Applications. DASFAA 2023 International Workshops*, LNCS 13922, https://doi.org/10.1007/978-3-031-35415-1_10

In the original version of this chapter funding program numbers in Acknowledgement section is not added. This has been corrected.

The updated original version of this chapter can be found at
https://doi.org/10.1007/978-3-031-35415-1_10

A. El Abbadi et al. (Eds.): DASFAA 2023 Workshops, LNCS 13922, p. C1, 2023.
https://doi.org/10.1007/978-3-031-35415-1_24

Author Index

A. El Abbadi et al. (Eds.): DASFAA 2023 Workshops, LNCS 13922, pp. 345–346, 2023.
https://doi.org/10.1007/978-3-031-35415-1

Printed in the United States
by Baker & Taylor Publisher Services

Printed in the United States
by Baker & Taylor Publisher Services